The Geometry of Higher-Order Hamilton Spaces

Fundamental Theories of Physics

An International Book Series on The Fundamental Theories of Physics: Their Clarification, Development and Application

Volume 132

The Geometry of Higher-Order Hamilton Spaces

Applications to Hamiltonian Mechanics

by

Radu Miron

Faculty of Mathematics,
Al. I. Cuza' University,
Iaşi, Romania

SPRINGER-SCIENCE+BUSINESS MEDIA, B.V.

A C.I.P. Catalogue record for this book is available from the Library of Congress.

ISBN 978-94-010-3995-6 ISBN 978-94-010-0070-3 (eBook)
DOI 10.1007/978-94-010-0070-3

Printed on acid-free paper

Contents

Preface

As is known, the Lagrange and Hamilton geometries have appeared relatively recently [76, 86]. Since 1980 these geometries have been intensively studied by mathematicians and physicists from Romania, Canada, Germany, Japan, Russia, Hungary, U.S.A. etc.

Prestigious scientific meetings devoted to Lagrange and Hamilton geometries and their applications have been organized in the above mentioned countries and a number of books and monographs have been published by specialists in the field: R. Miron [94, 95], R. Miron and M. Anastasiei [99, 100], R. Miron, D. Hrimiuc, H. Shimada and S. Sabău [115], P.L. Antonelli, R. Ingarden and M. Matsumoto [7]. Finsler spaces, which form a subclass of the class of Lagrange spaces, have been the subject of some excellent books, for example by M. Matsumoto [76], M. Abate and G. Patrizio [1], D. Bao, S.S. Chern and Z. Shen [17] and A. Bejancu and H.R.Farran [20]. Also, we would like to point out the monographs of M. Crampin [34], O.Krupkova [72] and D.Opriş, I. Butulescu [125], D.Saunders [144], which contain pertinent applications in analytical mechanics and in the theory of partial differential equations. Applications in mechanics, cosmology, theoretical physics and biology can be found in the well known books of P.L. Antonelli and T.Zawstaniak [11], G.S. Asanov [14], S. Ikeda [59], M. de Leone and P.Rodrigues [73].

The importance of Lagrange and Hamilton geometries consists of the fact that variational problems for important Lagrangians or Hamiltonians have numerous applications in various fields, such as mathematics, the theory of dynamical systems, optimal control, biology, and economy.

In this respect, P.L. Antonelli's remark is interesting:

"There is now strong evidence that the symplectic geometry of Hamiltonian dynamical systems is deeply connected to Cartan geometry, the dual of Finsler geometry", (see V.I.Arnold, I.M.Gelfand and V.S.Retach [13]).

The above mentioned applications have also imposed the introduction of the notions of higher order Lagrange spaces and, of course, higher order Hamilton spaces. The base manifolds of these spaces are bundles of accelerations of superior order. The methods used in the construction of these geometries are the natural extensions of the classical methods used in the edification of Lagrange and Hamilton geometries. These methods allow us to solve an old problem of differential geometry formulated by Bianchi and Bompiani [94] more than 100 years ago, namely the problem of prolongation of a Riemannian structure g defined on the base manifold M, to the tangent bundle $T^kM,\ k > 1$. By means of this solution of the previous problem, we can construct, for the first time, good examples of regular Lagrangians and Hamiltonians of higher order.

While higher order Lagrange geometry has developed sufficiently, we cannot say the same thing about higher order Hamilton geometry. A beginning was made for the case $k = 2$, in the book [115], as in the year 2001 we had not yet solved the variational problem for a Hamiltonian which depends on the higher order accelerations and momenta. However, this problem was solved in 2002, [98]. Another reason was the absence of a consistent theory of subspaces in the Hamilton space of order k, $k \geq 1$, which is indispensable for applications.

In the present book, we give the general geometrical theory of Hamilton spaces of order $k \geq 1$. This is not a simple generalization of the theory expounded for the case $k = 2$ in the monograph *The Geometry of Hamilton and Lagrange Spaces*, Kluwer Academic Publishers, Fundamental Theories of Physics series no. 118, written by the present author together with D. Hrimiuc, H. Shimada and V. S. Sabău, but it is a global picture of this new geometry, extremely useful in applications from Hamiltonian mechanics, quantum physics, optimal control and biology.

Consequently, this book must be considered a direct continuation of the monographs [94], [95], [99] and [115]. It contains new developments of subjects such as variational principles for higher order Hamiltonians; higher order energies; laws of conservations; Noether theorems; and the Hamilton subspaces of order k and their fundamental equations. Also, the Cartan spaces of order k are investigated in detail as dual of Finsler spaces of the same order.

In this respect, a more explicit argumentation is as follows.

The geometry of Lagrange space of order $k \geq 1$ is based on the geometrical edifice of the k-accelerations bundle (T^kM, π^k, M).

In analytical mechanics the manifold M is the space of configurations of a physical system. A point $x = (x^i)$, $(i = 1, ..., n = \dim M)$ in M is called a configuration. A mapping $c : t \in I \to (x^i(t)) \in U \subset M$ is a law of moving (evolution), t is time, a pair (t, x) is an event and the k-uple $\left(\frac{dx^i}{dt}, \cdots, \frac{1}{k!}\frac{d^kx^i}{dt^k}\right)$ gives the velocity and generalized accelerations of order 1, ..., $k-1$. The factors $\frac{1}{h!}$ $(h = 1, ..., k)$ are introduced here for the simplicity of calculus. In this book we omit the word 'generalized' and just use the shorter term the accelerations of order h, for $\frac{1}{h!}\frac{d^hx^i}{dt^h}$. A law of moving $c : t \in I \to c(t) \in U$ will be called a curve parametrized by time t.

A Lagrangian of order $k \geq 1$ is a real scalar function $L(x, y^{(1)}, ..., y^{(k)})$ on T^kM, where $y^{(h)i} = \frac{1}{h!}\frac{d^hx^i}{dt^h}$. This definition is for autonomous Lagrangians. A similar definition can be formulated for nonautonomous Lagrangians of order k, by

$$L : (t, x, y^{(1)}, ..., y^{(k)}) \in \mathbf{R} \times T^kM \to L(t, x, y^{(1)}, ..., y^{(k)}) \in \mathbf{R},$$

L being the scalar functions on the manifold $\mathbf{R} \times T^kM$.

The previous considerations can be done for the Hamiltonians of order k. Let $p_i = \frac{1}{2}\frac{\partial L}{\partial y^{(k)i}}$ be the 'momenta' determined by the Lagrangian L of order k. Then a scalar function

$$H : (x, y^{(1)}, ..., y^{(k-1)}, p) \in T^{*k}M \to H(x, y^{(1)}, ..., y^{(k-1)}, p) \in \mathbf{R},$$

is an autonomous Hamiltonian of order k. It is a function of the configurations x, accelerations $y^{(1)}$, ..., $y^{(k-1)}$ of order 1, ..., $k-1$ and momenta p.

A similar definition can be formulated for a nonautonomous Hamiltonian of order k.

For us it is preferable to study autonomous Lagrangians and Hamiltonians, because the notions of Lagrange space of order k or Hamilton space of order k are geometrical concepts and one can construct these geometries over the differentiable manifolds T^kM and $T^{*k}M$, respectively. Of course, the geometries of nonautonomous Lagrangians $L(t, x, y^{(1)}, ..., y^{(k)})$ and nonautonomous Hamiltonians $H(t, x, y^{(1)}, ..., y^{(k-1)}, p)$ can be constructed by means of the same methods. One obtains the rheonomic Lagrange spaces of order k and rheonomic Hamiltonian spaces of order k.

Now we know the usefulness of the geometry of higher order Lagrange space (see the book [94]). But why do we need a geometry of Hamilton spaces of order $k \geq 1$? Clearly, for the same reason as for the Lagrange case: For the determination of adequate geometrical models for the Hamiltonian mechanics of order k. This must be a natural extension of classical Hamiltonian mechanics, expounded by V.I. Arnold in the book [12] or R.M. Santilli in the book [139].

The problem is, why did we use the manifold $T^{*k}M$ as the background for the construction of the Hamilton geometry of order k. The answer is as follows. We need a 'dual' of the k -acceleration bundle (T^kM, π^k, M) denoted by $(T^{*k}M, \pi^{*k}, M)$ which must have the following properties:

1°. $T^{*1}M = T^*M$, ((T^*M, π^*, M) is the cotangent bundle).

2°. $\dim T^{*k}M = \dim T^kM = (k+1)n$.

3°. The manifold $T^{*k}M$ carries a natural presympletic structure.

4°. $T^{*k}M$ carries a natural Poisson structure.

5°. $T^{*k}M$ is local diffeomorphic to T^kM.

We solved this problem by considering the differentiable bundle $(T^{*k}M, \pi^{*k}, M)$ as the fibred bundle $(T^{k-1}M \times_M T^*M, \pi^{k-1} \times_M \pi^*, M)$. So we have

$$T^{*k}M = T^{k-1}M \times_M T^*M, \quad \pi^{*k} = \pi^{k-1} \times_M \pi^*.$$

A point $u \in T^{*k}M$ is of the form $u = (x, y^{(1)}, ..., y^{(k-1)}, p)$. It is determined by a configuration $x = (x^i)$, the accelerations $y^{(1)i} = \dfrac{dx^i}{dt}, ...,$ $y^{(k-1)i} = \dfrac{1}{(k-1)!}\dfrac{d^{k-1}x^i}{dt^{k-1}}$ and the momenta $p = (p_i)$.

All the previous five conditions are satisfied. These considerations imply the fact that the geometries of higher order Lagrange space and higher order Hamilton space are dual.

Duality is obtained via a Legendre transformation.

For a good understanding of the important concept of duality we had to make a short introduction to the geometrical theory of Lagrange and Finsler spaces of order k and then continue with the main subject of the book, the geometry of Hamilton and Cartan spaces of order k.

The Lagrange spaces of order k are defined as the pairs $L^{(k)n} = (M, L)$, where L is a regular Lagrangian of order k. By means of variational calculus the integral of action $I(c) = \int_0^1 L\left(x(t), \frac{dx}{dt}(t), ..., \frac{1}{k!}\frac{d^k t}{dt^k}\right) dt$ gives the Euler-Lagrange equations and the Craig-Synge equations. The last equations determine a canonical k-semispray S. The geometry of the space $L^{(k)n}$ can be developed by means of the fundamental function L, of the fundamental tensor $g_{ij} = \frac{1}{2}\frac{\partial^2 L}{\partial y^{(k)i}\partial y^{(k)j}}$ and of the canonical k-semispray S. The lifting of the previous geometrical edifice to the total space $T^k M$ will give us a metrical almost contact structure, canonically related to the Lagrange space of order k, $L^{(k)n}$. Of course, this structure involves the geometry of the space $L^{(k)n}$.

An important problem was to find some remarkable examples of spaces $L^{(k)n}$, for $k > 1$. By solving the problem of prolongations to $T^k M$ of a Riemannian structure g given on the base manifold M, we found interesting examples of Lagrange spaces of order k.

For the applications, the notions of energy of order $1, 2, ..., k$ are studied, and the law of conservation for the energy of order k and a Noether type theorem is proved.

The spaces $L^{(k)n}$ have two important particular cases.

The Finsler spaces of order k, $F^{(k)n}$, obtained when the fundamental function L is homogeneous with respect to accelerations $y^{(1)}$, ..., $y^{(k)}$, and the Riemann spaces of order k, $\mathcal{R}^{(k)n}$ are the spaces $L^{(k)n}$ for which the fundamental tensor g_{ij} does not depend on the accelerations $y^{(1)}$, ..., $y^{(k)}$.

Therefore we have the following sequence of inclusions, [94]:

$$(*) \qquad \{\mathcal{R}^{(k)n}\} \subset \{F^{(k)n}\} \subset \{L^{(k)n}\} \subset \{GL^{(k)n}\}.$$

In the case $k = 1$ this sequence admits a 'dual', which is obtained via Legendre transformation. In the book [115] we have introduced the "dual" of the sequence (*) for the case $k = 2$.

Now, the main problem for us is to define and study a 'dual' sequence of the inclusions (*) in the 'dual' Hamilton space of order k.

A Hamilton space of order k is a pair $H^{(k)n} = (M, H)$, where $H : (x, y^{(1)}, ..., y^{(k-1)}, p) \in T^{*k}M \to H(x, y^{(1)}, ..., y^{(k-1)}, p) \in \mathbf{R}$ is a regular Hamiltonian. Here, regularity means the Hessian of H, with respect to the momenta p_i, is not singular. The elements of the Hessian matrix are $g^{ij} = \frac{1}{2}\frac{\partial^2 H}{\partial p_i \partial p_j}$. Thus H is called the fundamental function and g^{ij} the fundamental tensor of the space $H^{(k)n}$. The geometry of the space $H^{(k)n}$ can be based on these two geometrical object fields: H and g^{ij}. In the case where $g^{ij} = g^{ij}(x)$ we have a particular class of Hamilton spaces $\mathcal{R}^{*(k)n}$ called Riemannian. If the fundamental function H is $2k$ -homogeneous on the fibres of the bundle $T^{*k}M$, the spaces $H^{(k)n}$ are called Cartan spaces of order k and denoted $\mathcal{C}^{(k)n}$.

Finally, a pair $GH^{(k)n} = (M, g^{ij})$, where $g^{ij}(x, y^{(1)}, ..., y^{(k-1)}, p)$ is a symmetric, nonsingular, distinguished tensor field which is called a generalized Hamilton space of order k.

Consequently, we obtain the sequence of inclusions

$$(**) \qquad \{\mathcal{R}^{*(k)n}\} \subset \{\mathcal{C}^{*(k)n}\} \subset \{H^{(k)n}\} \subset \{GH^{(k)n}\}.$$

This is the 'dual' sequence of the sequence (*) via the Legendre transformation.

The main goal of this book is to study the classes of spaces from the sequence (**). Chapters 4-11 of the book are devoted to this subject.

Therefore we begin with the geometry of the total space $T^{*k}M$ of the dual bundle $(T^{*k}M, \pi^{*k}, M)$ of the k-tangent bundle (T^kM, π^k, M) underline: vertical distributions; Liouville vector fields; Liouville 1-form $\omega = p_i dx^i$; the closed 2-form $\theta = d\omega$ which defines a natural presymplectic structure on $T^{*k}M$. In Chapter 5 a new theory of variational problem for the Hamiltonian H of order k is developed starting from the integral of action of H defined by

$$I(c) = \int_0^1 [p_i \frac{dx^i}{dt} - \frac{1}{2} H(x, \frac{dx}{dt}, ..., \frac{1}{(k-1)!}\frac{d^{k-1}x}{dt^{k-1}}, p)]dt$$

It is proved that the extremal curves are the solutions of the following Hamilton-Jacobi equations:

$$\frac{dx^i}{dt} = \frac{1}{2}\frac{\partial H}{\partial p_i},$$

$$\frac{dp_i}{dt} = -\frac{1}{2}[\frac{\partial H}{\partial x^i} - \frac{d}{dt}\frac{\partial H}{\partial y^{(1)i}} + ... + (-1)^{k-1}\frac{1}{(k-1)!}\frac{d^{k-1}}{dt^{k-1}}\frac{\partial H}{\partial y^{(k-1)i}}].$$

These equations are fundamental in the whole construction of the geometry of Hamiltonians of order k. They allow the introduction of the notion of energy of order $k-1,...,1$, $\mathcal{E}^{k-1}(H),...,\mathcal{E}^{1}(H)$ and prove a law of conservation for $\mathcal{E}^{k-1}(H)$ along the extremal curves. Now we can introduce in a natural way the Jacobi-Ostrogradski momenta and the Hamilton-Jacobi-Ostrogradski equations.

A theory of symmetries of the Hamiltonians H and the Noether type theorems are investigated, as well as a specific theory of tangent structure J and its adjoint J^*; canonical Poisson structure; the notion of dual semispray, which can be defined only by $k \geq 2$;the nonlinear connection N; the dual coefficients of N; the almost product structure $\mathbb{P}$, the almost contact structure $\mathbb{F}$ and Riemannian structure $\mathbb{G}$ are all studied.

We pay special attention to the theory of N-linear connections; curvatures and torsions; parallelism and structures equations.

Chapter 8 is devoted to the main subject of the book: Hamilton spaces of order k, $H^{(k)n} = (M, H(x, y^{(1)},.., y^{(k-1)}, p))$. To begin with, we prove the existence of these spaces and the existence of a natural presymplectic structure, as well as of a natural Poisson structure. Using the Legendre mapping from a Lagrange space of order k, $L^{(k)n} = (M, L)$ to the Hamilton space of order k, $H^{(k)n} = (M, H)$ it is proved that there is a local diffeomorphism between these spaces. As a direct consequence of previous results one can determine some important geometric object fields on the Hamilton spaces $H^{(k)n}$, namely the canonical nonlinear connection, the N-linear metrical connection given by generalized Christoffel symbols. The structure equations, curvatures and torsions of the above mentioned connections are pointed out. The Hamilton-Jacobi equations and an example from Electrodynamics end this chapter.

A theory of subspaces $\check{H}^{(k)m} = (\check{M}, \check{H})$ in the Hamilton spaces $H^{(k)n} = (M, H)$ appears for the first time in this book, in Chapter 9. Of course, it is absolutely necessary, especially for applications. But $\check{M}$ being a submanifold in the manifold M, the immersion $i : \check{M} \to M$ does not automatically imply an immersion of $T^{*k}\check{M}$ into the dual manifold $T^{*k}M$.

So, by means of an immersion of the cotangent bundle $T^*\check{M}$ into T^*M we construct $T^{*k}\check{M}$ as an immersed submanifold of the manifold $T^{*k}M$.

The Hamilton space $H^{(k)n} = (M, H)$ induces Hamilton subspaces $\check{H}^{(k)m} = (\check{M}, \check{H})$. So, we study the intrinsic geometrical object fields on $\check{H}^{(k)m}$ and the induced geometrical object fields, as well as the relations between them.

These problems are studied using the method of moving frame - suggested by the theory of subspaces in Lagrange spaces of order k. The Gauss-Weingarten formulae and the Gauss-Codazzi equations are important results.

In Chapter 10 we investigate the notion of Cartan space of order $k \geq 1$ as dual of that of Finsler space of same order. We point out the canonical linear connection, N-metrical connection, structure equations, the fundamental equations of Hamilton Jacobi and the Riemannian almost contact model of these spaces.

The last chapter, Chapter 11, is devoted to generalized Hamilton spaces $GH^{(k)n}$, generalized Cartan spaces $GC^{(k)n}$ and applications in the Hamiltonian relativistic optics.

The book can be divided into three parts: the Lagrange geometry of order k, presented in the first three chapters; the geometrical theory of the dual manifolds $T^{*k}M$ in chapters 4-7; and the geometry of Hamilton spaces of order k and their subspaces, contained in the last four chapters. They are studied directly and as 'dual' geometry, via Legendre transformation. More details for Lagrange geometry of order k can be found in the book [94]. Also, the particular case $k = 2$, of the geometry of Hamilton spaces $H^{(k)n}$ can be found in the book [115].

For these reasons, the book is accessible to readers from graduate students to researchers in mathematics, mechanics, physics, biology, informatics, etc.

Acknowledgments. I would like to express my gratitude to P.L. Antonelli, M. Anastasiei, A. Bejancu, M. Matsumoto, R.M. Santilli, P.S. Morey Jr. and Izu Vaisman for their continuous moral support and numerous valuable suggestions, as well as to my collaborators H. Shimada, D. Hrimiuc and V.S. Sabău, for the realization of the joint book [115].

Special thanks to I. Bucataru, M. Roman (University of Iaşi), to Ph. D. Assistants L. Popescu, F. Munteanu and to Professor P. Stavre (University of Craiova) who gave the manuscript a meticulous reading and typeset the manuscript into its final excellent form.

Finally, I should like to thank Kluwer Academic Publishers for their cooperation.

Chapter 1

Geometry of the k-Tangent Bundle T^kM

The notion of k- tangent bundle (or k-accelerations bundle or k- osculator bundle), (T^kM, π^k, M) is sufficiently known. It was presented in the book [115].

The manifold T^kM carries some geometrical object fields as the vertical distributions $V_1, ..., V_k$, the Liouville independent vector fields $\overset{1}{\Gamma}, ..., \overset{k}{\Gamma}$, with the properties $\overset{k}{\Gamma}$ belongs to V_1, $\overset{k-1}{\Gamma}$ belongs to V_2,... and $\overset{1}{\Gamma}$ belongs to the distribution V_k. On T^kM is defined a k- tangent structure J which maps $\overset{1}{\Gamma}$ on $\overset{2}{\Gamma}$, $\overset{2}{\Gamma}$ on $\overset{3}{\Gamma}$, $\overset{k-1}{\Gamma}$ on $\overset{k}{\Gamma}$ and $J\overset{k}{\Gamma}= 0$.

Besides these fundamental notion on T^kM we can introduce new concepts as the k- semisprays S, nonlinear connections N and the N- linear connections D. But for D we can get the curvatures, torsions, structure equations, geodesics, etc. The k- semispray S is defined by the conditions $JS = \overset{k}{\Gamma}$. It is important to remark that S is used for introducing those notions as nonlinear connection, or N- linear connection.

Concluding the geometry of k-accelerations bundle is basic for a geometrical theory of higher order Lagrange spaces or higher order Finsler spaces. In this book we need it for a theory of duality between higher order Lagrange spaces and higher order Hamilton spaces.

1.1 The Category of k-Accelerations Bundles

Let M be a real n-dimensional manifolds of C^∞ class and (T^kM, π^k, M) its bundle of accelerations of order k. It can be identified with the k-osculator bundles [94] or with the tangent bundle of order k. In the case $k = 1$, (T^1M, π^1, M) is the tangent bundle of the manifold M.

A point $u \in T^kM$ will be written as $u = (x, y^{(1)}, ..., y^{(k)})$ and $\pi^k(u) = x$, $x \in M$. The canonical coordinates of u are $(x^i, y^{(1)i}, ..., y^{(k)i})$, $i = \overline{1,n}$, $n = \dim M$.

These coordinates have a geometrical meaning. If $c : I \to M$ is a differentiable curve, $c(0) = x_0 \in M$, and $Im\ c \subset U$, U being a local chart of the base manifold M, and the mapping $c : I \to M$ is represented by $x^i = x^i(t)$, $t \in I$, then the osculating space of the curve c, in the point $x^i_0 = x^i(0)$ is characterized by the set of numbers:

$$(1.1.1) \qquad x^i_0 = x^i(0),\ y_0^{(1)i} = \frac{dx^i}{dt}(0), ..., y^{(k)i} = \frac{1}{k!}\frac{d^k x^i}{dt^k}(0)$$

Thus, the formulas (1.1.1) give us the canonical coordinates of a point $u_0 = (x_0, y_0^{(1)}, ..., y_0^{(k)})$ of the domain of the local chart $(\pi^k)^{-1}(U) \subset T^k M$.

Starting from (1.1.1) it is not difficult to see which are the changing rules of the local coordinates on $T^k M : (x^i, y^{(1)i}, ..., y^{(k)i}) \to (\widetilde{x}^i, \widetilde{y}^{(1)i}, ..., \widetilde{y}^{(k)i})$.

We deduce:

$$(1.1.2) \qquad \begin{cases} \widetilde{x}^i = \widetilde{x}^i(x^1, ..., x^n),\ rank \left\| \dfrac{\partial \widetilde{x}^i}{\partial x^j} \right\| = n \\ \widetilde{y}^{(1)i} = \dfrac{\partial \widetilde{x}^i}{\partial x^j} y^{(1)j} \\ 2\widetilde{y}^{(2)i} = \dfrac{\partial \widetilde{y}^{(1)i}}{\partial x^j} y^{(1)j} + 2\dfrac{\partial \widetilde{y}^{(1)i}}{\partial y^{(1)j}} y^{(2)j} \\ \dots\dots\dots\dots\dots\dots\dots\dots \\ k\widetilde{y}^{(k)i} = \dfrac{\partial \widetilde{y}^{(k-1)i}}{\partial x^j} y^{(1)j} + 2\dfrac{\partial \widetilde{y}^{(k-1)i}}{\partial y^{(1)j}} y^{(2)j} + ... + k\dfrac{\partial \widetilde{y}^{(k-1)i}}{\partial y^{(k-1)j}} y^{(k)j}. \end{cases}$$

But we must remark the following identities:

$$(1.1.3) \qquad \frac{\partial \widetilde{y}^{(\alpha)i}}{\partial x^j} = \frac{\partial \widetilde{y}^{(\alpha+1)i}}{\partial y^{(1)j}} = ... = \frac{\partial \widetilde{y}^{(k)i}}{\partial y^{(k-\alpha)j}},\ (\alpha = 0, ..., k-1;\ y^{(0)} = x).$$

In the following $T^0 M$ is canonically identified to M. Sometimes we employ the notations $y^{(0)i} = x^i$. The projections:

$$\pi^k_l : (x, y^{(1)}, ..., y^{(k)}) \in T^k M \to (x, y^{(1)}, ..., y^{(l)}) \in T^l M,\ (0 \leq l < k)$$

are submersions. Clearly $\pi^k_0 = \pi^k$.

A section $S : M \to T^k M$ of the projection π^k is a differentiable mapping with the property $\pi^k \circ S = 1_M$. It is a local section if $\pi^k \circ S\,|_U = 1_U$. Of course a section $S : M \to T^k M$ along a curve $c : I \to M$ has the property $\pi^k(S(c)) = c$.

If $c : I \to M$ is locally represent on $U \subset M$ by $x^i = x^i(t)$ then the mapping $\widetilde{c} : I \to T^k M$ given by:

$$(1.1.4) \qquad x^i = x^i(t),\ \ y^{(1)i} = \frac{1}{1!}\frac{dx^i}{dt}(t), ...,\ \ y^{(k)i} = \frac{1}{k!}\frac{dx^{(k)i}}{dt^k}(t),\ t \in I$$

is the extension of order k of c. Of course $\pi^k \circ \widetilde{c} = c$. So $\widetilde{c}$ is a section of π^k along c.

More general, if V is a vector field on the domain of a chart U and $c : I \to U$ is a curve, then the mapping
$S_V : c \to (\pi^k)^{-1}(U) \subset T^kM$, defined by

$$(1.1.5) \quad S_V : x^i = x^i(t),\ y^{(1)i} = V^i(x(t)), ...,\ y^{(k)i} = \frac{1}{k!}\frac{d^{k-1}V^i(x(t))}{dt^{k-1}},\ t \in I$$

is a section of the projection π^k along curve c.

Of course the notion of the section of π^k_l along T^lM can be defined, as in the previous case.

The following property hold:

Theorem 1.1.1 *If the differentiable manifold M is paracompact, then T^kM is a paracompact manifold.*

We can see, that

$$T^k : Man \to Man,$$

where Man is the category of differentiable manifolds, is a covariant functor.

Indeed we define:

$T^k : M \in ObMan \to T^kM \in ObMan$ and

$T^k : \{f : M \to M'\} \to \{T^kf : T^kM \to T^kM'\}$ as follows:

if $f(x)$ in the local coordinate of M is given by $x^{i'} = x^{i'}(x^1, ..., x^n)$, $i', j' = 1, ..., m = \dim M'$, then the morpfism $T^kf : T^kM \to T^kM'$ is defined by:

$$(1.1.6) \quad \begin{cases} x^{i'} = x^{i'}(x^1, ..., x^n), \\ y^{(1)i'} = \dfrac{\partial x^{i'}}{\partial x^j} y^{(1)j}, \\ 2y^{(2)i'} = \dfrac{\partial y^{(1)i'}}{\partial x^j} y^{(1)j} + 2\dfrac{\partial y^{(1)i'}}{\partial y^{(1)j}} y^{(2)j}, \\ \dots\dots\dots\dots\dots\dots\dots\dots\dots \\ ky^{(k)i'} = \dfrac{\partial y^{(k-1)i'}}{\partial x^j} y^{(1)j} + 2\dfrac{\partial y^{(k-1)i'}}{\partial y^{(1)j}} y^{(2)j} + ... + k\dfrac{\partial y^{(k-1)i'}}{\partial y^{(k-1)j}} y^{(k)j}. \end{cases}$$

Remarking that

$$(1.1.7) \quad \frac{\partial y^{(\alpha)i'}}{\partial x^j} = \frac{\partial y^{(\alpha+1)i'}}{\partial y^{(1)j}} = ... = \frac{\partial y^{(k)i'}}{\partial y^{(k-\alpha)j}},\ (\alpha = 0, ..., k-1;\ y^{(0)} = x),$$

we can prove without difficulties that T^k is a covariant functor.

1.2 Liouville Vector Fields. k-Semisprays

A local coordinate changing (1.1.2) transforms the natural basis of the tangent space $T_u(T^kM)$ by the following rule:

$$
\text{(1.2.1)}\quad
\begin{cases}
\dfrac{\partial}{\partial x^i} = \dfrac{\partial \widetilde{x}^j}{\partial x^i}\dfrac{\partial}{\partial \widetilde{x}^j} + \dfrac{\partial \widetilde{y}^{(1)j}}{\partial x^i}\dfrac{\partial}{\partial \widetilde{y}^{(1)j}} + \dots + \dfrac{\partial \widetilde{y}^{(k)j}}{\partial x^i}\dfrac{\partial}{\partial \widetilde{y}^{(k)j}} \\[2ex]
\dfrac{\partial}{\partial y^{(1)i}} = \dfrac{\partial \widetilde{y}^{(1)j}}{\partial y^{(1)i}}\dfrac{\partial}{\partial \widetilde{y}^{(1)j}} + \dots + \dfrac{\partial \widetilde{y}^{(k)j}}{\partial y^{(1)i}}\dfrac{\partial}{\partial \widetilde{y}^{(k)j}} \\
\dots\dots\dots\dots\dots\dots\dots\dots\dots\dots\dots\dots \\
\dfrac{\partial}{\partial y^{(k)i}} = \dfrac{\partial \widetilde{y}^{(k)j}}{\partial y^{(k)i}}\dfrac{\partial}{\partial \widetilde{y}^{(k)j}},
\end{cases}
$$

calculated at the point $u \in T^kM$.

These formulas imply the transformation of the natural cobasis at the point $u \in T^kM$ by the rule:

$$
\text{(1.2.1')}\quad
\begin{cases}
d\widetilde{x}^i = \dfrac{\partial \widetilde{x}^i}{\partial x^j}dx^j, \\[2ex]
d\widetilde{y}^{(1)i} = \dfrac{\partial \widetilde{y}^{(1)i}}{\partial x^j}dx^j + \dfrac{\partial \widetilde{y}^{(1)i}}{\partial y^{(1)j}}dy^{(1)j}, \\
\dots\dots\dots\dots\dots\dots\dots\dots\dots\dots\dots\dots \\
d\widetilde{y}^{(k)i} = \dfrac{\partial \widetilde{y}^{(k)i}}{\partial x^j}dx^j + \dfrac{\partial \widetilde{y}^{(k)i}}{\partial y^{(1)j}}dy^{(1)i} + \dots + \dfrac{\partial \widetilde{y}^{(k)i}}{\partial y^{(k)j}}dy^{(k)j}.
\end{cases}
$$

The matrix of coefficients of second member of (1.2.1) is the Jacobian matrix of the changing of coordinates (1.1.2). Since

$$
\frac{\partial \widetilde{x}^i}{\partial x^j} = \frac{\partial \widetilde{y}^{(1)i}}{\partial y^{(1)j}} = \dots = \frac{\partial \widetilde{y}^{(k)i}}{\partial y^{(k)j}},
$$

it follows.

Theorem 1.2.1 *If the number k is odd, then the manifold T^kM is orientable.*

Also the formulae (1.2.1), (1.2.1') allow to determine some important geometric object fields on the total space of accelerations bundle T^kM.

The distribution $V_1 : u \in T^kM \to V_{1,u} \subset T_u(T^kM)$ generated by the tangent vectors $\{\dfrac{\partial}{\partial y^{(1)i}}, \dots, \dfrac{\partial}{\partial y^{(k)i}}\}_u$, $\forall u \in T^kM$ is the vertical distribution on the bundle T^kM. Its dimension is kn.

$V_2 : u \in T^kM \to V_{2,u} \subset T_u(T^kM)$ generated by $\{\dfrac{\partial}{\partial y^{(2)i}}, \dots, \dfrac{\partial}{\partial y^{(k)i}}\}_u$, $\forall u \in T^kM$ is a subdistribution of V_1 of local dimension $(k-1)n$, ...,

$V_k : u \in T^kM \to V_{k,u} \subset T_u(T^kM)$ generated by $\{\frac{\partial}{\partial y^{(k)i}}\}_u$, $\forall u \in T^kM$ is a subdistribution of dimension n of the distribution V_{k-1}. All these subdistributions are integrable and the following sequence holds:

$$V_1 \supset V_2 \supset ... \supset V_k$$

Using again (1.2.1) we deduce:

Theorem 1.2.2 *The following operators in the algebra of functions $\mathcal{F}(T^kM)$:*

$$\tag{1.2.2} \begin{aligned} \overset{1}{\Gamma} &= y^{(1)i}\frac{\partial}{\partial y^{(k)i}}, \\ \overset{2}{\Gamma} &= y^{(1)i}\frac{\partial}{\partial y^{(k-1)i}} + 2y^{(2)i}\frac{\partial}{\partial y^{(k)i}}, \\ &\dots\dots\dots\dots \\ \overset{k}{\Gamma} &= y^{(1)i}\frac{\partial}{\partial y^{(1)i}} + 2y^{(2)i}\frac{\partial}{\partial y^{(2)i}} + ... + ky^{(k)i}\frac{\partial}{\partial y^{(k)i}} \end{aligned}$$

are vector fields globally defined on T^kM and independent on the manifold $\widetilde{T^kM} = T^kM \setminus \{0\}$, $\overset{1}{\Gamma}$ belongs to distribution V_k, $\overset{2}{\Gamma}$ belongs to distribution $V_{k-1}, ..., \overset{k}{\Gamma}$ belongs to distribution V_1.

Taking into account (1.2.1') it is not hard to prove:

Theorem 1.2.3 *For any differentiable function $L(x, y^{(1)}, ..., y^{(k)})$ on the manifold $\widetilde{T^kM}$, the following entries $d_0L, ..., d_kL$ are 1-form fields on $\widetilde{T^kM}$:*

$$\tag{1.2.2'} \begin{aligned} d_0L &= \frac{\partial L}{\partial y^{(k)i}}dx^i, \\ d_1L &= \frac{\partial L}{\partial y^{(k-1)i}}dx^i + \frac{\partial L}{\partial y^{(k)i}}dy^{(1)i}, \\ &\dots\dots\dots\dots \\ d_kL &= \frac{\partial L}{\partial x^i}dx^i + \frac{\partial L}{\partial y^{(1)i}}dy^{(1)i} + ... + \frac{\partial L}{\partial y^{(k)i}}dy^{(k)i}. \end{aligned}$$

Evidently, $d_kL = dL$ is the differential of the function L. But:
d_0L vanish on the distribution V_1,
d_1L vanish on the distribution V_2,
....
$d_{k-1}L$ vanish on the distribution V_k.
In applications we shall use also the following nonlinear operator

$$\tag{1.2.3} \Gamma = y^{(1)i}\frac{\partial}{\partial x^i} + 2y^{(2)i}\frac{\partial}{\partial y^{(1)i}} + ... + ky^{(k)i}\frac{\partial}{\partial y^{(k-1)i}}.$$

A k-tangent structure J on T^kM is defined as usual by the following $\mathcal{F}(T^kM)$-linear mapping $J : \mathcal{X}(T^kM) \to \mathcal{X}(T^kM)$:

$$J(\frac{\partial}{\partial x^i}) = \frac{\partial}{\partial y^{(1)i}}; \quad J(\frac{\partial}{\partial y^{(1)i}}) = \frac{\partial}{\partial y^{(2)i}}; ...;$$

(1.2.4)

$$J(\frac{\partial}{\partial y^{(k-1)i}}) = \frac{\partial}{\partial y^{(k)i}}; \quad J(\frac{\partial}{\partial y^{(k)i}}) = 0.$$

J is globally defined on T^kM. It is a tensor fields of type (1,1) on T^kM, locally expressed by

(1.2.4') $$J = \frac{\partial}{\partial y^{(1)i}} \otimes dx^i + \frac{\partial}{\partial y^{(2)i}} \otimes dy^{(1)i} + ... + \frac{\partial}{\partial y^{(k)i}} \otimes dy^{(k-1)i}.$$

The last form of J implies that J is an integrable structure. The k -tangent structure J has the following properties:

1°. $ImJ = V_1, \quad KerJ = V_k$.

2°. $rankJ = kn$.

3°. $J \overset{k}{\Gamma} = \overset{k-1}{\Gamma}, ..., J \overset{2}{\Gamma} = \overset{1}{\Gamma}, J \overset{1}{\Gamma} = 0$.

4°. $J \circ ... \circ J = 0$, $(k+1$ factors).

A k-semispray on the manifold T^kM is a vector field S on T^kM with the property

(1.2.5) $$JS = \overset{k}{\Gamma}.$$

The notion of local k-semispray is obvious.

Any k-semispray S can be uniquely written in the form

(1.2.6) $$S = y^{(1)i}\frac{\partial}{\partial x^i} + ... + ky^{(k)i}\frac{\partial}{\partial y^{(k-1)i}} - (k+1)G^i(x, y^{(1)}, ..., y^{(k)})\frac{\partial}{\partial y^{(k)i}},$$

or shortly

(1.2.6') $$S = \Gamma - (k+1)G^i(x, y^{(1)}, ..., y^{(k)})\frac{\partial}{\partial y^{(k)i}}.$$

The set of functions G^i is the set of *coefficients* of S. With respect to (1.1.2), G^i are transformed as following:

(1.2.6") $$(k+1)\widetilde{G}^i = (k+1)G^j\frac{\partial \widetilde{x}^i}{\partial x^j} - \Gamma \widetilde{y}^{(k)i}.$$

A curve $c : I \to M$ is called a *k-path* on M with respect to a k-semispray S if its extension $\widetilde{c}$ to T^kM is an integral curve of S. If S is given in the form (1.2.6), then its k-paths are characterized by the system of differential equations:

(1.2.7) $$\frac{d^{k+1}x^i}{dt^{k+1}} + (k+1)!G^i(x, \frac{dx}{dt}, ..., \frac{d^kx}{dt^k}) = 0.$$

Indeed, the solutions curves of S are given by the system of ordinary differential equations

$$\frac{dx^i}{dt} = y^{(1)i}, \quad \frac{dy^{(1)i}}{dt} = 2y^{(2)i}, ..., \quad \frac{dy^{(k-1)i}}{dt} = ky^{(k)i},$$

$$\frac{dy^{(k)i}}{dt} = -(k+1)G^i(x, y^{(1)}, ..., y^{(k)}).$$

If we eliminate $y^{(1)i}, ..., y^{(k)i}$ we obtain (1.2.7).

The previous theory will be used in the geometry of higher order Lagrange spaces, which is based on the regular Lagrangians of higher order.

Now, let us considered the *adjunct* k-tangent structure J^*. It is the endomorphism of the module $\mathcal{X}^*(T^kM)$, defined by:

$$J^*(dy^{(k)i}) = dy^{(k-1)i} = dy^{(k)i} \circ J,$$

$$J^*(dy^{(k-1)i}) = dy^{(k-2)i} = dy^{(k-1)i} \circ J,$$

$$\cdots\cdots\cdots\cdots\cdots\cdots\cdots\cdots\cdots\cdots$$

$$J^*(dy^{(2)i}) = dy^{(1)i} = dy^{(2)i} \circ J, \tag{1.2.8}$$

$$J^*(dy^{(1)i}) = dx^i = dy^{(1)i} \circ J,$$

$$J^*(dx^i) = 0.$$

By using the formula (1.2.1') is not difficult to prove:

Theorem 1.2.4 *J^* is globally defined on T^kM.*

If $\omega \in X^*(T^kM)$ is given by

$$\omega = \overset{(0)}{\omega}_i \, dx^i + \overset{(1)}{\omega}_i \, dy^{(1)i} + ... + \overset{(k)}{\omega}_i \, dy^{(k)i},$$

then

$$J^*\omega = \overset{(1)}{\omega}_i \, dx^i + ... + \overset{(k)}{\omega}_i \, dy^{(k-1)i}.$$

We put $J^*f = f$ for any function $f \in \mathcal{F}(T^kM)$ and observe that J^* is a tensor field, of type $(1,1)$ on T^kM. Namely, we have

$$J^* = dy^{(k-1)i} \otimes \frac{\partial}{\partial y^{(k)i}} + dy^{(k-2)i} \otimes \frac{\partial}{\partial y^{(k-1)i}} + ... + dx^i \otimes \frac{\partial}{\partial y^{(1)i}}. \tag{1.2.9}$$

The rank $\|J^*\| = kn$. J^* can be extended to an endomorphism of the exterior algebra $\wedge(T^kM)$, by putting

$$(J^*\omega)(X_1, ..., X_p) = \omega(JX_1, ..., JX_p), \forall \omega \in \wedge^p(T^kM). \tag{1.2.10}$$

The existence of J^* allow to introduce the so called *vertical* differential operators in the exterior algebra $\wedge(T^kM)$.

Indeed, let us consider the operators of differentiation d_k :

$$d_k = d = \frac{\partial}{\partial x^i}dx^i + \frac{\partial}{\partial y^{(1)i}}dy^{(1)i} + ... + \frac{\partial}{\partial y^{(k)i}}dy^{(k)i}.$$

We get:

$$(1.2.11)\qquad \begin{aligned} J^*d_k = d_{k-1} = &\quad \frac{\partial}{\partial y^{(1)i}}dx^i + ... + \frac{\partial}{\partial y^{(k)i}}dy^{(k-1)i}, \\ &\cdots\cdots\cdots\cdots\cdots\cdots \\ J^*d_1 = d_0 = &\quad \frac{\partial}{\partial y^{(k)i}}dx^i. \end{aligned}$$

It is not difficult to prove that the operators $d_k, d_{k-1}, ..., d_0$ do not depend on the changing of coordinates on the manifold T^kM. If $L(x, y^{(1)}, ..., y^{(k)})$ is a 1-form function on T^kM, then $d_kL, d_{k-1}L, ..., d_0L$ are differentiable given by (1.2.2').

But $d_0, d_1, ..., d_k$ can be extended to the exterior algebra $\wedge(TM)$ giving them restrictions to $\mathcal{F}(T^kM)$ and $\wedge^1(T^kM)$. So, we will take $d_0L, ..., d_kL$ expressed in (1.2.2') and

$$(1.2.12)\qquad d_\alpha(dy^{(\beta)i}) = 0, (\alpha, \beta = 0, 1, ...k; y^{(0)} = x).$$

Consequently $d_0, ..., d_k$ are the antiderivations of degree 1 in the exterior algebra $\wedge(T^kM)$. For instance, if $\omega \in \wedge^1(T^kM)$ and it is locally express by

$$\omega = \overset{(0)}{\omega}_i\, dx^i + \overset{(1)}{\omega}_i\, dy^{(1)i} + ... + \overset{(k)}{\omega}_i\, dy^{(k)i}$$

then

$$d_\alpha \omega = \sum_{\beta=0}^{k} d_\alpha \overset{(\beta)}{\omega}_i \wedge dy^{(\beta)i}, \ (\alpha = 0, ..., k).$$

It is not so difficult to see that the following properties hold:

$$(1.2.13)\qquad d_\alpha \circ d_\alpha = 0, \ (\alpha = 0, ..., k).$$

In the case $k = 1$, $\frac{1}{2}d_0L = \frac{1}{2}\frac{\partial L}{\partial y^i}dx^i = p_i dx^i$ is the Liouville 1- form and $\frac{1}{2}(d_1 \circ d_2)L = dp_i \wedge dx^i$ is the symplectic structure on TM.

1.3 Nonlinear Connections

The notion of nonlinear connection is also known, [94]. A subbundle HT^kM of the tangent bundle $(TT^kM, d\pi^k, T^kM)$ which is supplementary to the vertical subbundle V_1T^kM :

$$TT^kM = HT^kM \oplus V_1T^kM$$

is called a *nonlinear connection.*

The fibres of HT^kM, determine a horizontal distribution $N : u \in T^kM \to N_u = H_uT^kM \subset T_uT^kM, \forall u \in T^kM$, supplementary to the vertical distribution V_1, i.e:

$$T_uT^kM = N_u \oplus V_{1,u}, \quad \forall u \in T^kM. \tag{1.3.1}$$

If the base manifold M is paracompact then on T^kM there exist nonlinear connections. The dimension of the horizontal distribution N is $n = \dim M$.

Consider a nonlinear connection N on T^kM and denote by h and v the horizontal and vertical projectors with respect to the distributions N and V_1 :

$$h + v = I, h^2 = h, v^2 = v, hv = vh = 0.$$

As usual we denote

$$X^H = hX,\ X^V = vX,\ \forall X \in \mathcal{X}(T^kM).$$

An horizontal lift, with respect to N is a $\mathcal{F}(M)$−linear mapping $l_h : \mathcal{X}(M) \to \mathcal{X}(T^kM)$ which has the properties:

$$v \circ l_h = 0, d\pi^k \circ l_h = \mathrm{I_d}$$

There exists an unique local basis adapted to the horizontal distribution N. It is given by

$$\frac{\delta}{\delta x^i} = l_h(\frac{\partial}{\partial x^i}), (i = 1, ..., n). \tag{1.3.3}$$

The linearly independent vector fields $\dfrac{\delta}{\delta x^i}, (i = 1, ..., n)$ can be uniquely written in the form

$$\frac{\delta}{\delta x^i} = \frac{\partial}{\partial x^i} - \underset{(1)}{N}{}^j_i \frac{\partial}{\partial y^{(1)j}} - ... - \underset{(k)}{N}{}^j_i \frac{\partial}{\partial y^{(k)j}}. \tag{1.3.4}$$

The system of functions $(\underset{(1)}{N}{}^j_i, ..., \underset{(k)}{N}{}^j_i)$ gives the *coefficients* of the nonlinear connection N.

We remark that:

1) For $X = X^i(x)\dfrac{\partial}{\partial x^i} \in \mathcal{X}(M)$ the horizontal lifts is $l_h\ X = X^i\dfrac{\delta}{\delta x^i}$.

2) With respect to (1.1.2), we obtain

$$\frac{\delta}{\delta \widetilde{x}^i} = \frac{\partial x^j}{\partial \widetilde{x}^i}\frac{\delta}{\delta x^j}.$$

3) With respect to (1.1.2) the coefficients of N are transformed by the rule

$$\begin{aligned}
&\underset{(1)}{\widetilde{N}}{}^i_m \frac{\partial \widetilde{x}^m}{\partial x^j} = \underset{(1)}{N}{}^m_j \frac{\partial \widetilde{x}^i}{\partial x^m} - \frac{\partial \widetilde{y}^{(1)i}}{\partial x^j},\\
&\dots\dots\dots\dots\dots\dots\dots\dots\dots\dots\dots\dots\\
&\underset{(k)}{\widetilde{N}}{}^i_m \frac{\partial \widetilde{x}^m}{\partial x^j} = \underset{(k)}{N}{}^m_j \frac{\partial \widetilde{x}^i}{\partial x^m} + ... + \underset{(1)}{N}{}^m_j \frac{\partial \widetilde{y}^{(k-1)i}}{\partial x^m} - \frac{\partial \widetilde{y}^{(k)i}}{\partial x^j}.
\end{aligned} \tag{1.3.5}$$

The k-tangent structure J, defined in (1.2.4) applies the horizontal distribution N into a vertical distribution $N_1 \subset V_1$ of dimension n, supplementary to the distribution V_2. Then it applies the distribution N_1 in a distribution $N_2 \subset V_2$, supplementary to the distribution V_3 and so on. Of course, we have $\dim N_0 = \dim N_1 = \cdots = \dim N_{k-1} = n$.

Setting $N_0 = N$, we can write

$$N_1 = J(N_0), N_2 = J^2(N_0), ..., N_{k-1} = J^{k-1}(N_0) \tag{1.3.6}$$

and we obtain the following direct decomposition:

$$T_u T^k M = N_{0,u} \oplus N_{1,u} \oplus ... \oplus N_{k-1,u} \oplus V_{k,u}\ , \qquad \forall u \in T^k M. \tag{1.3.7}$$

An adapted basis to the distributions N_0, N_1, ..., N_{k-1}, V_k at a point $u \in T^k M$ is given by:

$$\frac{\delta}{\delta x^i}, \frac{\delta}{\delta y^{(1)i}}, \dots, \frac{\delta}{\delta y^{(k)i}}, \tag{1.3.8}$$

where

$$\frac{\delta}{\delta y^{(1)i}} = J\left(\frac{\delta}{\delta x^i}\right), \frac{\delta}{\delta y^{(2)i}} = J^2\left(\frac{\delta}{\delta x^i}\right), \dots, \frac{\delta}{\delta y^{(k)i}} = J^k\left(\frac{\delta}{\delta x^i}\right).$$

Therefore, using (1.2.4) and (1.3.4) we get:

$$\begin{aligned}
&\frac{\delta}{\delta x^i} = \frac{\partial}{\partial x^i} - \underset{(1)}{N}{}^j_i \frac{\partial}{\partial y^{(1)j}} - \cdots - \underset{(k)}{N}{}^j_i \frac{\partial}{\partial y^{(k)j}},\\
&\frac{\delta}{\delta y^{(1)i}} = \frac{\partial}{\partial y^{(1)i}} - \underset{(1)}{N}{}^j_i \frac{\partial}{\partial y^{(2)j}} - \cdots - \underset{(k-1)}{N}{}^j_i \frac{\partial}{\partial y^{(k)j}},\\
&\dots\dots\dots\dots\dots\dots\dots\dots\dots\dots\dots\dots\\
&\frac{\delta}{\delta y^{(k-1)i}} = \frac{\partial}{\partial y^{(k-1)i}} - \underset{(1)}{N}{}^j_i \frac{\partial}{\partial y^{(k)j}},\\
&\frac{\delta}{\delta y^{(k)i}} = \frac{\partial}{\partial y^{(k)i}}.
\end{aligned} \tag{1.3.9}$$

With respect to (1.1.2) we have:

$$\frac{\delta}{\delta y^{(\alpha)i}} = \frac{\partial \widetilde{x}^j}{\partial x^i} \frac{\delta}{\delta \widetilde{y}^{(\alpha)j}}, \quad (\alpha = 0, ..., k; y^{(0)} = x). \tag{1.3.10}$$

Taking into account the direct sum (1.3.1) and (1.3.7), it follows that the vertical distribution V_1 at a point u gives rise to the direct decomposition:

$$V_{1,u} = N_{1,u} \oplus ... \oplus N_{k-1,u} \oplus V_{k,u}\ , \qquad \forall u \in T^k M. \tag{1.3.11}$$

Let $h, v_1, ..., v_k$ be the projectors determined by (1.3.7):

$$h + v_1 + \cdots + v_k = I, h^2 = h, v_\alpha v_\alpha = v_\alpha, hv_\alpha = v_\alpha h = 0, (\alpha = 1, ..., k),$$

$$v_\alpha v_\beta = v_\beta v_\alpha = 0, \quad (\alpha \neq \beta; \alpha, \beta = 1, ..., k).$$

If we denote

$$X^H = hX,\ X^{V_\alpha} = v_\alpha X,\ \forall X \in \mathcal{X}(T^kM) \tag{1.3.12}$$

we have, uniquely,

$$X = X^H + X^{V_1} + ... + X^{V_k}. \tag{1.3.13}$$

In the adapted basis (1.3.8) we can write:

$$X^H = X^{(0)i}\frac{\delta}{\delta x^i}, X^{V_\alpha} = X^{(\alpha)i}\frac{\delta}{\delta y^{(\alpha)i}}, (\alpha = 1, ..., k).$$

The following properties are important in applications.
1) The distribution $N = N_0$ is integrable if, and only if

$$[X^H, Y^H]^{V_\alpha} = 0, (\alpha = 1, ..., k).$$

2) The distribution N_α is integrable, if and only if:

$$[X^{V_\alpha}, Y^{V_\alpha}]^H = 0,\ [X^{V_\alpha}, Y^{V_\alpha}]^{V_\beta} = 0,\ (\alpha \neq \beta\ , \alpha, \beta = 1, ..., k).$$

The notion of h- or v_α- lift of a vector fields $X = \mathcal{X}(M)$, $X = X^i(x)\dfrac{\partial}{\partial x^i}$ is obvious. We have:

$$l_h(X) = X^i(x)\frac{\delta}{\delta x^i},\ l_{v_\alpha}(X) = X^i(x)\frac{\delta}{\delta y^{(\alpha)i}}, (\alpha = 1, ..., k). \tag{1.3.14}$$

1.4 The Dual Coefficients of a Nonlinear Connection

Consider a nonlinear connection N, having the coefficients $(\underset{(1)}{N^i_j}, ..., \underset{(k)}{N^i_j})$. The adapted basis $(\dfrac{\delta}{\delta x^i}, \dfrac{\delta}{\delta y^{(1)i}}, ..., \dfrac{\delta}{\delta y^{(k)i}})$ to the direct decomposition (1.3.7) is expressed in the formulae (1.3.9). Its dual basis, denoted by

$$\delta x^i, \delta y^{(1)i}, ..., \delta y^{(k)i}, \tag{1.4.1}$$

can be uniquely written in the form:

$$\begin{aligned}&\delta x^i = dx^i,\\ &\delta y^{(1)i} = dy^{(1)i} + \underset{(1)}{M^i_j} dx^j,\\ &\dots\dots\dots\dots\dots\dots\dots\dots\dots\dots\\ &\delta y^{(k)i} = dy^{(k)i} + \underset{(1)}{M^i_j} dy^{(k-1)j} + \dots + \underset{(k-1)}{M^i_j} dy^{(1)j} + \underset{(k)}{M^i_j} dx^j,\end{aligned} \tag{1.4.2}$$

where

$$\begin{aligned}&\underset{(1)}{M^i_j} = \underset{(1)}{N^i_j},\ \underset{(2)}{M^i_j} = \underset{(2)}{N^i_j} + \underset{(1)}{N^m_j}\underset{(1)}{M^i_m}, \dots,\\ &\underset{(k)}{M^i_j} = \underset{(k)}{N^i_j} + \underset{(k-1)}{N^m_j}\underset{(1)}{M^i_m} + \dots + \underset{(1)}{N^m_j}\underset{(k-1)}{M^i_m}.\end{aligned} \tag{1.4.3}$$

The system of functions $(\underset{(1)}{M^i_j}, \dots, \underset{(k)}{M^i_j})$ is called the system of dual coefficients of the nonlinear connection N. They are determined entirely by means of the coefficients $(\underset{(1)}{N^i_j}, \dots, \underset{(k)}{N^i_j},)$.

Conversely, if the dual coefficients are given, then the coefficients $(\underset{(1)}{N^i_j}, \dots, \underset{(k)}{N^i_j},)$ are expressed by:

$$\begin{aligned}&\underset{(1)}{N^i_j} = \underset{(1)}{M^i_j},\ \underset{(2)}{N^i_j} = \underset{(2)}{M^i_j} - \underset{(1)}{M^i_m}\underset{(1)}{N^m_j}, \dots,\\ &\underset{(k)}{N^i_j} = \underset{(k)}{M^i_j} - \underset{(1)}{M^i_m}\underset{(k-1)}{N^m_j} - \dots - \underset{(k-1)}{M^i_m}\underset{(1)}{N^m_j}.\end{aligned} \tag{1.4.3'}$$

It follows, without difficulties, that with respect to (1.1.2) we have:

$$\delta\widetilde{x}^i = \frac{\partial\widetilde{x}^i}{\partial x^j}\delta x^j,\ \delta\widetilde{y}^{(\alpha)i} = \frac{\partial\widetilde{x}^i}{\partial x^j}\delta y^{(\alpha)j},\ (\alpha = 1, \dots, k). \tag{1.4.3''}$$

Also, with respect to (1.1.2) the dual coefficients are transformed by the rule:

(1.4.4)

$$\begin{cases}\underset{(1)}{M^m_j}\dfrac{\partial\widetilde{x}^i}{\partial x^m} = \underset{(1)}{\widetilde{M}^i_m}\dfrac{\partial\widetilde{x}^m}{\partial x^j} + \dfrac{\partial\widetilde{y}^{(1)i}}{\partial x^j},\\ \dots\dots\dots\dots\dots\dots\dots\dots\dots\dots\\ \underset{(k)}{M^m_j}\dfrac{\partial\widetilde{x}^i}{\partial x^m} = \underset{(k)}{\widetilde{M}^i_m}\dfrac{\partial\widetilde{x}^m}{\partial x^j} + \underset{(k-1)}{\widetilde{M}^i_m}\dfrac{\partial\widetilde{y}^{(1)m}}{\partial x^j} + \cdots + \underset{(1)}{\widetilde{M}^i_m}\dfrac{\partial\widetilde{y}^{(k-1)m}}{\partial x^j} + \dfrac{\partial\widetilde{y}^{(k)i}}{\partial x^j}.\end{cases}$$

The relations between the natural basis on the manifold T^kM and the adapted basis are immediately:

$$(1.4.5)\qquad \begin{aligned} &\frac{\partial}{\partial x^i} = \frac{\delta}{\delta x^i} + \underset{(1)}{M_i^j}\frac{\delta}{\delta y^{(1)j}} + \cdots + \underset{(k)}{M_i^j}\frac{\delta}{\delta y^{(k)j}},\\ &\frac{\partial}{\partial y^{(1)i}} = \frac{\delta}{\delta y^{(1)i}} + \underset{(1)}{M_i^j}\frac{\delta}{\delta y^{(2)j}} + \cdots + \underset{(k-1)}{M_i^j}\frac{\delta}{\delta y^{(k)j}},\\ &\cdots\cdots\cdots\cdots\cdots\cdots\cdots\cdots\\ &\frac{\partial}{\partial y^{(k)i}} = \frac{\delta}{\delta y^{(k)i}}. \end{aligned}$$

Similarly, we have:

$$(1.4.5')\qquad \begin{aligned} &dx^i = \delta x^i,\\ &dy^{(1)i} = \delta y^{(1)i} - \underset{(1)}{N_j^i}\delta x^j,\\ &\cdots\cdots\cdots\cdots\cdots\cdots\cdots\cdots\\ &dy^{(k)i} = \delta y^{(k)i} - \underset{(1)}{N_j^i}\delta y^{(k-1)j} - \cdots - \underset{(k-1)}{N_j^i}\delta y^{(1)j} - \underset{(k)}{N_j^i}\delta x^j. \end{aligned}$$

As an application we can prove:

Theorem 1.4.1 *1) The Liouville vector fields $\overset{1}{\Gamma}, \ldots, \overset{k}{\Gamma}$ can be expressed in the adapted basis (1.3.8) in the form*

$$(1.4.6)\qquad \begin{aligned} &\overset{1}{\Gamma} = z^{(1)i}\frac{\delta}{\delta y^{(k)i}},\\ &\overset{2}{\Gamma} = z^{(1)i}\frac{\delta}{\delta y^{(k-1)i}} + 2z^{(2)i}\frac{\delta}{\delta y^{(k)i}},\\ &\cdots\cdots\cdots\cdots\cdots\cdots\cdots\cdots\\ &\overset{k}{\Gamma} = z^{(1)i}\frac{\delta}{\delta y^{(1)i}} + 2z^{(2)i}\frac{\delta}{\delta y^{(2)i}} + \cdots + kz^{(k)i}\frac{\delta}{\delta y^{(k)i}}, \end{aligned}$$

where

$$(1.4.7)\qquad \begin{aligned} &z^{(1)i} = y^{(1)i},\ 2z^{(2)i} = 2y^{(2)i} + \underset{(1)}{M_m^i}y^{(1)m},\ \ldots,\\ &kz^{(k)i} = ky^{(k)i} + (k-1)\underset{(1)}{M_m^i}y^{(k-1)m} + \cdots + \underset{(k-1)}{M_m^i}y^{(1)m} \end{aligned}$$

2) With respect to (1.1.2) we have:

$$(1.4.7')\qquad \widetilde{z}^{(\alpha)i} = \frac{\partial\widetilde{x}^i}{\partial x^j}z^{(\alpha)j},\ (\alpha = 1, \ldots, k).$$

We note that the formulas (1.4.7') express the geometrical meaning of each entry $z^{(1)i}$, ...,$z^{(k)i}$. So, we call them *the Liouville distinguished vector fields* (shortly, *d-vector fields*). These vectors are important in the geometry of higher order Lagrangians.

A field of 1-form $\omega \in \mathcal{X}^*(T^kM)$ can be uniquely written as

$$\omega = \omega^H + \omega^{V_1} + \cdots + \omega^{V_k} \tag{1.4.8}$$

where

$$\omega^H = \omega \circ h,\ \omega^{V_\alpha} = \omega \circ v_\alpha,\ (\alpha = 1, ..., k). \tag{1.4.8'}$$

In the adapted cobasis (1.4.1) we get:

$$\omega^H = \omega_i^{(o)} \delta x^i,\ \omega^{V_\alpha} = \omega_i^{(\alpha)} \delta y^{(\alpha)i},\ (\alpha = 1, ..., k). \tag{1.4.8"}$$

For any function $f \in \mathcal{F}(T^kM)$, the 1-form df has the components:

$$df = (df)^H + (df)^{V_1} + \cdots + (df)^{V_k}.$$

Using (1.4.8") we obtain:

$$(df)^H = \frac{\delta f}{\delta x^i}\delta x^i,\ (df)^{V_\alpha} = \frac{\delta f}{\delta y^{(\alpha)i}}\delta y^{(\alpha)i},\ (\alpha = 1, ..., k). \tag{1.4.8"'}$$

Let $\gamma : I \to T^kM$ be a parametrized curve, locally expressed by

$$x^i = x^i(t),\ y^{(\alpha)i} = y^{(\alpha)i}(t),\ t \in I,\ (\alpha = 1, ..., k). \tag{1.4.9}$$

The tangent vector field can be expressed as:

$$\frac{d\gamma}{dt} = \left(\frac{d\gamma}{dt}\right)^H + \cdots + \left(\frac{d\gamma}{dt}\right)^{V_k} =$$

$$= \frac{dx^i}{dt}\frac{\delta}{\delta x^i} + \frac{\delta y^{(1)i}}{dt}\frac{\delta}{\delta y^{(1)i}} + \cdots + \frac{\delta y^{(k)i}}{dt}\frac{\delta}{\delta y^{(k)i}},$$

where, by means of (1.4.2) one can write:

$$\begin{aligned} &\frac{\delta y^{(1)i}}{dt} = \frac{dy^{(1)i}}{dt} + \underset{(1)}{M^i_j}\frac{dx^j}{dt}, \ldots, \\ &\ldots\ldots\ldots\ldots\ldots\ldots\ldots\ldots\ldots \\ &\frac{\delta y^{(k)i}}{dt} = \frac{dy^{(k)i}}{dt} + \underset{(1)}{M^i_j}\frac{dy^{(k-1)j}}{dt} + \cdots + \underset{(k)}{M^i_j}\frac{dx^j}{dt}. \end{aligned} \tag{1.4.10}$$

A parametrized curve γ is called *horizontal* if $\left(\frac{d\gamma}{dt}\right)^{V_\alpha} = 0$, $(\alpha = 1, .., k)$. It is characterized by the system of differential equations:

$$\frac{\delta y^{(1)i}}{dt} = \cdots = \frac{\delta y^{(k)i}}{dt} = 0. \tag{1.4.11}$$

A parametrized curve $c : I \to M$ on the base manifold M, analytically given by $x^i = x^i(t)$, $t \in I$, has its extension $\widetilde{c} : I \to T^kM$, given by:

$$x^i = x^i(t), y^{(1)i} = \frac{dx^i}{dt}, ..., y^{(k)i} = \frac{1}{k!}\frac{d^kx^i}{dt^k}.$$

A horizontal curve $\widetilde{c}$ is called an *autoparallel curve* of the nonlinear connection N.

Theorem 1.4.2 *The autoparallel curves of the nonlinear connection N are characterized by the system of differential equations*

$$\begin{aligned} &y^{(1)i} = \frac{dx^i}{dt}, \cdots, y^{(k)i} = \frac{1}{k!}\frac{d^kx^i}{dt^k}, \\ &\frac{\delta y^{(1)i}}{dt} = 0, \cdots, \frac{\delta y^{(k)i}}{dt} = 0. \end{aligned} \tag{1.4.12}$$

1.5 The Determination of a Nonlinear Connection

A nonlinear connection N on the manifold of accelerations of order k, T^kM can be determined by a k-semispray S or a Riemannian structure $g(x)$ defined on the base manifold, or by a Finslerian or Lagrangian structure over the manifold M.

A first result is as follows:

Theorem 1.5.1 *(R. Miron and Gh. Atanasiu,[94]) If a k-semispray S, with the coefficients $G^i(x, y^{(1)}, ..., y^{(k)})$ is given on T^kM, then the set of functions:*

$$\begin{aligned} &\underset{(1)}{M^i_j} = \frac{\partial G^i}{\partial y^{(k)j}}, \ \underset{(2)}{M^i_j} = \frac{1}{2}\left(S\underset{(1)}{M^i_j} + \underset{(1)}{M^i_m}\underset{(1)}{M^m_j}\right), ..., \\ &\underset{(k)}{M^i_j} = \frac{1}{k}\left(S\ \underset{(k-1)}{M^i_j} + \underset{(1)}{M^i_m}\underset{(k-1)}{M^m_j}\right) \end{aligned} \tag{1.5.1}$$

is the set of dual coefficients of a nonlinear connection N determined only by the k-semispray S.

The proof can be find in the book [94].

Other result obtained by I. Bucataru ([26,27]) is given in:

Theorem 1.5.2 *If S is a k-semispray with the coefficients G^i, then the following set of functions*

$$\underset{(1)}{M^{*i}_j} = \frac{\partial G^i}{\partial y^{(k)j}}, \underset{(2)}{M^{*i}_j} = \frac{\partial G^i}{\partial y^{(k-1)j}}, ..., \underset{(k)}{M^{*i}_j} = \frac{\partial G^i}{\partial y^{(1)j}} \tag{1.5.2}$$

define the dual coefficients of a nonlinear connection N^* determined only by the k-semispray S.

The problem is to determine a nonlinear connection N on T^kM from a Riemannian structure $g_{ij}(x)$, given on the base manifold M.

Let us consider $\gamma^i_{jk}(x)$ the Christoffel symbols of the tensor $g_{ij}(x)$. Then, we obtain, ([94]), without difficulties:

Theorem 1.5.3 *The following set of functions*

$$\begin{aligned} &\underset{(1)}{M^i_j}(x, y^{(1)}) = \gamma^i_{jm}(x) y^{(1)m}, \\ &\underset{(2)}{M^i_j}(x, y^{(1)}, y^{(2)}) = \frac{1}{2}\left(\Gamma \underset{(1)}{M^i_j} + \underset{(1)}{M^i_m}\underset{(1)}{M^m_j}\right), \\ &\cdots\cdots\cdots\cdots\cdots\cdots\cdots\cdots \\ &\underset{(k)}{M^i_j}(x, y^{(1)}, ..., y^{(k)}) = \frac{1}{k}\left(\Gamma \underset{(k-1)}{M^i_j} + \underset{(1)}{M^i_m}\underset{(k-1)}{M^m_j}\right), \end{aligned} \tag{1.5.3}$$

where Γ is the operator (1.2.3), has the properties:

1° It defines the dual coefficients of a nonlinear connection N on $\widetilde{T^kM}$, determined only by the Riemannian structure $g_{ij}(x)$.

2° $\underset{(1)}{M^i_j}(x, y^{(1)})$ depend linearly on $y^{(1)i}$, ..., $\underset{(k)}{M^i_j}(x, y^{(1)}, ..., y^{(k)})$ depend linearly on $y^{(k)i}$.

In the same manner we can determine a nonlinear connection on $\widetilde{T^kM}$ by means of a Finsler space. Namely, we can prove:

Theorem 1.5.4 *Let $N^i_j(x, y^{(1)})$ be the Cartan nonlinear connection of a Finsler space $F^n = (M, F(x, y^{(1)}))$. Then, the following set of functions*

$$\begin{aligned} &\underset{(1)}{M^i_j}(x, y^{(1)}) = N^i_j(x, y^{(1)}), \\ &\underset{(2)}{M^i_j}(x, y^{(1)}, y^{(2)}) = \frac{1}{2}\left(\Gamma \underset{(1)}{M^i_j} + \underset{(1)}{M^i_m}\underset{(1)}{M^m_j}\right), \\ &\cdots\cdots\cdots\cdots\cdots\cdots\cdots\cdots \\ &\underset{(k)}{M^i_j}(x, y^{(1)}, ..., y^{(k)}) = \frac{1}{k}\left(\Gamma \underset{(k-1)}{M^i_j} + \underset{(1)}{M^i_m}\underset{(k-1)}{M^m_j}\right), \end{aligned} \tag{1.5.4}$$

has the properties:

1° It defines the dual coefficients of a nonlinear connection N on $\widetilde{T^kM}$, determined only by the fundamental function $F(x, y^{(1)})$ of the Finsler space F^n.

2° $\underset{(2)}{M^i_j}$ depend linearly on $y^{(2)i}$, ..., $\underset{(k)}{M^i_j}$ depend linearly on $y^{(k)i}$. Γ being the operator (1.2.3).

Remark 1.5.1 *Theorems 1.5.3 and 1.5.4 prove the existence of the nonlinear connections on $\widetilde{T^kM}$ in the case when the base manifold is paracompact.*

Let us consider the adapted basis (1.3.8) and adapted cobasis (1.4.1) corresponding to the nonlinear connection with the dual coefficients (1.5.3).

Let $X^i(x)$ be a vector field on the manifold M. We obtain

$$
\begin{aligned}
&l_h X = X^i(x)\frac{\delta}{\delta x^i},\ l_{v_\alpha} X = X^i(x)\frac{\delta}{\delta y^{(\alpha)i}},\\
&(\alpha = 1, ..., n),\ \forall X = X^i(x)\frac{\partial}{\partial x^i}.
\end{aligned}
\tag{1.5.5}
$$

Of course $l_h X$, $l_{v_\alpha} X$ are h- and respectively v_α-lifts of the vector field $X = X^i(x)\dfrac{\partial}{\partial x^i}$.

Theorem 1.5.5 *If $g_{ij}(x)$ is a Riemannian structure on the base manifold M and N is a nonlinear connection with the dual coefficients (1.5.3) determined by $g_{ij}(x)$, then*

$$
\mathbb{G} = g_{ij}(x)dx^i \otimes dx^j + g_{ij}(x)\delta y^{(1)i} \otimes \delta y^{(1)j} + \cdots + g_{ij}(x)\delta y^{(k)i} \otimes \delta y^{(k)j} \tag{1.5.6}
$$

is a Riemannian structure on $\widetilde{T^kM}$ induced only by $g_{ij}(x)$.

The proof is not difficult. The Riemannian structure (1.5.6) is *the prolongation* to T^kM of the Riemannian structure $g_{ij}(x)$, [94].

Using the same nonlinear connection N with the dual coefficients (1.5.3) we can consider the $\mathcal{F}(T^kM)$-linear mapping

$$
\mathbb{F}\left(\frac{\delta}{\delta x^i}\right) = -\frac{\delta}{\delta y^{(k)i}}, \mathbb{F}\left(\frac{\delta}{\delta y^{(\alpha)i}}\right) = 0, (\alpha = 1, .., k), \mathbb{F}\left(\frac{\delta}{\delta y^{(k)i}}\right) = \frac{\delta}{\delta y^{(k)i}}
\tag{1.5.7}
$$

$$(i = 1, ..., n).$$

One proves:

1° $\mathbb{F}$ is globally defined on the manifold $\widetilde{T^kM}$.

2° $\mathbb{F}$ is a tensor field of type $(1,1)$:

$$\mathbb{F} = -\frac{\delta}{\delta y^{(k)i}} \otimes dx^i + \frac{\delta}{\delta x^i} \otimes \delta y^{(k)i}. \tag{1.5.8}$$

3° $Ker\ \mathbb{F} = N_1 \oplus \cdots \oplus N_{k-1}$, $Im\ \mathbb{F} = N_0 \oplus V_k$.

4° $rank\ \mathbb{F} = 2n$.

5° $\mathbb{F}^3 + \mathbb{F} = 0$.

6° $\mathbb{F}$ is determined only by $g_{ij}(x)$.

Concluding we have:

Theorem 1.5.6 *A Riemannian structure $g_{ij}(x)$ given on the base manifold determine an almost $(k-1)n$-contact structure $\mathbb{F}$ on $\widetilde{T^kM}$ depending only on $g_{ij}(x)$.*

Let $\left(\underset{(1)}{\xi_a}, ..., \underset{(k-1)}{\xi_a}\right)$, $(a = 1, ..., n)$ be a local basis adapted to the direct decomposition $N_1 \oplus \cdots \oplus N_{k-1}$ and $\left(\overset{(1)}{\eta^a}, ..., \overset{(k-1)}{\eta^a}\right)$, $(a = 1, ..., n)$ its dual.

In the book [94] is proved the result:

Theorem 1.5.7 *The set $\left(\mathbb{G}, \mathbb{F}, \underset{(1)}{\xi_a}, ..., \underset{(k-1)}{\xi_a}, \overset{(1)}{\eta^a}, ..., \overset{(k-1)}{\eta^a}\right)$ is a Riemannian almost $(k-1)n$-contact structure on the manifold $\widetilde{T^kM}$ determined only by the Riemannian structure $g_{ij}(x)$ defined on the base manifold M.*

A similar theory we can study for a Finsler space $F^n = (M, F(x, y^{(1)}))$ using the nonlinear connection (1.5.4) defined only by the fundamental function $F(x, y^{(1)})$.

Also, we can investigate the problem of determination of a nonlinear connection on T^kM by means of a Riemannian structure $\mathbb{G}$ given on the manifold T^kM.

One shows that: A Riemann structure $\mathbb{G}$ on T^kM determine a Riemannian almost $(k-1)n$-contact structure depending only on $\mathbb{G}$ (K. Matsumoto and R. Miron, [111]).

1.6 *d*-Tensor Fields. *N*-Linear Connections

Let N be a nonlinear connection on the manifold T^kM. Then, the direct decomposition (1.3.7) holds. With respect to (1.3.7) a vector field X on T^kM and a 1-form ω on T^kM can be uniquely written in the form (1.3.13) and (1.4.8), respectively.

A distinguished tensor field (shortly d-tensor) T on T^kM of type (r,s) is a tensor field T with the property

$$T\left(\overset{(1)}{\omega},...,\overset{(r)}{\omega},\underset{(1)}{X},...,\underset{(s)}{X}\right)=T\left(\overset{(1)}{\omega}{}^{H},...,\overset{(r)}{\omega}{}^{V_k},\underset{(1)}{X}{}^{H},...,\underset{(s)}{X}{}^{V_k}\right)$$

for any 1-forms $\overset{(1)}{\omega},...,\overset{(r)}{\omega}$ from $\mathcal{X}^*(T^kM)$ and any vector fields $\underset{(1)}{X},...,\underset{(s)}{X}$ from $\mathcal{X}(T^kM)$.

In the adapted cobasis (1.3.8) and its dual basis (1.4.1) a d-tensor field T has the components:

$$T^{i_1\ldots i_r}_{j_1\ldots j_s}(x,y^{(1)},...,y^{(k)})=T\left(\delta x^{i_1},...,\delta y^{(k)i_r},\frac{\delta}{\delta x^{j_1}},...,\frac{\delta}{\delta y^{(k)j_s}}\right).$$

Using the formulas (1.3.10) and (1.4.3")we deduce:

A transformation of coordinates (1.2) on T^kM implies a transformation of the components of the d-tensor field T by the classical rule

$$\widetilde{T}^{i_1\ldots i_r}_{j_1\ldots j_s}=\frac{\partial\widetilde{x}^{i_1}}{\partial x^{h_1}}\cdots\frac{\partial\widetilde{x}^{i_r}}{\partial x^{h_r}}\frac{\partial x^{k_1}}{\partial\widetilde{x}^{j_1}}\cdots\frac{\partial x^{k_s}}{\partial\widetilde{x}^{j_s}}T^{h_1\ldots h_r}_{k_1\ldots k_s}. \tag{1.6.1}$$

But, this fact is possible only for the components of a d-tensor in the adapted basis. The components of T in the natural basis $\left(\frac{\partial}{\partial x^i},\frac{\partial}{\partial y^{(1)i}},\cdots,\frac{\partial}{\partial y^{(k)i}}\right)$ and natural cobasis $\left(dx^i,dy^{(1)i},...,dy^{(k)i}\right)$ have very complicated rule of transformations with respect to the changing of local coordinates (1.1.2) on the manifold T^kM.

If $X\in\mathcal{X}(T^kM)$, its projections X^H, X^{V_1}, ..., X^{V_k} are d -vector fields and them components $X^{(0)i}$, $X^{(\alpha)i}$, $(\alpha=1,...,k)$ are called also d-vector fields.

As an example, we have that $z^{(1)i}$, ..., $z^{(k)i}$ from (1.4.6), are d-vector fields. They are called the Liouville d-vector fields.

For a 1-form ω, ω^H, ω^{V_α} from (1.4.8) are d-1-forms and them components $\omega^{(0)i}$, ..., $\omega^{(\alpha)i}$ from (1.4.8") are d-covector fields.

Of course, the set of d-tensor fields with respect to ordinary operation of addition and tensor product determines a tensor algebra on the ring of functions $\mathcal{F}(T^kM)$.

An N-linear connection on the total space of k-accelerations bundle T^kM is a linear connection D on T^kM with the properties:

(1) D preserves by parallelism the horizontal distribution N;

(2) The k-tangent structure J is absolutely parallel with respect to D.

As a consequence of this definition, the following theorems hold, [94]:

Theorem 1.6.1 *A linear connection D on the manifold T^kM is a N-linear connection if and only if the following properties hold:*

$$\begin{aligned}&\left(D_XY^H\right)^{V_\alpha}=0,\ \left(D_XY^{V_\alpha}\right)^H=0,\\&\left(D_XY^{V_\alpha}\right)^{V_\beta}=0,\ (\alpha\neq\beta,\alpha,\beta=1,...,k),\end{aligned} \tag{1.6.2}$$

(1.6.2') $$D_X(JY^H) = JD_XY^H, D_X(JY^{V_\alpha}) = JD_XY^{V_\alpha}, (\alpha = 1, ..., k),$$

for any $X, Y \in \mathcal{X}(T^kM)$.

If $X \in \mathcal{X}(T^kM)$ is written in the form (1.3.13), for any $Y \in \mathcal{X}(T^kM)$ and D an N-linear connection, we have:

(1.6.3) $$D_XY = D_{X^H}Y + D_{X^{V_1}}Y + \cdots + D_{X^{V_k}}Y.$$

Let us denote:

$$D_X^H = D_{X^H}, D_X^{V_1} = D_{X^{V_1}}, ..., D_X^{V_k} = D_{X^{V_k}}.$$

Consequently, we can write:

(1.6.3') $$D_XY = D_X^HY + D_X^{V_1}Y + \cdots + D_X^{V_k}Y.$$

The operators D_X^H, ..., $D_X^{V_\alpha}$ are not covariant derivations in the d-tensor algebra, since $D_X^Hf = X^Hf \neq Xf$.

But they have similar properties with the covariant derivative.

So, if T is a tensor field of type (r, s) we have:

(1.6.4)

$$(D_X^HT)\left(\overset{(1)}{\omega}, ..., \overset{(r)}{\omega}, \underset{(1)}{X}, ..., \underset{(s)}{X}\right) = X^HT\left(\overset{(1)}{\omega}, ..., \overset{(r)}{\omega}, \underset{(1)}{X}, ..., \underset{(s)}{X}\right) -$$
$$-T\left(D_X^H\overset{(1)}{\omega}, ..., \overset{(r)}{\omega}, \underset{(1)}{X}, ..., \underset{(s)}{X}\right) - ... - T\left(\overset{(1)}{\omega}, ..., D_X^H\overset{(r)}{\omega}, \underset{(1)}{X}, ..., \underset{(s)}{X}\right) -$$
$$-\left(\overset{(1)}{\omega}, ..., \overset{(r)}{\omega}, D_X^H\underset{(1)}{X}, ..., \underset{(s)}{X}\right) - ... - T\left(\overset{(1)}{\omega}, ..., \overset{(r)}{\omega}, \underset{(1)}{X}, ..., D_X^H\underset{(s)}{X}\right),$$

$$(D_X^{V_\alpha}T)\left(\overset{(1)}{\omega}, ..., \overset{(r)}{\omega}, \underset{(1)}{X}, ..., \underset{(s)}{X}\right) = X^{V_\alpha}T\left(\overset{(1)}{\omega}, ..., \overset{(r)}{\omega}, \underset{(1)}{X}, ..., \underset{(s)}{X}\right) -$$
$$-T\left(D_X^{V_\alpha}\overset{(1)}{\omega}, ..., \overset{(r)}{\omega}, \underset{(1)}{X}, ..., \underset{(s)}{X}\right) - ... - T\left(\overset{(1)}{\omega}, ..., D_X^{V_\alpha}\overset{(r)}{\omega}, \underset{(1)}{X}, ..., \underset{(s)}{X}\right) -$$
$$-\left(\overset{(1)}{\omega}, ..., \overset{(r)}{\omega}, D_X^{V_\alpha}\underset{(1)}{X}, ..., \underset{(s)}{X}\right) - ... - T\left(\overset{(1)}{\omega}, ..., \overset{(r)}{\omega}, \underset{(1)}{X}, ..., D_X^{V_\alpha}\underset{(s)}{X}\right).$$

For instance, the formula (1.6.4) for a 1-form $\omega \in \mathcal{X}^*(T^kM)$ leads to the following expressions of $D_X^H\omega$ and $D_X^{V_\alpha}\omega$:

(1.6.5)
$$(D_X^H\omega)(Y) = X^H\omega(Y) - \omega(D_X^HY),$$
$$(D_X^{V_\alpha}\omega)(Y) = X^{V_\alpha}\omega(Y) - \omega(D_X^{V_\alpha}Y), \ (\alpha = 1, ..., k).$$

In the adapted basis (1.3.13) an N-linear connection D can be uniquely represented in the form, [94]:

(1.6.6)

$$\begin{cases} D_{\frac{\delta}{\delta x^j}} \dfrac{\delta}{\delta x^i} = L^m_{ij} \dfrac{\delta}{\delta x^m}, \ D_{\frac{\delta}{\delta x^j}} \dfrac{\delta}{\delta y^{(\alpha)i}} = L^m_{ij} \dfrac{\delta}{\delta y^{(\alpha)m}}, (\alpha = \overline{1,k}), \\ D_{\frac{\delta}{\delta y^{(\beta)j}}} \dfrac{\delta}{\delta x^i} = \underset{(\beta)}{C^m_{ij}} \dfrac{\delta}{\delta x^m}, \ D_{\frac{\delta}{\delta y^{(\beta)j}}} \dfrac{\delta}{\delta y^{(\alpha)i}} = \underset{(\beta)}{C^m_{ij}} \dfrac{\delta}{\delta y^{(\alpha)m}}, (\alpha, \beta = \overline{1,k}). \end{cases}$$

The system of functions $(L^m_{ij}, \underset{(1)}{C^m_{ij}}, ..., \underset{(k)}{C^m_{ij}})$ represents *the coefficients* of D.

With respect to (1.1.2), L^m_{ij} are transformed by the same rule as the coefficients of a linear connection defined on the base manifold M. Others coefficients $\underset{(\alpha)}{C^m_{ij}}$, $\alpha = 1, .., k$ are transformed like d -tensors of type $(1, 2)$.

If T is a d-tensor field of type (r, s), represented in adapted basis in the form

$$T = T^{i_1 \ldots i_r}_{j_1 \ldots j_s} \frac{\delta}{\delta x^{i_1}} \otimes \cdots \otimes \frac{\delta}{\delta y^{(k)i_r}} \otimes dx^{j_1} \otimes \cdots \otimes \delta y^{(k)j_s} \tag{1.6.7}$$

and $X = X^H = X^i(x, y^{(1)}, ..., y^{(k)}) \dfrac{\delta}{\delta x^i}$, then the h-covariant derivative $D^H_X T$, by means of (1.6.6), is as follows

$$D^H_X T = X^m T^{i_1 \ldots i_r}_{j_1 \ldots j_s | m} \frac{\delta}{\delta x^{i_1}} \otimes \cdots \otimes \frac{\delta}{\delta y^{(k)i_r}} \otimes dx^{j_1} \otimes \cdots \otimes \delta y^{(k)j_s}, \tag{1.6.8}$$

where

$$T^{i_1 \ldots i_r}_{j_1 \ldots j_s | m} = \frac{\delta T^{i_1 \ldots i_r}_{j_1 \ldots j_s}}{\delta x^m} + L^{i_1}_{hm} T^{h i_2 \ldots i_r}_{j_1 \ldots j_s} + \cdots - L^h_{j_s m} T^{i_1 \ldots i_r}_{j_1 \ldots h}. \tag{1.6.9}$$

The operator "$|$" will be called *the h-covariant derivative*, with the same denomination as D^H_X .

Consider the v_α-covariant derivatives $D^{V_\alpha}_X$, for $X = \overset{(\alpha)}{X^i} \dfrac{\delta}{\delta y^{(\alpha)i}}$, $(\alpha = 1, ..., k)$. Then, using (1.6.6) and (1.6.7) we have:

$$D^{V_\alpha}_X T = \overset{(\alpha)}{X^m} T^{i_1 \ldots i_r}_{j_1 \ldots j_s} \overset{(\alpha)}{|}_m \frac{\delta}{\delta x^{i_1}} \otimes \cdots \otimes \frac{\delta}{\delta y^{(k)i_r}} \otimes dx^{j_1} \otimes \cdots \otimes \delta y^{(k)j_s}, \tag{1.6.8'}$$

where

$$T^{i_1 \ldots i_r}_{j_1 \ldots j_s} \overset{(\alpha)}{|}_m = \frac{\delta T^{i_1 \ldots i_r}_{j_1 \ldots j_s}}{\delta y^{(\alpha)m}} + \underset{(\alpha)}{C^{i_1}_{hm}} T^{h \ldots i_r}_{j_1 \ldots j_s} + \cdots - \underset{(\alpha)}{C^h_{j_s m}} T^{i_1 \ldots i_r}_{j_1 \ldots h}, \ (\alpha = 1, .., k). \tag{1.6.9'}$$

The operators "$\overset{(\alpha)}{|}$", in number of k are called *v_α-covariant derivatives*, with the same denominations as $D^{V_\alpha}_X$. Each of them has similar properties to those of h -covariant derivative. For instance

$$(fT^{\cdots}_{\cdots})\overset{(\alpha)}{|}_m = \frac{\delta f}{\delta y^{(\alpha)m}}T^{\cdots}_{\cdots} + fT^{\cdots}_{\cdots}\overset{(\alpha)}{|}_m,$$

$$\left(\overset{1}{T}{}^{\cdots}_{\cdots}\otimes\overset{2}{T}{}^{\cdots}_{\cdots}\right)\overset{(\alpha)}{|}_m = \overset{1}{T}{}^{\cdots}_{\cdots}\overset{(\alpha)}{|}_m\otimes\overset{2}{T}{}^{\cdots}_{\cdots} + \overset{1}{T}{}^{\cdots}_{\cdots}\otimes\overset{2}{T}{}^{\cdots}_{\cdots}\overset{(\alpha)}{|}_m.$$

The operators $_|$ and $\overset{(\alpha)}{|}$ commute with the operation of contraction, etc.

These operators applied to the Liouville d-vector fields $z^{(1)i}$, ..., $z^{(k)i}$ determine 'the deflection tensors' of D:

$$\overset{(\alpha)}{D}{}^i_j = z^{(\alpha)i}{}_{|j}; \quad \overset{(\beta\alpha)}{d}{}^i_j = z^{(\beta)i}\overset{(\alpha)}{|}_j. \tag{1.6.10}$$

1.7 Torsion and Curvature

The tensor of torsion $\mathbb{T}$ of an N-linear connection D:

$$\mathbb{T}(X,Y) = D_X Y - D_Y X - [X,Y], \quad \forall X,Y \in \mathcal{X}(T^k M), \tag{1.7.1}$$

for $X = X^H + X^{V_1} + \cdots + X^{V_k}$, $Y = Y^H + Y^{V_1} + \cdots + Y^{V_k}$, uniquely determines the following vector fields:

$$\mathbb{T}(X^H,Y^H), \mathbb{T}(X^H,Y^{V_\alpha}), \mathbb{T}(X^{V_\alpha},Y^{V_\beta}), (\alpha,\beta = 1,...,k). \tag{1.7.2}$$

Therefore:

The tensor of torsion $\mathbb{T}$ of the N-linear connection D is well determined by the following components, which are d-tensor fields of type $(1,2)$:

$$\begin{aligned} \mathbb{T}(X^H,Y^H) &= h\mathbb{T}(X^H,Y^H) + \sum_{\alpha=1}^{k} v_\alpha \mathbb{T}(X^H,Y^H), \\ \mathbb{T}(X^H,Y^{V_\beta}) &= h\mathbb{T}(X^H,Y^{V_\beta}) + \sum_{\alpha=1}^{k} v_\alpha \mathbb{T}(X^H,Y^{V_\beta}), \\ \mathbb{T}(X^{V_\alpha},Y^{V_\beta}) &= \sum_{\gamma=1}^{k} v_\gamma \mathbb{T}(X^{V_\alpha},Y^{V_\beta}). \end{aligned} \tag{1.7.3}$$

It is not difficult to prove that $h\mathbb{T}(X^{V_\alpha},Y^{V_\beta}) = 0$, $\forall\alpha,\beta = 1$, ..., k. The expressions of d-tensors of torsion are the following:

$$
(1.7.4)\qquad
\begin{cases}
h\mathbb{T}(X^H,Y^H) = D^H_X Y^H - D^H_Y X^H - [X^H,Y^H]^H,\\
v_\alpha\mathbb{T}(X^H,Y^H) = [X^H,Y^H]^{V_\alpha},
\end{cases}
\qquad
\begin{cases}
h\mathbb{T}(X^H,Y^{V_\beta}) = -\nabla_{Y^{V_\beta}}X^H - [X^H,Y^{V_\beta}]^H,\\
v_\alpha\mathbb{T}(X^H,Y^{V_\beta}) = \left(\nabla^H_X Y^{V_\beta}\right)^{V_\alpha} - [X^H,Y^{V_\beta}]^{V_\alpha},\\
v_\gamma\mathbb{T}(X^{V_\alpha},Y^{V_\beta}) = v_\gamma\left(\nabla^{V_\alpha}_X Y^{V_\beta} - \nabla_{Y^{V_\beta}}X^{V_\gamma} - [X^{V_\alpha},Y^{V_\beta}]\right).
\end{cases}
$$

In the adapted basis we set:

$$
h\mathbb{T}\left(\frac{\delta}{\delta x^h},\frac{\delta}{\delta x^j}\right) = \underset{(00)}{T^i_{jh}}\frac{\delta}{\delta x^i},\ v_\alpha\mathbb{T}\left(\frac{\delta}{\delta x^h},\frac{\delta}{\delta x^j}\right) = \underset{(0\alpha)}{T^i_{jh}}\frac{\delta}{\delta y^{(\alpha)i}},
$$

$$
(1.7.5)\qquad h\mathbb{T}\left(\frac{\delta}{\delta x^h},\frac{\delta}{\delta y^{(\beta)j}}\right) = \underset{(\beta 0)}{T^i_{jh}}\frac{\delta}{\delta x^i},\ v_\alpha\mathbb{T}\left(\frac{\delta}{\delta x^h},\frac{\delta}{\delta y^{(\beta)j}}\right) = \underset{(\beta\alpha)}{T^i_{jh}}\frac{\delta}{\delta y^{(\alpha)i}},
$$

$$
h\mathbb{T}\left(\frac{\delta}{\delta y^{(\alpha)h}},\frac{\delta}{\delta y^{(\beta)j}}\right) = 0,\ v_\gamma\mathbb{T}\left(\frac{\delta}{\delta y^{(\alpha)h}},\frac{\delta}{\delta y^{(\beta)j}}\right) = \underset{(\beta\alpha)}{\overset{\gamma}{T^i_{jh}}}\frac{\delta}{\delta y^{(\gamma)i}}.
$$

Using (1.7.4) and (1.6.6) as well as the following expressions of the Lie brackets, we can calculate the components of d-tensors of torsion.

The Lie brackets of the vector fields of the adapted basis are given by:

$$
\left[\frac{\delta}{\delta x^j},\frac{\delta}{\delta x^h}\right] = \underset{(01)}{R^i_{jh}}\frac{\delta}{\delta y^{(1)i}} + \cdots + \underset{(0k)}{R^i_{jh}}\frac{\delta}{\delta y^{(k)i}},
$$

$$
(1.7.6)\qquad \left[\frac{\delta}{\delta x^j},\frac{\delta}{\delta y^{(\alpha)h}}\right] = \underset{(\alpha 1)}{B^i_{jh}}\frac{\delta}{\delta y^{(1)i}} + \cdots + \underset{(\alpha k)}{B^i_{jh}}\frac{\delta}{\delta y^{(k)i}},
$$

$$
\left[\frac{\delta}{\delta y^{(\alpha)j}},\frac{\delta}{\delta y^{(\beta)h}}\right] = \underset{(\alpha\beta)}{\overset{(1)}{C^i_{jh}}}\frac{\delta}{\delta y^{(1)i}} + \cdots + \underset{(\alpha\beta)}{\overset{(k)}{C^i_{jh}}}\frac{\delta}{\delta y^{(k)i}},
$$

where the coefficients R, B, C can be calculated by means of the coefficients of the nonlinear connection N, [94].

We remark the following d-tensor of torsion

$$
(1.7.7)\qquad \underset{(00)}{T^i_{jk}} = L^i_{jk} - L^i_{kj},\quad \underset{(\alpha\alpha)}{\overset{\alpha}{T^i_{jk}}} = \underset{(\alpha)}{C^i_{jk}} - \underset{(\alpha)}{C^i_{kj}}\quad (\alpha = 1,\dots,k).
$$

For simplicity we denote

$$\underset{(00)}{T^i_{jk}} = \underset{(0)}{T^i_{jk}}, \quad \overset{\alpha}{\underset{(\alpha\alpha)}{T^i_{jk}}} = \underset{(\alpha)}{S^i_{jk}}. \tag{1.7.7'}$$

The notion of curvature can be investigated by the same method.

The curvature tensor $\mathbb{R}$ of the N-linear connection D is given by

$$\mathbb{R}(X,Y)Z = (D_X D_Y - D_Y D_X)\,Z - D_{[X,Y]}Z, \quad \forall X,Y,Z \in \mathcal{X}(T^kM). \tag{1.7.8}$$

Using the formula (1.6.2) we obtain

$$J\,(\mathbb{R}(X,Y)Z) = \mathbb{R}(X,Y)JZ, \dots, J^k\,(\mathbb{R}(X,Y)Z) = \mathbb{R}(X,Y)J^kZ. \tag{1.7.9}$$

Setting $Z^{V_\alpha} = J^\alpha Z^H$ we have

$$\mathbb{R}(X,Y)Z^{V_\alpha} = J^\alpha\left(\mathbb{R}(X,Y)Z^H\right), \quad (\alpha = 1, ..., k).$$

The essential components of the curvature tensor field $\mathbb{R}$ are $\mathbb{R}(X,Y)Z^H$. It has the properties:

$$v_\beta\left(\mathbb{R}(X,Y)Z^H\right) = 0, \quad h\left(\mathbb{R}(X,Y)Z^{V_\beta}\right) = 0,$$

$$v_\beta\left(\mathbb{R}(X,Y)Z^{V_\alpha}\right) = 0, \quad (\alpha \neq \beta, \alpha, \beta = 1, ..., k).$$

Thus, the curvature tensor $\mathbb{R}$ of the N-linear connection D gives rise to the d-vector fields:

$$\mathbb{R}(X^H,Y^H)Z^H = [D^H_X, D^H_Y]Z^H - D^H_{[X^H,Y^H]}Z^H - \sum_{\gamma=1}^{k} D^{V_\gamma}_{[X^H,Y^H]}Z^H,$$

$$\mathbb{R}(X^{V_\alpha},Y^H)Z^H = [D^{V_\alpha}_X, D^H_Y]Z^H - D^H_{[X^{V_\alpha},Y^H]}Z^H - \sum_{\gamma=1}^{k} D^{V_\gamma}_{[X^{V_\alpha},Y^H]}Z^H,$$

$$\mathbb{R}(X^{V_\beta},Y^{V_\alpha})Z^H = [D^{V_\beta}_X, D^{V_\alpha}_Y]Z^H - \sum_{\gamma=1}^{k} D^{V_\gamma}_{[X^{V_\beta},Y^{V_\alpha}]}Z^H, \tag{1.7.10}$$
$$(\alpha, \beta = 1, ..., k;\ \beta \leq \alpha).$$

The d-tensor fields (1.7.10) are obtained applying the operators J, J^2, ..., J^k and taking into account $J^\gamma Z^H = Z^{V_\gamma}$, $(\gamma = 1, ..., k)$.

In the applications it is suitable to consider the equalities (1.7.10) as Ricci identities, [94].

The local expressions of d-tensors of curvature in adapted basis are:

$$\begin{aligned}
&\mathbb{R}\left(\frac{\delta}{\delta x^m}, \frac{\delta}{\delta x^j}\right)\frac{\delta}{\delta x^h} = R^{\,i}_{h\ jm}\frac{\delta}{\delta x^i},\\
&\mathbb{R}\left(\frac{\delta}{\delta y^{(\alpha)m}}, \frac{\delta}{\delta x^j}\right)\frac{\delta}{\delta x^h} = \underset{(\alpha)}{P}{}^{\,i}_{h\ jm}\frac{\delta}{\delta x^i},\\
&\mathbb{R}\left(\frac{\delta}{\delta y^{(\beta)m}}, \frac{\delta}{\delta y^{(\alpha)j}}\right)\frac{\delta}{\delta x^h} = \underset{(\beta\alpha)}{S}{}^{\,i}_{h\ jm}\frac{\delta}{\delta x^i}.
\end{aligned} \tag{1.7.11}$$

Note that the other d-tensors of curvature (1.7.10) have the same components. The d-tensors (1.7.11) are called also *d-tensors of curvature.* They have the expressions:

$$R_{h\ jm}^{\ i} = \frac{\delta L^i_{hj}}{\delta x^m} - \frac{\delta L^i_{hm}}{\delta x^j} + L^p_{hj}L^i_{pm} - L^p_{hm}L^i_{pj} + \sum_{\gamma=1}^{k} \underset{(\gamma)}{C^i_{hp}} \underset{(0\gamma)}{R^p_{jm}}, \tag{1.7.12}$$

$$\underset{(\alpha)}{P}{}_{h\ jm}^{\ i} = \frac{\delta L^i_{hj}}{\delta y^{(\alpha)m}} - \frac{\delta \underset{(\alpha)}{C^i_{hm}}}{\delta x^j} + L^p_{hj}\underset{(\alpha)}{C^i_{pm}} - \underset{(\alpha)}{C^p_{hm}}L^i_{pj} + \sum_{\gamma=1}^{k} \underset{(\gamma)}{C^i_{hp}} \underset{(\alpha\gamma)}{B^p_{jm}},$$

$$\underset{(\beta\alpha)}{S}{}_{h\ jm}^{\ i} = \frac{\delta \underset{(\alpha)}{C^i_{hj}}}{\delta y^{(\beta)m}} - \frac{\delta \underset{(\beta)}{C^i_{hm}}}{\delta y^{(\alpha)j}} + \underset{(\alpha)}{C^p_{hj}}\underset{(\beta)}{C^i_{pm}} - \underset{(\beta)}{C^p_{hm}}\underset{(\alpha)}{C^i_{pj}} + \sum_{\gamma=1}^{k} \underset{(\gamma)}{C^i_{hp}} \underset{(\alpha\beta)}{\overset{\gamma}{C}{}^p_{jm}}$$

The connection 1-forms of the N-linear connection D given in ch. 7, §7 of the book [94] are:

$$\omega^i{}_j = L^i_{jh}dx^h + \underset{(1)}{C^i_{jh}}\delta y^{(1)h} + \cdots + \underset{(k)}{C^i_{jh}}\delta y^{(k)h}\ . \tag{1.7.13}$$

Therefore, the structure equations of D are given by the following theorem, [94]:

Theorem 1.7.1 *The structure equations of an N-linear connection D on the manifold T^kM are given by:*

$$\begin{aligned} &d(dx^i) - dx^m \wedge \omega^i{}_m = -\overset{(0)}{\Omega^i}\ , \\ &d(\delta y^{(\alpha)i}) - \delta y^{(\alpha)m} \wedge \omega^i{}_m = -\overset{(\alpha)}{\Omega^i}\ , \\ &d\omega^i{}_j - \omega^m_j \wedge \omega^i{}_m = -\Omega^i{}_j\ , \end{aligned} \tag{1.7.14}$$

where the 2-forms of torsion $\overset{(0)}{\Omega^i}$, $\overset{(\alpha)}{\Omega^i}$ *are*

$$\overset{(0)}{\Omega^i} = \frac{1}{2}\underset{(0)}{T^i_{jh}} dx^j \wedge dx^h + \underset{(1)}{C^i_{jh}} dx^j \wedge \delta y^{(1)h} + \cdots$$

$$+\underset{(k)}{C^i_{jh}} dx^j \wedge \delta x^{(k)h} ,$$

$$\overset{(\alpha)}{\Omega^i} = \frac{1}{2}\underset{(0\alpha)}{R^i_{jh}} dx^j \wedge dx^h + \sum_{\gamma=1}^{k} \underset{(\gamma\alpha)}{B^i_{jm}} dx^j \wedge \delta y^{(\gamma)m} + \tag{1.7.15}$$

$$+\sum_{\gamma=1}^{k} \overset{\beta}{\underset{(\alpha\gamma)}{C^i_{jm}}} \delta y^{(\beta)j} \wedge \delta y^{(\gamma)m} -$$

$$-\left[L^i_{jm} dx^j + \sum_{\gamma=1}^{k} \underset{(\gamma)}{C^i_{jm}} \delta y^{(\gamma)j} \wedge \delta y^{(\alpha)m}\right].$$

and where the 2-forms of curvature $\Omega^i{}_j$ *are:*

$$\Omega^i{}_j = \frac{1}{2}R^i_{j\ pq} dx^p \wedge dx^q +$$

$$+\sum_{\alpha=1}^{k} \underset{(\alpha)}{P}{}^{\ i}_{j\ pq} dx^p \wedge \delta y^{(\alpha)q} + \sum_{\alpha,\beta=1}^{k} \underset{(\alpha\beta)}{S}{}^{\ i}_{j\ pq} \delta y^{(\alpha)p} \wedge \delta y^{(\beta)q} . \tag{1.7.16}$$

The Bianchi identities of D can be derived from (1.7.14) applying the operator of exterior differentiation and calculating $d\,\overset{(0)}{\Omega^i}$, $d\,\overset{(\alpha)}{\Omega^i}$, $d\Omega^i{}_j$ from (1.7.15) and (1.7.16) modulo the system (1.7.14).

Chapter 2

Lagrange Spaces of Higher Order

We define the notion of higher order Lagrangian and the notion of integral of action. We investigate the variational problem for autonomous Lagrangians deriving the Euler-Lagrange equations. The notion of Lagrange space of order k is introduced by means of regular nondegenerate Lagrangian defined on the total space of the k-accelerations bundle T^kM. In this case the Craig-Synge equations determine a k-semispray, which depend only on the considered Lagrangian. Therefore the geometry of the

Lagrange space of order k is based on the mentioned k-semispray, [94].

2.1 Lagrangians of Order k

Let us consider the k-accelerations bundle (T^kM, π^k, M). In Analytical Mechanics M is the space of configurations of a physical system. A point $x = (x^i) \in U \subset M$ is called a *configuration*, a mapping $c : t \in I \to (x^i(t)) \in U$ is a *law of moving (evolution)*, t is called *time*, a couple (t, x) is an *event* and $\left(\frac{dx^i}{dt}, \frac{1}{2!}\frac{d^2x^i}{dt^2}, \dots, \frac{1}{k!}\frac{d^kx^i}{dt^k}\right)$ are the velocity and generalized accelerations of order 1, 2,..., $k-1$ (respectively). The factors $\frac{1}{h!}$ $(h = 1, ..., k)$ are introduced for convenience.

Throughout this book we omit the word *generalized* and say shortly, accelerations of order h for $\frac{1}{h!}\frac{d^hx^i}{dt^h}$, $h = 1, ..., k$. A law a moving $c : t \in I \to (x^i(t)) \in U \subset M$ will be called a *curve* parametrized by time t. As usual the curve $\widetilde{c} : t \in I \to \widetilde{c}(t) \in (\pi^k)^{-1}(U) \subset T^kM$,

$$\widetilde{c} : t \in I \to \left(x^i(t), \frac{dx^i}{dt}, \frac{1}{2!}\frac{d^2x^i}{dt^2}, \dots, \frac{1}{k!}\frac{d^kx^i}{dt^k}\right) \tag{2.1.1}$$

is the extension of the curve c to the total space T^kM of the k -acceleration bundle.

Definition 2.1.1 *A Lagrangian of order k, ($k \in \mathbf{N}^*$) is a mapping $L : T^kM \to \mathbf{R}$.*

This means that L is a real function $L(x, y^{(1)}, ..., y^{(k)})$ on T^kM. In the other words, with respect to a change of local coordinates on T^kM (1.1.2), we have

$$\widetilde{L}(\widetilde{x}, \widetilde{y}^{(1)}, ..., \widetilde{y}^{(k)}) = L(x, y^{(1)}, ..., y^{(k)}). \tag{2.1.2}$$

The previous definition is given for *autonomous Lagrangians*. A similar definition we have for *nonautonomous Lagrangians*. They are mappings $L : (t, x, y^{(1)}, ..., y^{(k)}) \in \mathbf{R}\times T^kM \to L(t, x, y^{(1)}, ..., y^{(k)}) \in \mathbf{R}$, which are real functions with respect to a change of coordinates on $\mathbf{R}\times T^kM$, $(t, x, y^{(1)}, ..., y^{(k)}) \to (\widetilde{t}, \widetilde{x}, \widetilde{y}^{(1)}, ..., \widetilde{y}^{(k)})$, where $t = \widetilde{t}$.

For us it is preferable to study the autonomous Lagrangians, because the notion of Lagrange space of order k is a geometrical one. But, one sees that the nonautonomous Lagrangians can be geometrized by means of the notion of

rheonomic Lagrange space of order k. Such kind of geometry can by constructed by the same methods as in the autonomous case.

In the following we investigate the Lagrangians $L(x, y^{(1)}, ..., y^{(k)})$ which have the property (2.1.2).

A Lagrangian of order k, $L : T^kM \to \mathbf{R}$ is called *differentiable* if it is of C^∞-class on $\widetilde{T^kM}$ and continuous on the null section of the projection $\pi^k : T^kM \to M$.

The Hessian of a differentiable Lagrangian L, with respect to the variables $y^{(k)i}$ on $\widetilde{T^kM}$ is the matrix $||g_{ij}||$, where

$$g_{ij} = \frac{1}{2}\frac{\partial^2 L}{\partial y^{(k)i}\partial y^{(k)j}}. \tag{2.1.3}$$

One can prove, [94] that g_{ij} is a d-tensor field, on the manifold $\widetilde{T^kM}$. This is covariant of order 2 and symmetric.

This property determines the geometrical covariance of the Hessian of L.

If

$$rank\ ||g_{ij}|| = n, \quad on\ \widetilde{T^kM} \tag{2.1.4}$$

we say that $L(x, y^{(1)}, ..., y^{(k)})$ is a *regular* (or *nondegenerate*) Lagrangian.

The existence of the regular Lagrangians of order k is assured by the following example.

Let $\gamma_{ij}(x)$ a Riemannian tensor field on the base manifold M and $z^{(k)i}$ the Liouville vector field on $\widetilde{T^kM}$ determined by the prolongation of order k of the Riemannian spaces $\mathcal{R}^n = (M, \gamma_{ij}(x))$, an arbitrary covector field b_i on $T^{k-1}M$ and a function b on $T^{k-1}M$. Then the Lagrangian of order k

$$\begin{aligned} L(x, y^{(1)}, ..., y^{(k)}) &= \gamma_{ij}(x) z^{(k)i} z^{(k)j} + b_i(x, y^{(1)}, ..., y^{(k-1)}) z^{(k)i} + \\ &+ b(x, y^{(1)}, ..., y^{(k-1)}) \end{aligned} \tag{2.1.5}$$

is a regular Lagrangian on T^kM. We have $g_{ij}(x, y^{(1)}, ..., y^{(k)}) = \gamma_{ij}(x)$.

Let us consider the scalar fields

$$I^1(L) = \mathcal{L}_{\overset{1}{\Gamma}} L, ..., I^k(L) = \mathcal{L}_{\overset{k}{\Gamma}} L, \tag{2.1.6}$$

where $\overset{1}{\Gamma}, ..., \overset{k}{\Gamma}$ are the Liouville vector fields on T^kM and $\mathcal{L}$ is the Lie operator of derivation. $I^1(L), ..., I^k(L)$ will be called the *main invariants* of the Lagrangian L.

For a smooth parametrized curve $c : [0, 1] \to M$ represented in a domain of a local chart by $x^i = x^i(t)$, $t \in [0, 1]$. The parameter t is called *time* and its extension to $\widetilde{T^kM}$ is $\widetilde{c}$, given by (2.1.1).

The integral of action for the differentiable Lagrangian $L(x, y^{(1)}, ..., y^{(k)})$ along curve c is defined by

$$I(c) = \int_0^1 L\left(x(t), \frac{dx}{dt}, \frac{1}{2!}\frac{d^2x}{dt^2}, \cdots, \frac{1}{k!}\frac{d^kx}{dt^k}\right) dt. \tag{2.1.7}$$

Ones proves, [94] the following important results:

Theorem 2.1.1 *The necessary conditions for the integral of action be independent on the parametrization of the curve c are*

$$I^1(L) = \cdots = I^{k-1}(L) = 0, I^k(L) = L. \tag{2.1.8}$$

The conditions (2.1.8) are called the *Zermello conditions*, [69, 94].

Theorem 2.1.2 *If the differentiable Lagrangian L of order k, $k > 1$, satisfies the Zermello conditions (2.1.8), then it is degenerate (singular), i.e.*

$$rank\ ||g_{ij}(x, y^{(1)}, ..., y^{(k)})|| < n, \text{ on } \widetilde{T^kM}. \tag{2.1.9}$$

2.2 Variational Problem

The variational problem concerning the functional $I(c)$ from (2.1.7) was studied in the book [94], ch. 8. So we shall present here only the corresponding results.

Theorem 2.2.1 *In order for the curve $c : t \in [0, 1] \to (x^i(t)) \in U \subset M$ to be an extremal curve for the integral of action $I(c)$ it is necessary that the following Euler-Lagrange equations hold:*

$$\begin{aligned} &\overset{\circ}{E}_i(L) := \frac{\partial L}{\partial x^i} - \frac{d}{dt}\frac{\partial L}{\partial y^{(1)i}} + \cdots + (-1)^k \frac{1}{k!}\frac{d^k}{dt^k}\frac{\partial L}{\partial y^{(k)i}} = 0, \\ &y^{(1)i} = \frac{dx^i}{dt}, \cdots, y^{(k)i} = \frac{1}{k!}\frac{d^kx^i}{dt^k}. \end{aligned} \tag{2.2.1}$$

An important remark regard the geometrical meaning of the system of functions $\overset{\circ}{E}_i(L)$. The following property holds:

Every $\overset{\circ}{E}_i(L)$ from (2.2.1) determines a d-covector field along curve c.

This means that, with respect to a change of local coordinates on the manifold T^kM, $\overset{\circ}{E}_i(L)$ obeys the transformation

$$\overset{\circ}{\widetilde{E}}_i(\widetilde{L})\frac{\partial \widetilde{x}^i}{\partial x^j} = \overset{\circ}{E}_j(L). \tag{2.2.2}$$

Let $c : t \in [0,1] \to (x^i(t)) \in U \subset M$ be a smooth curve (or a law of moving) parametrized by the time t and $V^i(x(t))$ a differentiable vector field on c.

The mapping $S_V : (x(t)) \in U \to S_V(x(t)) \in T^kM$, defined by

$$\begin{cases} x^i = x^i(t), \ t \in [0,1], \\ y^{(1)i} = V^i(x(t)), \ 2y^{(2)i} = \dfrac{1}{1!}\dfrac{dV^i}{dt}, ..., ky^{(k)i} = \dfrac{1}{(k-1)!}\dfrac{d^{k-1}V^i}{dt^{k-1}}, \end{cases} \tag{2.2.3}$$

is a section of the projection $\pi^k : T^kM \to M$.

It is not difficult to see that the operator

$$\frac{d_V}{dt} = V^i\frac{\partial}{\partial x^i} + \frac{dV^i}{dt}\frac{\partial}{\partial y^{(1)i}} + \cdots + \frac{1}{k!}\frac{d^kV^i}{dt^k}\frac{\partial}{\partial y^{(k)i}} \tag{2.2.4}$$

is invariant with respect to the coordinate transformations (1.1.2).

If $V^i = \dfrac{dx^i}{dt}$, then $\dfrac{d_V}{dt} = \dfrac{d}{dt}$ and for any Lagrangian $L(x, y^{(1)}, ..., y^{(k)})$, $\dfrac{d_VL}{dt}$ is a scalar field.

The action of the k-tangent structure J (cf. §2.4, ch. 1) on the operator $\dfrac{d_V}{dt}$ leads to k new operators:

$$I_V^k = J\left(\frac{d_V}{dt}\right), \ I_V^{k-1} = J\left(I_V^k\right), ..., I_V^1 = J\left(I_V^2\right), \ 0 = J\left(I_V^1\right), \tag{2.2.5}$$

where

$$I_V^k = V^i\frac{\partial}{\partial y^{(1)i}} + \frac{dV^i}{dt}\frac{\partial}{\partial y^{(2)i}} + \cdots + \frac{1}{(k-1)!}\frac{d^{k-1}V^i}{dt^{k-1}}\frac{\partial}{\partial y^{(k)i}}. \tag{2.2.5'}$$

But the previous operators are vector fields. Consequently, $I_V^1(L)$, ..., $I_V^k(L)$ are scalar fields. For $V^i = \dfrac{dx^i}{dt}$ they coincide with the main invariants $I^1(L)$, ..., $I^k(L)$.

The relations between the operators $\dfrac{d_V}{dt}$, I_V^1,..., I_V^k are expressed cf.[94] by:

Theorem 2.2.2 *The following identities hold:*

$$\begin{aligned} \frac{d_VL}{dt} &= V^i \overset{\circ}{E}_i(L) + \frac{d}{dt}I_V^k(L) - \frac{1}{2!}\frac{d^2}{dt^2}I_V^{k-1}(L) + \cdots + \\ &+ (-1)^{k-1}\frac{1}{k!}\frac{d^k}{dt^k}I_V^1(L). \end{aligned} \tag{2.2.6}$$

Besides of the covector field $\overset{\circ}{E_i}(L)$, Craig and Synge introduced other covectors, important in the variational problem of the integral of action. They are denoted by $\overset{1}{E_i}, .., \overset{k}{E_i}$ and are provided by:

Lemma 2.2.1

For any differentiable Lagrangian $L(x, y^{(1)}, ..., y^{(k)})$ and any differentiable function $\Phi(t)$ we have:

$$\overset{\circ}{E_i}(\Phi L) = \Phi \overset{\circ}{E_i}(L) + \frac{d\Phi}{dt}\overset{1}{E_i}(L) + \cdots + \frac{d^k\Phi}{dt^k}\overset{k}{E_i}(L), \tag{2.2.7}$$

where $\overset{1}{E_i}(L)$, ..., $\overset{k}{E_i}(L)$ are d-covector fields. They are expressed by the actions on L of the following operators:

$$\overset{\circ}{E_i} = \frac{\partial}{\partial x^i} - \frac{d}{dt}\frac{\partial}{\partial y^{(1)i}} + \cdots + (-1)^k \frac{1}{k!}\frac{d^k}{dt^k}\frac{\partial}{\partial y^{(k)i}},$$

$$(2.2.8)\quad \begin{aligned} \overset{1}{E_i} &= \sum_{\alpha=1}^{k} (-1)^{\alpha}\frac{1}{\alpha!}\binom{\alpha}{\alpha-1}\frac{d^{k-1}}{dt^{k-1}}\frac{\partial}{\partial y^{(\alpha)i}},\\ \overset{2}{E_i} &= \sum_{\alpha=2}^{k} (-1)^{\alpha}\frac{1}{\alpha!}\binom{\alpha}{\alpha-2}\frac{d^{k-2}}{dt^{k-2}}\frac{\partial}{\partial y^{(\alpha)i}},\\ &\cdots\cdots\cdots\cdots\cdots\cdots\\ \overset{k}{E_i} &= (-1)^k\frac{1}{k!}\frac{\partial}{\partial y^{(k)i}}. \end{aligned}$$

In the light of the above results we get:

Theorem 2.2.3 *For any differentiable Lagrangians $L(x, y^{(1)}, ..., y^{(k)})$ and any function $F(x, y^{(1)}, ..., y^{(k-1)})$ the following properties hold:*

$$\overset{0}{E_i}\left(L + \frac{dF}{dt}\right) = \overset{0}{E_i}(L), \tag{2.2.9}$$

$$\overset{0}{E_i}\left(\frac{dF}{dt}\right) = 0,\ \overset{1}{E_i}\left(\frac{dF}{dt}\right) = -\overset{0}{E_i}(F),\ ...,\ \overset{k}{E_i}\left(\frac{dF}{dt}\right) = -\overset{k-1}{E}_i(F). \tag{2.2.10}$$

A consequence of the property (2.2.9) is as follows.

Theorem 2.2.4 *The integral of action*

$$I(c) = \int_0^1 L dt,\ I'(c) = \int_0^1 \left(L + \frac{dF}{dt}\right) dt \tag{2.2.11}$$

determine the same Euler-Lagrange equations, (F depending on $(x, y^{(1)}, ..., y^{(k-1)})$).

2.3 Higher Order Energies

In the book [94] we introduce the notion of higher order energies of a Lagrangian $L(x, y^{(1)}, ..., y^{(k)})$.

Definition 2.3.1 *We call energies of order k, $k-1$, ..., 1 of the differentiable Lagrangian $L(x, y^{(1)}, ..., y^{(k)})$ the following invariants:*

$$\mathcal{E}^k(L) = I^k(L) - \frac{1}{2!}\frac{dI^{k-1}(L)}{dt} + \cdots + (-1)^{k-1}\frac{1}{k!}\frac{d^{k-1}I^1(L)}{dt^{k-1}} - L,$$

$$\mathcal{E}^{k-1}(L) = -\frac{1}{2!}I^{k-1}(L) + \frac{1}{3!}\frac{dI^{k-2}(L)}{dt}\cdots + (-1)^{k-1}\frac{1}{k!}\frac{d^{k-2}I^1(L)}{dt^{k-2}}, \tag{2.3.1}$$

....

$$\mathcal{E}^1(L) = (-1)^{k-1}\frac{1}{k!}I^1(L).$$

As we shall see, the energies $\mathcal{E}^k(L)$, ..., $\mathcal{E}^1(L)$ are involved in a Nöther theory of symmetries of the higher order Lagrangians.

The next theorem is well known:

Theorem 2.3.1 *For any differentiable Lagrangian $L(x, y^{(1)}, ..., y^{(k)})$ the following identity holds*

$$\frac{d\mathcal{E}^k(L)}{dt} = -\overset{0}{E_i}(L)\frac{dx^i}{dt}. \tag{2.3.2}$$

An immediate consequence of the previous theorem is the following *law of conservation*:

Theorem 2.3.2 *For any differentiable Lagrangian $L(x, y^{(1)}, ..., y^{(k)})$ the energy of order k, $\mathcal{E}^k(L)$ is conserved along every extremal curve of the Euler-Lagrange equations $\overset{0}{E_i}(L) = 0$.*

Theorem 2.2.4 says that the integrals of actions $I(c)$ and $I'(c)$ from (2.2.11) determine the same Euler-Lagrange equations, if F is a Lagrangian with the property $\frac{\partial F}{\partial y^{(k)i}} = 0$.

Definition 2.3.2 *A symmetry of a differentiable Lagrangian $L(x, y^{(1)}, ..., y^{(k)})$ is a C^∞-diffeomorphism $\varphi : M \times \mathbf{R} \to M \times \mathbf{R}$ which preserves the variational principle of the integral of action $I(c)$ from (2.1.7).*

One can consider the notion of local symmetry of the Lagrangian L, taking φ as local diffeomorphism. If $U \times (a, b)$ is a domain of a local chart on the manifold $M \times \mathbf{R}$, then we can express an infinitesimal diffeomorphism φ in the form

$$\begin{aligned} x'^i &= x^i + \varepsilon V^i(x, t), \quad (i = 1, ..., n) \\ t' &= t + \varepsilon\tau(x, t), \end{aligned} \tag{2.3.3}$$

where (V^i, τ) is a vector field on $U \times (a,b)$ defined along curve $c : t \in [0,1] \to (x^i(t), t) \in U \times (a,b)$ and ε is a real number, sufficiently small in absolute value, so that $Im\ \varphi \subset U \times (a,b)$.

The infinitesimal transformation (1.3.3) is a symmetry for the differentiable Lagrangian L if and only if for any differentiable function $F(x, y^{(1)}, ..., y^{(k-1)})$ the following equation holds:

$$L\left(x', \frac{dx'}{dt'}, \cdots, \frac{1}{k!}\frac{d^k x'}{dt'^k}\right) dt' = \left\{L\left(x, \frac{dx}{dt}, \cdots, \frac{1}{k!}\frac{d^k x}{dt^k}\right) + \right.$$
$$\left. + \quad F\left(x, \frac{dx}{dt}, \cdots, \frac{1}{(k-1)!}\frac{d^{k-1}x}{dt^{k-1}}\right)\right\} dt$$

One proves the following Nöther theorem, [94]:

Theorem 2.3.3 *For any infinitesimal symmetry (1.3.3) of a Lagrangian $L(x, y^{(1)}, ..., y^{(k)})$ and for any function $\Phi(x, y^{(1)}, ..., y^{(k-1)})$, the function:*

$$\mathcal{F}^k(L, \Phi) = I_V^k(L) - \frac{1}{2!}\frac{d}{dt}I_V^{k-1}(L) + \cdots + (-1)^{k-1}\frac{1}{k!}\frac{d^{k-1}}{dt^{k-1}}I_V^1(L) -$$
$$- \quad \tau\mathcal{E}^k(L) + \frac{d\tau}{dt}\mathcal{E}^{k-1}(L) + \cdots + (-1)^k\frac{d^{k-1}\tau}{dt^{k-1}}\mathcal{E}^1(L) - \Phi.$$

is conserved along extremal curves of the Euler-Lagrange equations $\overset{\circ}{E}_i(L) = 0$.

In particular, if the Zermello conditions (2.1.8) are satisfied, then the energies $\mathcal{E}^1(L)$, ..., $\mathcal{E}^k(L)$ vanish and the previous theorem has a simpler form.

2.4 Jacobi-Ostrogradski Momenta

We introduce the Jacobi-Ostrogradski momenta and the Hamilton - Jacobi - Ostrogradski equations. The main results on these are given without proofs, [94].

Consider the energy of order k, $\mathcal{E}^k(L)$ of the Lagrangian L from (2.3.1). Noticing that $\mathcal{E}^k(L)$ is a polynomial function of degree one in $\frac{dx^i}{dt}$, ..., $\frac{d^k x^i}{dt^k}$ we can write:

(2.4.1) $$\mathcal{E}^k(L) = p_{(1)i}\frac{dx^i}{dt} + p_{(2)i}\frac{d^2x^i}{dt^2} + \cdots + p_{(k)i}\frac{d^k x^i}{dt^k} - L,$$

where

$$p_{(1)i} = \frac{\partial L}{\partial y^{(1)i}} - \frac{1}{2!}\frac{d}{dt}\frac{\partial L}{\partial y^{(2)i}} + \cdots + (-1)^{k-1}\frac{1}{k!}\frac{d^{k-1}}{dt^{k-1}}\frac{\partial L}{\partial y^{(k)i}},$$

(2.4.2) $$p_{(2)i} = \frac{1}{2!}\frac{\partial L}{\partial y^{(2)i}} - \frac{1}{3!}\frac{d}{dt}\frac{\partial L}{\partial y^{(3)i}} + \cdots + (-1)^{k-2}\frac{1}{k!}\frac{d^{k-2}}{dt^{k-2}}\frac{\partial L}{\partial y^{(k)i}},$$

....

$$p_{(k)i} = \frac{1}{k!}\frac{\partial L}{\partial y^{(k)i}}.$$

$p_{(1)i}, ..., p_{(k)i}$ are called *the Jacobi-Ostrogradski momenta.*

M. de Léon and others, [73, 74] established the following important result.

Theorem 2.4.1 *Along extremal curves of Euler-Lagrange equations,* $\overset{\circ}{E}_i (L) = 0$, *the following Hamilton-Jacobi-Ostrogradski equations hold:*

$$\frac{\partial \mathcal{E}^k(L)}{\partial p_{(\alpha)i}} = \frac{d^\alpha x^i}{dt^\alpha}, \quad (\alpha = 1, ..., k),$$

$$\frac{\partial \mathcal{E}^k(L)}{\partial x^i} = -\frac{dp_{(1)i}}{dt}, \tag{2.4.3}$$

$$\frac{\partial \mathcal{E}^k(L)}{\partial y^{(\alpha)i}} = -\alpha!\frac{dp_{(\alpha+1)i}}{dt}, \quad (\alpha = 1, ..., k-1).$$

The Jacobi-Ostrogradski momenta $p_{(1)i}, ..., p_{(k)i}$ allow the introduction of the 1-forms:

$$\begin{aligned} p_{(1)} &= p_{(1)i}dx^i + p_{(2)i}dy^{(1)i} + \cdots + p_{(k)i}dy^{(k-1)i}, \\ p_{(2)} &= p_{(2)i}dx^i + p_{(3)i}dy^{(1)i} + \cdots + p_{(k)i}dy^{(k-2)i}, \\ &\cdots\cdots\cdots\cdots\cdots\cdots\cdots\cdots\cdots \\ p_{(k)} &= p_{(k)i}dx^i. \end{aligned} \tag{2.4.4}$$

The following properties hold:

$$J^*p_{(1)} = p_{(2)}, \;\; ...., \;\; J^*p_{(k-1)} = p_{(k)}.$$

2.5 Higher Order Lagrange Spaces

The notion of Lagrange space of order k is a natural extension of that of classical Lagrange space $L^n = (M, L(x,y))$. It was introduced by the author of this monograph, in the papers quoted in his book [94]. It was introduced by the author in the paper quoted in his book *The Geometry of Higher Order Lagrange Spaces. Applications to Mechanics and Physics*, (Kluwer, FTPH no. 82) entirely devoted to this subject.

Here we give, without prooofs, the main results from the geometry of higher order Lagrange spaces, [94].

We call a *Lagrange space of order* k a pair $L^{(k)n} = (M, L)$ formed by a real n-dimensional manifold M and a differentiable Lagrangian of order k, $L : (x, y^{(1)}, ..., y^{(k)}) \in T^kM \to L(x, y^{(1)}, ..., y^{(k)}) \in \mathbf{R}$ for which the Hessian with the entries

$$g_{ij}(x, y^{(1)}, ..., y^{(k)}) = \frac{1}{2}\frac{\partial^2 L}{\partial y^{(k)i}\partial y^{(k)j}}, \;\; on \;\widetilde{T^kM},$$

has the property

$$rank \; ||g_{ij}|| = n$$

and the quadratic form $\Psi = g_{ij}\xi^i\xi^j$ on $\widetilde{T^kM}$ has constant signature.

L is called the *fundamental function* and g_{ij} the *fundamental (or metric) tensor field* of the space $L^{(k)n}$.

The geometry of the pair $(T^kM, L(x, y^{(1)}, ..., y^{(k)}))$ is called the geometry of the space $L^{(k)} = (M, L)$.

Notice that the geometry of Lagrange space of order k is not coincident with the geometrization of the Lagrangians of order k, $L(x, y^{(1)}, ..., y^{(k)})$, which could be degenerate, i.e. *rank* $||g_{ij}|| < n$.

We shall study this geometry using the methods suggested by the Higher Order Lagrangian Mechanics [35] and by the geometry of higher order Finsler spaces [95]. Consequently, we determine a canonical semispray S and derive the main geometrical object field of the space $L^{(k)n}$ by means of S.

Consider the integral of action of the Lagrangian L, the fundamental function of the Lagrange space of order k, $L^{(k)n} = (M, L)$ and determine the covector fields $\overset{0}{E}_i(L)$, ..., $\overset{k-1}{E_i}(L)$, $\overset{k}{E}_i(L)$.

Then $\overset{0}{E}_i(L) = 0$ are just the Euler-Lagrange equations. Thus, $\mathcal{E}^k(L)$, ..., $\mathcal{E}^1(L)$ from (2.3.1) give us the *energies* of the space $L^{(k)n}$. Theorems 2.3.1 and 2.3.2 can be applied and Theorem 2.3.3 gives the infinitesimal symmetries of the considered space.

The following result is known, [94]:

Theorem 2.5.1 *The equations* $g^{ij}\overset{k-1}{E}_i(L) = 0$ *determine the k-semispray*

$$S = y^{(1)i}\frac{\partial}{\partial x^i} + 2y^{(2)i}\frac{\partial}{\partial y^{(1)i}} + \cdots + ky^{(k)}\frac{\partial}{\partial y^{(k-1)i}} - (k+1)G^i\frac{\partial}{\partial y^{(k)i}} \tag{2.5.1}$$

with the coefficients

$$(k+1)G^i = \frac{1}{2}g^{ij}\left\{\Gamma\left(\frac{\partial L}{\partial y^{(k)j}}\right) - \frac{\partial L}{\partial y^{(k-1)j}}\right\}, \tag{2.5.2}$$

Γ *being the operator (1.2.3).*

The k-semispray S depends on the fundamental function L of the space $L^{(k)n}$. It will be called canonical. If L is globally defined on T^kM, then S has the same property on $\widetilde{T^kM}$.

Taking into account Theorem 1.5.1 we find, [94]:

Theorem 2.5.2 *The set of functions*

$$\begin{gathered}\underset{(1)}{M^i_j} = \frac{\partial G^i}{\partial y^{(k)j}},\ \underset{(2)}{M^i_j} = \frac{1}{2}\left(S\underset{(1)}{M^i_j} + \underset{(1)}{M^i_m}\underset{(1)}{M^m_j}\right),\ \ldots, \\ \underset{(k)}{M^i_j} = \frac{1}{k}\left(S\underset{(k-1)}{M^i_j} + \underset{(1)}{M^i_m}\underset{(k-1)}{M^m_j}\right)\end{gathered} \tag{2.5.3}$$

are the dual coefficients of a nonlinear connection N determined only by the canonical semispray S.

N is called canonical, too.

Theorem 1.5.2 give us new dual coefficients $\underset{(1)}{M^{*i}_{j}}, ..., \underset{(k)}{M^{*i}_{j}}$, (I. Bucataru, [26]).

The coefficients $\underset{(1)}{N^i_j}, ..., \underset{(k)}{N^i_j}$ of N are obtained from (1.4.3').

The existence of the spaces $L^{(k)n} = (M, L)$, when M is a paracompact manifold is assured by the following examples.

Example 5.1 Let $g_{ij}(x)$ be a Riemannian tensor on M and the nonlinear connection N with the dual coefficients $\underset{(1)}{M^i_j}(x, y^{(1)}), ...\underset{(k)}{M^i_j}(x, y^{(1)}, ...y^{(k)})$ given by (1.5.3). Consider the Liouville d-vector field, $z^{(k)i}$ (Th. 21.4.1):

$$kz^{(k)i} = ky^{(k)i} + (k-1)\underset{(1)}{M^i_m}y^{(k-1)m} + ... + \underset{(k-1)}{M^i_m}y^{(1)m} \tag{2.5.4}$$

and remark that $z^{(k)i}$ is a *d*-vector field, linear in the vertical variables $y^{(k)i}$. Consequence, the function

$$L(x, y^{(1)}, ...y^{(k)}) = g_{ij}(x)z^{(k)i}z^{(k)j} \tag{2.5.5}$$

is a fundamental function of a Lagrange space $L^{(k)n}$. Its fundamental tensor field is exactly $g_{ij}(x)$.

Example 5.2 Let $\overset{\circ}{L}(x, y^{(1)})$ be the Lagrangian of electrodynamics

$$\overset{\circ}{L}(x, y^{(1)}) = mc\gamma_{ij}(x)y^{(1)i}y^{(1)j} + \frac{2e}{m}b_i(x)y^{(1)i} \tag{2.5.6}$$

m, c, e being the well known physical constants, $\gamma_{ij}(x)$ the gravitational potentials and $b_i(x)$ the electromagnetic potentials.

Let us consider the nonlinear connection N determined like in previous example. Then

$$L(x, y^{(1)}, ..., y^{(k)}) = mc\gamma_{ij}(x)z^{(k)i}z^{(k)j} + \frac{2e}{m}b_i(x)z^{(k)i} \tag{2.5.7}$$

is a fundamental function of a Lagrange space of order k, $L^{(k)n} = (M, L)$.

Evidently L from (2.5.7) is a particular case of the Lagrangian from (2.1.5). It will be called the *prolongation* of order k of the electrodynamic Lagrangian in (2.5.6).

In the end of this chapter we will study the geometry of the Lagrange space $L^{(k)n}$, with fundamental function (2.5.7).

The adapted basis $\left\{\frac{\delta}{\delta x^i}, \frac{\delta}{\delta y^{(1)i}}, ..., \frac{\delta}{\delta y^{(k)i}}\right\}$ determined by the canonical nonlinear connection N is given by (2.3.9) and its dual basis $\{\delta x^i, \delta y^{(1)i}, ..., \delta y^{(k)i}\}$ is expressed in the formulae (2.4.2).

The horizontal curves of the space $L^{(k)n}$ with respect to canonical nonlinear connection N are characterized by the system of differential equations

$$\frac{\delta y^{(1)i}}{dt} = \cdots = \frac{\delta y^{(k)i}}{dt} = 0.$$

In the light of these properties we determine the autoparallel curves of N by adding the conditions to the previous differential equations

$$y^{(1)i} = \frac{dx^i}{dt}, ..., y^{(k)i} = \frac{1}{k!}\frac{d^k x^i}{dt^k}.$$

Using the results from chapter 1 one can proves:

The canonical nonlinear connection N is integrable if, and only if the following equations hold

$$\underset{(01)}{R^i_{jk}} = \cdots = \underset{(0k)}{R^i_{jk}} = 0.$$

2.6 Canonical Metrical N-Connections

Consider the canonical nonlinear connection N of the Lagrange space of order k, $L^{(k)n} = (M, L)$. A linear connection D on $T^k M$ is called an N-linear connection if:

1) D preserves by parallelism the horizontal distribution N,

2) $DJ = 0$.

The coefficients of D with respect to the adapted basis denoted by $D\Gamma(N) = (L^i_{jh}, \underset{(1)}{C^i_{jh}}, ..., \underset{(k)}{C^i_{jh}})$, can be uniquely determined if D is compatible with the Riemannian (or pseudoriemannian) metric $\mathbb{G}$ on $T^k M$, determined by the fundamental tensor field g_{ij} of $L^{(k)n}$.

Namely, $\mathbb{G}$ is expressed in the adapted basis by

$$\mathbb{G} = g_{ij} dx^i \otimes dx^j + g_{ij}\delta y^{(1)i} \otimes \delta y^{(1)j} + \cdots + g_{ij}\delta y^{(k)i} \otimes \delta y^{(k)j}. \tag{2.6.1}$$

D is compatible with $\mathbb{G}$ if

$$D_X \mathbb{G} = 0, \forall X \in \mathcal{X}(\widetilde{T^k M}).$$

Theorem 2.6.1 *The following properties hold:*

1) There exists a unique N-linear connection D on $\widetilde{T^k M}$ verifying the axioms:

$$g_{ij|h} = 0, \ g_{ij} \overset{(1)}{|}_h = \cdots = g_{ij} \overset{(k)}{|}_h = 0, \tag{2.6.2}$$

$$L^i_{jh} = L^i_{hj}, \ \underset{(\alpha)}{C^i_{jh}} = \underset{(\alpha)}{C^i_{hj}}, \ (\alpha = 1, ...k). \tag{2.6.2'}$$

2) The coefficients $C\Gamma(N) = (L^i_{jh}, \underset{(1)}{C}{}^i_{jh}, ..., \underset{(k)}{C}{}^i_{jh})$ *of* D *are given by the generalized Christoffel symbols:*

$$(2.6.3)\qquad \begin{cases} L^m_{ij} = \dfrac{1}{2} g^{ms}\left(\dfrac{\delta g_{is}}{\delta x^j} + \dfrac{\delta g_{sj}}{\delta x^i} - \dfrac{\delta g_{ij}}{\delta x^s}\right), \\ \underset{(\alpha)}{C}{}^m_{ij} = \dfrac{1}{2} g^{ms}\left(\dfrac{\delta g_{is}}{\delta y^{(\alpha)j}} + \dfrac{\delta g_{sj}}{\delta y^{(\alpha)i}} - \dfrac{\delta g_{ij}}{\delta y^{(\alpha)s}}\right), \ (\alpha = 1, ..., k). \end{cases}$$

3) D *depends only on the fundamental function* L *of the space* $L^{(k)n}$.

The connection D from the previous theorem is called the *canonical metrical* N-connection of the Lagrange space of order k, $L^{(k)n}$.

The d-tensors of curvature of D satisfy the following identities:

$$g_{sj} R_i{}^s{}_{hm} + g_{is} R_j{}^s{}_{hm} = 0,$$

$$g_{sj} \underset{(\alpha)}{P}{}_i{}^s{}_{hm} + g_{is} \underset{(\alpha)}{P}{}_j{}^s{}_{hm} = 0,$$

$$g_{sj} \underset{(\alpha\beta)}{S}{}_i{}^s{}_{hm} + g_{is} \underset{(\alpha\beta)}{S}{}_j{}^s{}_{hm} = 0.$$

The connection 1-forms of D are

$$(2.6.4)\qquad \omega^i_j = L^i_{jh} dx^h + \sum_{\alpha=1}^{k} \underset{(\alpha)}{C}{}^i_{jh} \delta y^{(\alpha)h}.$$

Finally, the structure equations of the canonical metrical N-connection D are given by Theorem 1.7.1 in the conditions (2.6.3).

The Bianchi identities of D are obtained from the structure equations applying the operator of exterior differentiation and taking into account the same equations of structure.

Now let us consider the tensor field

$$(2.6.5)\qquad \mathbb{F} = -\frac{\delta}{\delta y^{(k)i}} \otimes dx^i + \frac{\delta}{\delta x^i} \otimes \delta y^{(k)i}.$$

We can see that $\mathbb{F}$ is globally defined on $\widetilde{T^k M}$ and $\mathbb{F}^3 + \mathbb{F} = 0$. The pair $(\mathbb{G}, \mathbb{F})$ determines a Riemannian almost $(k-1)n$-contact structure on $\widetilde{T^k M}$ depending only on then fundamental function L of the space $L^{(k)n}$.

Examples.

1°. $\mathcal{R}^{(k)n} = \mathrm{Prol}^k \mathcal{R}^n$.

Let N be the nonlinear connection with the dual coefficients (2.5.3). It is determined only by $g_{ij}(x)$. If $\{\delta x^i, \delta y^{(1)i}, ..., \delta y^{(k)i}\}$ is the adapted cobasis to N and V_1, then the formula (2.6.1) gives us a Riemannian structure $\mathbb{G}$ on

$\widetilde{T^kM}$ which depend on $g_{ij}(x)$, only. The space $\mathcal{R}^{(k)n} = (\widetilde{T^kM}, \mathbb{G})$ is called the prolongation to $\widetilde{T^kM}$ of the Riemannian structure $(M, g_{ij}(x))$.

Consider the k-Liouville vector field $z^{(k)i}$ from (2.5.4) constructed by means of (2.5.3).

In this case the pair $L^{(k)n} = (M, L(x, y^{(1)}, ..., y^{(k)})$ is a Lagrange space of order k, where $L = g_{ij}(x)z^{(k)i}z^{(k)j}$.

The fundamental tensor field of $L^{(k)n}$ is exactly $g_{ij}(x)$, because $z^{(k)i}$ is of the form $z^{(k)i} = y^{(k)i} + \lambda^i(x, y^{(1)}, ..., y^{(k-1)})$. The canonical metrical N-connection has the coefficients

$$L^i_{jh} = \gamma^i_{jh}(x), \ \underset{(\alpha)}{C^i_{jh}} = 0, \ (\alpha = 1, ..., k).$$

2°. $\mathcal{F}^{(k)n} = \text{Prol}^k F^n$. The prolongation of order k of a Finsler space $F^n = (M, F(x, y^{(1)}))$ leads to a second example of Lagrange space of order k.

Consider N the nonlinear connection with the dual coefficients (1.5.4). It is determined only by F^n. The N-lift of the fundamental tensor $g_{ij}(x, y^{(1)})$ of the space F^n is given by (2.6.1). It is a Riemannian metric on $\widetilde{T^kM}$ determined by F^n only. Thus,

$$L = g_{ij}(x, y^{(1)})z^{(k)i}z^{(k)j},$$

$z^{(k)i}$ being the k-Liouville vector fields corresponding to N, is the fundamental function of a space $L^{(k)n}$.

The fundamental tensor of space is $g_{ij}(x, y^{(1)i})$ and the canonical metrical N-connection has the coefficients

$$L^i_{jk} = F^i_{jk}(x, y^{(1)}), \ \underset{(1)}{C^i_{jk}} = C^i_{jk}, \ \underset{(\alpha)}{C^i_{jk}} = 0, \ (\alpha = 2, ..., k-1).$$

2.7 Generalized Lagrange Spaces of Order k

The notion of generalized Lagrange spaces of order k is a natural extension of that of space $L^{(k)n}$. It was used, [94], in the geometrical theory of the higher order relativistic optics.

Definition 2.7.1 *A generalized Lagrange spaces of order k is a pair $GL^{(k)n} = (M, g_{ij}(x, y^{(1)}, ..., y^{(k)}))$ formed by a real n -dimensional differentiable manifold M and a differentiable symmetric d-tensor field g_{ij} defined on $\widetilde{T^kM}$, having two properties:*

a. g_{ij} has a constant signature on $\widetilde{T^kM}$;

b. rank $||g_{ij}|| = n$ on $\widetilde{T^kM}$.

Of course, $GL^{(k)n}$ can be defined locally, in the case when g_{ij} is given on an open set $(\pi^k)^{-1}(U)$, $U \subset M$.

We say that g_{ij} is the *fundamental tensor* of the space $GL^{(k)n}$.

Notice that any Lagrange space of order k, $L^{(k)n} = (M, L)$ is a generalized Lagrange space $GL^{(k)n} = (M, g_{ij})$ with

$$g_{ij} = \frac{1}{2}\frac{\partial^2 L}{\partial y^{(k)i}\partial y^{(k)j}}. \tag{2.7.1}$$

But not and conversely. Indeed, is possible that the system of differential partial equations (2.7.1) does not admit any solution $L(x, y^{(1)}, ..., y^{(k)})$ when the d-tensor field $g_{ij}(x, y^{(1)}, ..., y^{(k)})$ is apriori given.

Let us consider the tensor field:

$$\underset{(k)}{C}{}_{ijh} = \frac{1}{2}\frac{\partial g_{ij}}{\partial y^{(k)h}}. \tag{2.7.2}$$

It is not hard to see that $\underset{(k)}{C}{}_{ijh}$ is a covariant of order three d-tensor field .

Easily follows:

Proposition 2.7.1

A necessary condition so that the system of differential partial equation (2.7.1) admits a solution $L(x, y^{(1)}, ..., y^{(k)})$ is that the d-tensor field $\underset{(k)}{C}{}_{ijh}$ be completely symmetric.

If the tensor $\underset{(k)}{C}{}_{ijh}$ is not completely symmetric we say that the space $GL^{(k)n}$ is not reducible to a Lagrange space of order k.

Example 2.7.1 Let $\mathcal{R}^n = (M, \gamma_{ij}(x))$ be a Riemannian space and $\sigma(x, y^{(1)}, ..., y^{(k)})$ a smooth function on T^kM with the property $\frac{\partial\sigma}{\partial y^{(k)i}} \neq 0$.

Consider the d-tensor field on T^kM:

$$g_{ij}(x, y^{(1)}, ..., y^{(k)}) = e^{2\sigma(x,y^{(1)},...,y^{(k)})}(\gamma_{ij} \circ \pi^k)(x, y^{(1)}, ..., y^{(k)}). \tag{2.7.3}$$

We can prove that the pair $GL^{(k)n} = (M, g_{ij})$ with g_{ij} from (2.7.3) is a generalized Lagrange space of order k, which is not reducible to a Lagrange space.

This example proves the existence of spaces $GL^{(k)n}$ on the paracompact manifolds. For $k = 1$, this space was introduced by R. Miron and R. Tavakol [100], [114].

Example 2.7.2 Consider again $\mathcal{R}^n = (M, \gamma_{ij}(x))$ and the Liouville d-vector field $z^{(k)i}$ of the space $Prol^k\mathcal{R}^n$. It is expressed by

$$kz^{(k)i} = ky^{(k)i} + (k-1)\underset{(1)}{M}{}^i_m y^{(k-1)m} + \cdots + \underset{(k-1)}{M}{}^i_m y^{(1)m}, \tag{2.7.4}$$

where the dual coefficients $\underset{(1)}{M}{}^i_j$, ..., $\underset{(k-1)}{M}{}^i_j$ are given by the formulae (2.5.3).

Evidently $z^{(k)i}$ linearly depends on $y^{(k)i}$. So that its covariant $z^{(k)}_i = \gamma_{ij}z^{(k)j}$ linearly depends on $y^{(k)i}$.

Assuming that there exists a function $n(x, y^{(1)}, ..., y^{(k)}) \geq 1$ on $T^k M$, we can construct the following d-tensor field

$$g_{ij}(x, y^{(1)}, ..., y^{(k)}) = \gamma_{ij}(x) + \left(1 - \frac{1}{n^2(x, y^{(1)}, ..., y^{(k)})}\right) z_i^{(k)} z_j^{(k)}. \tag{2.7.5}$$

One proves that the pair $GL^{(k)n} = (M, g_{ij})$, with g_{ij} from (2.7.5), is a generalized Lagrange space of order k, which is not reducible to a Lagrange space of order k.

In the case $k = 1$, this space was studied by the author [91, 100] and applied, together with T. Kawaguchi [104] in the relativistic geometrical optics.

In the previous two examples we have a natural canonical nonlinear connection, with the dual coefficients (2.5.3), determined by the space $\mathrm{Prol}^k \mathcal{R}^n$.

Returning to the generalized Lagrange spaces of order k, $GL^{(k)n}$, we remark the difficulty to find a nonlinear connection derived only from the fundamental tensor g_{ij} of the space.

Therefore we assume that a nonlinear connection N on $\widetilde{T^k M}$ is apriori given. Thus we, shall study the pair $(N, GL^{(k)n})$ using the same methods like in the geometry of the space $L^{(k)n}$.

Indeed, considering the adapted basis $\left(\frac{\delta}{\delta x^i}, \frac{\delta}{\delta y^{(1)i}}, \cdots, \frac{\delta}{\delta y^{(k)i}}\right)$ and its dual basis $\left(\delta x^i, \delta y^{(1)i}, ..., \delta y^{(k)i}\right)$, determined by the nonlinear connection, we define the N-lift of the fundamental tensor g_{ij}:

$$\mathbb{G} = g_{ij} dx^i \otimes dx^j + g_{ij} \delta y^{(1)i} \otimes \delta y^{(1)j} + \cdots + g_{ij} \delta y^{(k)i} \otimes \delta y^{(k)j}. \tag{2.7.6}$$

A N-linear connection compatible to G is furnished by the following theorem:

Theorem 2.7.1 *1^0 There exists an unique N-linear connection D for which*

$$\begin{aligned} &g_{ij|h} = 0, \quad g_{ij} \overset{(\alpha)}{|}_h = 0, \quad (\alpha = 1, ..., k), \\ &L^i_{jh} = L^i_{hj}, \quad \underset{(\alpha)}{C}{}^i_{jh} = \underset{(\alpha)}{C}{}^i_{hj}, \quad (\alpha = 1, ..., k). \end{aligned} \tag{2.7.7}$$

2^0 The coefficients $C\Gamma(N) = \left(L^i_{jh}, \underset{(1)}{C}{}^i_{jh}, ..., \underset{(k)}{C}{}^i_{jh}\right)$ of D are given by the generalized Christoffel symbols:

$$\begin{aligned} &L^m_{ij} = \frac{1}{2} g^{ms} \left(\frac{\delta g_{is}}{\delta x^j} + \frac{\delta g_{sj}}{\delta x^i} - \frac{\delta g_{ij}}{\delta x^s}\right), \\ &\underset{(\alpha)}{C}{}^m_{ij} = \frac{1}{2} g^{ms} \left(\frac{\delta g_{is}}{\delta y^{(\alpha)j}} + \frac{\delta g_{sj}}{\delta y^{(\alpha)i}} - \frac{\delta g_{ij}}{\delta y^{(\alpha)s}}\right), (\alpha = 1, ..., k). \end{aligned} \tag{2.7.7'}$$

The previous N-linear connection of the space $GL^{(k)n}$ is called metrical canonical N-linear connection.

The structure equations can be written, exactly as in the chapter 1 (theorem 1.7.1).

The tensor $\mathbb{F}$ on $\widetilde{T^kM}$:

$$\mathbb{F} = -\frac{\delta}{\delta y^{(k)i}} \otimes dx^i + \frac{\delta}{\delta x^i} \otimes \delta y^{(k)i}, \tag{2.7.8}$$

together with the metric tensor $\mathbb{G}$ from (2.7.6) determine a Riemannian $(k-1)n$-almost contact structure on $\widetilde{T^kM}$. It is 'the geometrical model' of the generalized Lagrange space of order k, $GL^{(k)n} = (M, g_{ij})$.

Chapter 3

Finsler Spaces of Order k

The geometry of Finsler spaces of order k, introduced by the author and presented in his book 'The geometry of Higher- Order Finsler Spaces' Hadronic Press, 1998 is a natural extension to T^kM of the classical theory of Finsler Spaces. The impact of this geometry in Differential Geometry, Variational Calculus, Analytical Mechanics or Theoretical Physics is decisive. Finsler spaces play a role in applications to Biology, Engineering, Physics or Optimal Control. Also the introduction of the notion of Finsler space of order k is demanded by the solution of problem of prolongation to T^kM of the Riemannian or Finslerian structures defined on the base manifold M.

In the present chapter we will develop the geometrical theory of the Finsler spaces of order k, based on the geometry of Lagrange spaces of the same order. Such that the Finsler spaces $F^{(k)n}$ form a subclass of the class of spaces $L^{(k)n}$. Consequently we obtain an extension of the sequence

$\{R^n\} \subset \{F^n\} \subset \{L^n\} \subset \{GL^n\}$ to the following sequence of the spaces of order k :

$\{R^{(k)n}\} \subset \{F^{(k)n}\} \subset \{L^{(k)n}\} \subset \{GL^{(k)n}\}$.

In the next chapter we shall investigate the dual of this sequence, made by the Hamilton spaces of order k.

3.1 Spaces $F^{(k)n}$

In order to introduce the notion of Finsler space of order k there are necessary some preliminaries. A functions $f : T^kM \to R$, of C^∞ class on $\widetilde{T^kM}$ and continuous on the null sections of the mapping $\pi^k : T^kM \to M$ is called homogeneous of degree $r \in Z$ on the fibres on T^kM (briefly r-homogeneous) if for any positive constant a, we have:

$$f(x, ay^{(1)}, a^2y^{(2)}, ..., a^ky^{(k)}) = a^r f(x, y^{(1)}, ..., y^{(k)}). \tag{3.1.1}$$

An Euler theorem holds:

A function $f \in \mathcal{F}(T^kM)$, differentiable on $\widetilde{T^kM}$ and continuous on the null section of π^k is $r-$homogeneous if and only if:

$$\mathcal{L}_{\overset{k}{\Gamma}} f = rf, \tag{3.1.2}$$

$\mathcal{L}_{\overset{k}{\Gamma}}$ being the Lie operator of derivation with respect to the Lioville vector field $\overset{k}{\Gamma}$, i.e.:

$$\mathcal{L}_{\overset{k}{\Gamma}} f = \overset{k}{\Gamma} f = y^{(1)i}\frac{\partial f}{\partial y^{(1)i}} + ... + ky^{(k)i}\frac{\partial f}{\partial y^{(k)i}}. \tag{3.1.2'}$$

Notice the following property:

If the function $f \in \mathcal{F}(T^kM)$ is differentiable on T^kM (inclusive in the points $(x, 0, ..., 0)$) and $k-$homogeneous then f is a polynom of order k in the variable $y^{(1)i}, y^{(2)i}, ..., y^{(k)i}$.

The notion of homogeneity can be extended to the vector fields and $1-$form fields on T^kM. One proves:

A vector field $X \in \mathcal{X}(T^kM)$ is r-homogeneous if and only if

$$\mathcal{L}_{\overset{k}{\Gamma}} X = (r-1)X. \tag{3.1.3}$$

Of course, $\mathcal{L}_{\overset{k}{\Gamma}} X = [\overset{k}{\Gamma}, X]$.

For instance, the vector fields $\dfrac{\partial}{\partial x^i}, \dfrac{\partial}{\partial y^{(1)i}}, ..., \dfrac{\partial}{\partial y^{(k)i}}$ are $1, 0, ..., 1-k$ homogeneous, respectively.

An homogeneous k- semispray S is called a $k-$ spray. The following sentences hold:

1°. A $k-$ spray S given by

$$S = y^{(1)i}\frac{\partial f}{\partial x^i} + ... + ky^{(k)i}\frac{\partial f}{\partial y^{(k-1)i}} - (k+1)G^i\frac{\partial}{\partial y^{(k)i}} \tag{3.1.4}$$

is 2-homogeneous if and only if its coefficients G^i are $k+1$- homogeneous.

2°. The dual coefficients $\underset{(1)}{M^i_j}, \underset{(2)}{M^i_j}, ..., \underset{(k)}{M^i_j}$ of the nonlinear connection N, (according to Theorem 2.5.2, ch.2) determined by a 2 -homogeneous spray S are $1, 2, ..., k$-homogeneous, respectively.

3°. The same property have the coefficients $\underset{(1)}{N^i_j}, ..., \underset{(k)}{N^i_j}$ of the nonlinear connection N.

4°. The local adapted basis to N, $\dfrac{\delta}{\delta x^i}, \dfrac{\delta}{\delta y^{(1)i}}, ..., \dfrac{\delta}{\delta y^{(k)i}}$ has the following degree of homogeneity 1,0, ...,1- k, respectively.

5°. If $X \in \mathcal{X}$ $(\widetilde{T^kM})$ is r-homogeneous and $f \in \mathcal{F}(\widetilde{T^kM})$ is s-homogeneous, then $Xf$is $s+r-1$-homogeneous.

Evidently the 1-forms $dx^i, dy^{(1)i}, ..., dy^{(k)i}$ are homogeneous of degree $0, 1, ..., k$, respectively. The same degree of homogeneity have $\delta x^i, \delta y^{(1)i}, ..., \delta y^{(k)i}$ (the dual basis adapted to N).

A q-form $\omega \in \Lambda^q(\widetilde{T^kM})$ is s-homogeneous, if and only if

$$\mathcal{L}_{\underset{\Gamma}{k}}\omega = s\omega. \tag{3.1.5}$$

Evidently, if ω is s - homogeneous and ω' is s'- homogeneous then $\omega \wedge \omega'$ is s+s'- homogeneous.

For any s-homogeneous $\omega \in \Lambda^q(\widetilde{T^kM})$ and any r-homogeneous $\widetilde{T^kM})$ the function $\omega(\underset{1}{X}, .., \underset{q}{X})$ is $r + (s-1)q$ homogeneous.

Now we can formulate:

Definition 3.1.1 *A Finsler space of order k, $k \geq 1$, is a pair $F^{(k)n} = (M, F)$ determined by a real differentiable manifold M of dimension n and a function $F : T^kM \to R$ having the following properties:*

1°. F is differentiable on $\widetilde{T^kM}$ and continuous on the null section $0 : M \to T^kM$.

2°. F is positive.

3°. F is k-homogeneous on the fibres of the bundle T^kM,

4°. The Hessian of F^2 with the entries

$$g_{ij} = \frac{1}{2}\frac{\partial^2 F^2}{\partial y^{(k)i}\partial y^{(k)j}} \tag{3.1.6}$$

is positively defined on $\widetilde{T^kM}$.

The k-homogeneous means *positively k- homogeneous* since $F(x, ay^{(1)}, ..., a^k y^{(k)}) = a^k F(x, y^{(1)}, ..., y^{(k)})$ holds for any $a > 0$.

It is not difficult to see that this definition has a geometrical meaning, g_{ij} from (3.1.6) being a d-tensor field on $\widetilde{T^kM}$. Such that the function F is called the fundamental or metric function and the d-tensor field g_{ij} is called fundamental or metric tensor of the Finsler space of order k, $F^{(k)n}$. Of course, the axiom 4° implies:

$$rank\,\|g_{ij}\| = n \tag{3.1.7}$$

Clearly, in the case $k = 1, F^{(1)n} = (M, F)$ is a Finsler space [115].

We can see, without difficulties that the following theorem holds:

Theorem 3.1.1 *The pair $L^{(k)n} = (M, F^2(x, y^{(1)}, ..., y^{(k)}))$ is a Lagrange space of order k. Conversely, if $L^{(k)n} = (M, L(x, y^{(1)}, ..., y^{(k)})$ is a Lagrange space of order k, having the fundamental function L positively, 2k- homogeneous and the fundamental tensor g_{ij} positively defined, then the pair $F^{(k)n} = (M, \sqrt{L})$ is a Finsler space of order k.*

Consequently, the class of spaces $F^{(k)n}$ is a subclass of spaces $L^{(k)n}$.

Taking into account the k-homogeneity of the fundamental function F and $2k$-homogeneity of F^2 we get:

1°. p_i given by

$$p_i = \frac{1}{2}\frac{\partial F^2}{\partial y^{(k)i}} \tag{3.1.7'}$$

is a k-homogeneous d-covector fields.

2°. $p = p_i dx^i$ is a k-homogeneous 1-form.

3°. The fundamental tensor g_{ij} is 0-homogeneous. Its contravariant g^{ij} is 0-homogeneous ($g_{ij}g^{jh} = \delta_i^h$), too.

These homogeneities imply:

$$\mathcal{L}_{\overset{k}{\Gamma}} F^2 = 2kF^2, \mathcal{L}_{\overset{k}{\Gamma}} \frac{\partial F^2}{\partial y^{(k)}} = k\frac{\partial F^2}{\partial y^k}, (or\ \mathcal{L}_{\overset{k}{\Gamma}} p_i = kp_i), \tag{3.1.8}$$

$$\mathcal{L}_{\overset{k}{\Gamma}} g_{ij} = 0, \mathcal{L}_{\overset{k}{\Gamma}} \underset{(k)}{C_{ijh}} = -k\underset{(k)}{C_{ijh}}, \tag{3.1.8'}$$

where

$$\underset{(k)}{C_{ijh}} = \frac{1}{2}\frac{\partial g_{ij}}{\partial y^{(k)h}} = \frac{1}{4}\frac{\partial^3 F^2}{\partial y^{(k)i}\partial y^{(k)j}\partial y^{(k)h}}. \tag{3.1.8''}$$

Of course, the equation $\mathcal{L}_{\overset{k}{\Gamma}} g_{ij} = 0$ can be written as follows:

$$\mathcal{L}_{\overset{k}{\Gamma}} g_{rs} = (y^{(1)i}\frac{\partial}{\partial y^{(1)i}} + ... + y^{(k)i}\frac{\partial}{\partial y^{(k)i}})g_{rs} = 0. \tag{3.1.8'''}$$

The integral of action of F along a parametrized curve c :

$$I(c) = \int_0^1 F(x, \frac{dx}{dt}, ..., \frac{1}{k!}\frac{d^k x}{dt^k})dt \tag{3.1.9}$$

can be considered for determining *the length* of $\widetilde{c} : [0,1] \to \widetilde{T^k M}$ in the given parametrization.

Theorem 2.1.1 shows that the necessary conditions for the integral of action $I(c)$, (3.1.9), to be independent on the parametrization of a curve c are given by the Zermelo conditions

$$\mathcal{L}_{\overset{1}{\Gamma}} F = ... = \mathcal{L}_{\overset{k-1}{\Gamma}} F = 0,\ \mathcal{L}_{\overset{k}{\Gamma}} F = F.$$

But $\mathcal{L}_{\overset{k}{\Gamma}} F = F$ and $\mathcal{L}_{\overset{k}{\Gamma}} F = kF$, for $k > 1$ are contradictory.

Consequently, in a Finsler space of order k, $k > 1$, the integral of action (3.1.9) essentially depend on the parametrization of a curve c.

Example 3.5.1 Let $F^n = (M, F(x, y^{(1)}))$ *be a Finsler space having* $g_{ij}(x, y^1)$ as the fundamental tensor and $\underset{(1)}{M^i_j}(x, y^1)$ as coefficients of the Cartan nonlinear connection. Then, theorem 1.5.4 gives us the dual coefficients on T^kM :

$$\underset{(1)}{M_j^i}(x, y^{(1)}), \underset{(2)}{M_j^i}(x, y^{(1)}, y^{(2)}), ..., \underset{(k)}{M_j^i}(x, y^{(1)}, ..., y^{(k)})$$

of a nonlinear connection, which depends only on the fundamental function $F(x, y^{(1)})$ of the Finsler space F^n.

It is not hard to see that these coefficients are homogeneous of degree $1, 2, ..., k$ respectively. This property implies that the d-Liouville vector field $z^{(k)i}$:

$$kz^{(k)i} = ky^{(k)i} + (k-1)\underset{(1)}{M_j^i} y^{(k-1)j} + ... + \underset{(k-1)}{M_j^i}\, y^{(1)j} \tag{3.1.10}$$

is linear in the variables $y^{(k)i}$ and it is k-homogeneous.

Consider the function

$$F(x, y^{(1)}, ..., y^{(k)}) = \{g_{ij}(x, y^{(1)}) z^{(k)i} z^{(k)j}\}^{1/2}, \tag{3.1.11}$$

$g_{ij}(x, y^{(1)})$ being a d-tensor positively defined. It follows that F from (3.1.11) is a positive differentiable function on $\widetilde{T^kM}$ and continuous on the null section of π^k. It is k -homogeneous and has $g_{ij}(x, y^{(1)})$ as the fundamental tensor.

Consequently, the pair $F^{(k)n} = (M, F(x, y^{(1)}, ..., y^{(k)})$ for F from (3.1.11) is a Finsler space of order k.

Concluding, we have:

Theorem 3.1.2 *If the base manifold is paracompact then there exist a Finsler space of order k,* $F^{(k)n}$.

The spaces $F^{(k)n}$ constructed in example (3.1.1) is called the Prolongation of order k of the Finsler space F^n. It is denoted by $\text{Prol}^k F^n$.

In order to determine the geodesics of the space $F^{(k)n}$ we take the integral of action of the regular Lagrangian F^2. The variational problem leads to the Euler - Lagrange equations.

$$\overset{\circ}{E}_i (F^2) = \frac{\partial F^2}{\partial x^i} - \frac{d}{dt}\frac{\partial F^2}{\partial y^{(1)i}} + ... + (-1)^k \frac{1}{k!}\frac{d^k}{dt^k}\frac{\partial F^2}{\partial y^{(k)i}} = 0. \tag{3.1.12}$$

The integral curves of the previous equations are called the geodesics of the space $F^{(k)n}$. Applying the theory from the section 2, ch. 2 we can determine the infinitesimal symmetries of the spaces $F^{(k)n}$.

The energies of order $k, k-1, ..., 1$ of the Finsler space of order k, $F^{(k)n} = (M, F(x, y^{(1)}, ..., y^{(k)}))$ are given by the formulae (2.3.1), for $L = F^2$.

In particular, Theorem 2.3.2 can be applied in order to obtain

Theorem 3.1.3 *The energy of order k,* $\mathcal{E}^k(F^2)$ *of the Finsler space* $F^{(k)n}$ *is conserved along every geodesic of this space.*

Let us consider

$$\overset{\vee}{p}_{(k)i}= \frac{\partial F^2}{\partial y^{(k)i}}, \overset{\vee}{p}_{(k-1)i}= \frac{\partial F^2}{\partial y^{(k-1)i}}, ..., \overset{\vee}{p}_{(1)i}= \frac{\partial F^2}{\partial y^{(1)i}}, \overset{\vee}{p}_{(0)i}= \frac{\partial F^2}{\partial x^i}. \tag{3.1.13}$$

Then we have

Theorem 3.1.4 *1°. The Cartan differential 1-forms are the followings*

$$\begin{aligned} d_0F^2 &= \overset{\vee}{p}_{(k)i}\, dx^i, \\ d_1F^2 &= \overset{\vee}{p}_{(k-1)i}\, dx^i + \overset{\vee}{p}_{(k)i}\, dy^{(k)i}, \\ &\ldots\ldots\ldots\ldots\ldots\ldots, \\ d_kF^2 &= \overset{\vee}{p}_{(0)i}\, dx^i + \overset{\vee}{p}_{(1)i}\, dy^{(1)i} + ... + \overset{\vee}{p}_{(k)i}\, dy^{(k)i}. \end{aligned} \tag{3.1.14}$$

2°. And the Poincare 2-forms are given by

$dd_0F^2 = d\,\overset{\vee}{p}_{(k)i} \wedge dx^i,$

$\ldots\ldots\ldots\ldots\ldots\ldots,$

$dd_{(k-1)}F^2 = d\,\overset{\vee}{p}_{(1)i} \wedge dx^i + ... + d\,\overset{\vee}{p}_{(k)i}\, dy^{(k)i}.$

Here we have $dd_kF^2 = d^2F^2 = 0$.

3°. The 1-forms $d_0F^2, d_1F^2, ..., d_kF^2$ *are* $k, k+1, ..., 2k$ *homogeneous, respectively.*

3.2 Cartan Nonlinear Connection in $F^{(k)n}$

The considerations made in the previous chapter allow us to introduce in a Finsler space of order k, $F^{(k)n} = (M, F)$ the main geometrical object fields as: canonical k-spray, Cartan nonlinear connection, canonical N-linear connection etc. Canonical mean here that all these object fields depend only on the fundamental function F.

Taking into account the operators $\overset{0}{E_i}, \overset{1}{E_i}, ..., \overset{k}{E_i}$ given by (2.2.8) we construct the system of d -covector fields

$$\overset{0}{E_i}(F^2), \overset{1}{E_i}(F^2), ..., \overset{k}{E_i}(F^2) \tag{3.2.1}$$

All equations

$$\overset{0}{E_i}(F^2) = 0, \overset{1}{E_i}(F^2) = 0, ..., \overset{k-1}{E_i}(F^2) = 0$$

have geometrical meanings. The equation

$$\overset{k-1}{E_i}(F^2) = 0$$

is important for us. It will be called the *Craig-Synge equation.* Using (2.2.8), this equation is expressed as follows

$$\frac{\partial F^2}{\partial y^{(k-1)i}} - \frac{d}{dt}\frac{\partial F^2}{\partial y^{(k)i}} = 0. \tag{3.2.2}$$

But $\frac{d}{dt}\frac{\partial F^2}{\partial y^{(k)i}} = \Gamma\frac{\partial F^2}{\partial y^{(k)i}} + \frac{2}{k!}g_{ij}\frac{d^{k+1}x^i}{dt^{k+1}}$, where Γ is the operator (1.2.3).

Consequently, the Craig-Synge equations (3.2.2) is equivalent to the following equations

$$g^{ij}\overset{k-1}{E_j}(F^2)= 0, y^{(1)i} = \frac{dx^i}{dt}, ..., y^{(k)i} = \frac{1}{k!}\frac{d^kx^i}{dt^k}$$

or

$$\frac{d^{k+1}x^i}{dt^{k+1}} + (k+1)G^i(x, \frac{dx}{dt}, ..., \frac{1}{k!}\frac{d^kx}{dt^k}) = 0, \tag{3.2.3}$$

where

$$(k+1)G^i(x, y^{(1)}, ..., y^{(k)}) = \frac{1}{2}g^{ij}\{\Gamma(\frac{\partial F^2}{\partial y^{(k)j}}) - \frac{\partial F^2}{\partial y^{(k-1)j}}\}. \tag{3.2.4}$$

Applying the Theorem 2.5.1 one obtains:

Theorem 3.2.1 *The Craig-Synge equations (3.2.3) determines a canonical k-spray S :*
(3.2.5)

$$S = y^{(1)i}\frac{\partial}{\partial x^i} + 2y^{(2)i}\frac{\partial}{\partial y^{(1)i}} ... + ky^{(k)i}\frac{\partial}{\partial y^{(k-1)i}} - (k+1)G^i(x, y^{(1)}, ..., y^{(k)})\frac{\partial}{\partial y^{(k)i}}$$

with the coefficients G^i from (3.2.4).

Note that S depend only on the fundamental function F of the space $F^{(k)n}$.
It is a k-spray, since it is a 2-homogeneous vector field. The paths of S are given by the differential equations (3.2.3).
By means of the Theorems 1.5.1 and 3.2.1 the dual coefficients of the nonlinear connection N determined by the canonical k-spray S are given by:

Theorem 3.2.2 *In a Finsler space of order k, $F^{(k)n} = (M, F)$ there exist nonlinear connections, depending only on the fundamental function F. One of these, denoted by N, has the dual coefficients:*

$$\underset{(1)}{M^i_j} = \frac{1}{2(k+1)}\frac{\partial}{\partial y^{(k)j}}\{g^{im}[\Gamma\frac{\partial F^2}{\partial y^{(k)m}} - \frac{\partial F^2}{\partial y^{(k-1)m}}]\},$$

$$\underset{(2)}{M^i_j} = \frac{1}{2}\{S\underset{(1)}{M^j_i} + \underset{(1)}{M^i_m}\underset{(1)}{M^m_j}]\}, \tag{3.2.6}$$

....

$$\underset{(k)}{M^i_j} = \frac{1}{k}\{S\ \underset{(k-1)}{M^i_j} + \underset{(1)}{M^i_m}\ \underset{(k-1)}{M^m_j}]\},$$

where S is the canonical k-spray of the space $F^{(k)n}$.

N is called the Cartan nonlinear connection of the space $F^{(k)n}$.

In the case $k = 1$, N reduces to the classical Cartan nonlinear connection of the Finsler space $F^n = (M, F(x, y^{(1)}))$.

Some properties of N.

1°. The Cartan nonlinear connection N is globally defined on $\widetilde{T^kM}$ (if $F(x, y^{(1)}, ..., y^{(k)})$ has this property)

2°. The dual coefficients (3.2.6) of N are homogeneous of degree $1, 2, ..., k$ i.e:

$$\mathcal{L}_{\overset{k}{\Gamma}} \underset{(\alpha)}{M^i_j} = \alpha \underset{(\alpha)}{M^i_j}, \alpha = 1, ..., k. \tag{3.2.7}$$

The coefficients $\underset{(1)}{N^i_j}, ..., \underset{(k)}{N^i_j}$ of N, (ch.1) are expressed by

$$\begin{aligned} &\underset{(1)}{N^i_j} = \underset{(1)}{M^i_j}, \\ &\underset{(2)}{N^i_j} = \underset{(2)}{M^i_j} - \underset{(1)}{N^i_m} \underset{(1)}{M^m_j}, \\ &\dots\dots\dots\dots\dots\dots\dots\dots \\ &\underset{(k)}{N^i_j} = \underset{(k)}{M^i_j} - \underset{(k-1)}{N^i_m} \underset{(1)}{M^m_j} - ... - \underset{(1)}{N^i_m} \underset{(k-1)}{M^m_j}. \end{aligned} \tag{3.2.8}$$

These coefficients are homogeneous functions of degree $1, ..., k$ respectively, i.e

$$\mathcal{L}_{\overset{k}{\Gamma}} \underset{(\alpha)}{N^i_j} = \alpha \underset{(\alpha)}{N^i_j}, \quad (\alpha = 1, ..., k). \tag{3.2.7'}$$

The Cartan nonlinear connection N gives rise to a distribution $N_u \subset T_u(\widetilde{T^kM})$ supplementary to the vertical distribution $V_u \subset T_u(\widetilde{T^kM})$, $\forall u \in \widetilde{T^kM}$ with the property:

$$T_u(\widetilde{T^kM}) = N_u \oplus V_u \ , \ \forall u \in \widetilde{T^kM}.$$

If we consider the distributions

$$N_0 = N, N_1 = J(N_0), ..., N_{k-1} = J(N_{k-2}), V_k = J(N_{k-1}),$$

then according to the general theory we obtain the direct decomposition of the vector spaces:

$$T_u(\widetilde{T^kM}) = N_{0,u} \oplus N_{1,u} \oplus ... \oplus N_{k-1,u} \oplus V_{k,u}, \forall u \in \widetilde{T^kM}. \tag{3.2.9}$$

The local adapted basis to the direct decomposition (3.2.9) is

$$\left(\frac{\delta}{\delta x^i}, \frac{\delta}{\delta y^{(1)i}}, ..., \frac{\delta}{\delta y^{(k)i}}\right), \tag{3.2.10}$$

where

$$(3.2.11)\qquad \begin{aligned} &\frac{\delta}{\delta x^i} = \frac{\partial}{\partial x^i} - \underset{(1)}{N_i^j}\frac{\partial}{\partial y^{(1)j}} - \dots - \underset{(k)}{N_i^j}\frac{\partial}{\partial y^{(k)j}},\\ &\frac{\delta}{\delta y^{(1)i}} = \frac{\partial}{\partial y^{(1)i}} - \underset{(1)}{N_i^j}\frac{\partial}{\partial y^{(2)j}} - \dots - \underset{(k-1)}{N_i^j}\frac{\partial}{\partial y^{(k)j}},\\ &\dots\dots\dots\dots\dots\dots\dots\dots\dots\dots\\ &\frac{\delta}{\delta y^{(k)i}} = \frac{\partial}{\partial y^{(k)i}}, \end{aligned}$$

$\underset{(1)}{N_i^j}, \dots, \underset{(k)}{N_i^j}$ being the coefficients (3.2.8) of the Cartan nonlinear connection.

Of course we have

$$\frac{\delta}{\delta y^{(1)i}} = J\left(\frac{\delta}{\delta x^i}\right), \dots, \frac{\delta}{\delta y^{(k)i}} = J\left(\frac{\delta}{\delta y^{(k-1)i}}\right), 0 = J\left(\frac{\delta}{\delta y^{(k)i}}\right).$$

Taking into account section 1.4 ch.1, the dual (adapted) cobasis, of the basis (3.2.10) is:

$$(3.2.12)\qquad (\delta x^i, \delta y^{(1)i}, \dots, \delta y^{(k)i})$$

where

$$(3.2.12')\qquad \begin{aligned} &\delta x^i = dx^i,\\ &\delta y^{(1)i} = dy^{(1)i} + \underset{(1)}{M_j^i}dx^j,\\ &\dots\dots\dots\dots\dots\dots\dots\dots\\ &\delta y^{(k)i} = dy^{(k)i} + \underset{(1)}{M_j^i}dy^{(k-1)i} + \dots + \underset{(k)}{M_j^i}dx^j, \end{aligned}$$

$\underset{(1)}{M_j^i}, \dots, \underset{(k)}{M_j^i}$, being the dual coefficients (3.2.6) of the Cartan nonlinear connection N.

It is not difficult to see that the following identities hold:

$$(3.2.13)\qquad \begin{cases} J^*(\delta y^{(k)i}) = \delta y^{(k-1)i},\ J^*(\delta y^{(k-1)i}) = \delta y^{(k-2)i}, \dots,\\ J^*(\delta y^{(1)i}) = \delta x^i,\ J^*(dx^i) = 0. \end{cases}$$

J^* being the adjoint of the k-structure J.

Now we can determine the differential operators $d_k, d_{k-1}, \dots, d_0$ defined in (1.2.11), using the expression of the operator of differentiation $d_k = d$ in the adapted basis:

$$(3.2.14)\qquad d_k = \frac{\delta}{\delta x^i}\delta x^i + \frac{\delta}{\delta y^{(1)i}}\delta y^{(1)i} + \dots + \frac{\delta}{\delta y^{(k)i}}\delta y^{(k)i}.$$

Applying (3.2.13), we get d_k in (3.2.14), and

$$d_{k-1} = J^* d_k, d_{k-2} = J^* d_{k-1}, ..., d_0 = J^* d_1.$$

One obtains:

$$\begin{aligned} d_{k-1} &= \frac{\delta}{\delta y^{(1)i}}\delta x^i + \frac{\delta}{\delta y^{(2)i}}\delta y^{(1)i} + ... + \frac{\delta}{\delta y^{(k)i}}\delta y^{(k-1)i}, \\ d_{k-2} &= \frac{\delta}{\delta y^{(2)i}}\delta x^i + \frac{\delta}{\delta y^{(3)i}}\delta y^{(1)i} + ... + \frac{\delta}{\delta y^{(k)i}}\delta y^{(k-2)i}, \\ &\cdots\cdots\cdots\cdots\cdots\cdots \\ d_1 &= \frac{\delta}{\delta y^{(k-1)i}}\delta x^i + \frac{\delta}{\delta y^{(k)i}}\delta y^{(1)i}, \\ d_0 &= \frac{\delta}{\delta y^{(k)i}}\delta x^i. \end{aligned} \tag{3.2.15}$$

Consequently, we get:

Theorem 3.2.3 *With respect to the direct decomposition (3.2.9), in adapted basis (3.2.11), (3.2.12), the Cartan 1-forms $d_0F^2, d_1F^2, ..., d_kF^2$ of a Finsler space of order k, $F^{(k)n} = (M, F)$ can be expressed as follows:*

$$\begin{aligned} d_0F^2 &= (d_0F^2)^H, \\ d_1F^2 &= (d_1F^2)^H + (d_1F^2)^{V_1}, \\ &\cdots\cdots\cdots\cdots\cdots\cdots \\ d_kF^2 &= (d_kF^2)^H + (d_kF^2)^{V_1} + ... + (d_kF^2)^{V_k}. \end{aligned} \tag{3.2.16}$$

Equivalently,

$$\begin{aligned} d_0F^2 &= \frac{\delta F^2}{\delta y^{(k)i}}\delta x^i, \\ d_1F^2 &= \frac{\delta F^2}{\delta y^{(k-1)i}}\delta x^i + \frac{\delta F^2}{\delta y^{(k)i}}\delta y^{(1)i}, \\ &\cdots\cdots\cdots\cdots\cdots\cdots \\ d_kF^2 &= \frac{\delta F^2}{\delta x^i}\delta x^i + \frac{\delta F^2}{\delta y^{(1)i}}\delta y^{(1)i} + ... + \frac{\delta F^2}{\delta y^{(k)i}}\delta y^{(k)i}. \end{aligned} \tag{3.2.17}$$

In the previous expressions every term is an 1-form field on $\widetilde{T^kM}$. So we have the following main 1-form fields

$$
(3.2.18)\quad
\begin{aligned}
\theta_0 &= (d_0F^2)^H = \frac{\delta F^2}{\delta y^{(k)i}}\delta x^i = \overset{\vee}{p}_{(k)i}\,\delta x^i,\\
\theta_1 &= (d_1F^2)^{V_1} = \frac{\delta F^2}{\delta y^{(k)i}}\delta y^{(k)i} = \overset{\vee}{p}_{(k)i}\,\delta y^{(1)i},\\
&\cdots\cdots\cdots\cdots\cdots\cdots\cdots\cdots\\
\theta_{k-1} &= (d_{k-1}F^2)^{V_{k-1}} = \frac{\delta F^2}{\delta y^{(k)i}}\delta y^{(k-1)i} = \overset{\vee}{p}_{(k)i}\,\delta y^{(k-1)i},\\
\theta_k &= (d_kF^2)^{V_k} = \frac{\delta F^2}{\delta y^{(k)i}}\delta y^{(k)i} = \overset{\vee}{p}_{(k)i}\,\delta y^{(k)i}.
\end{aligned}
$$

Theorem 3.2.4 *1°. The 1-form fields $\theta_0, ..., \theta_k$ depend only on the fundamental function F of the Finsler space $F^{(k)n}$.*

2°. The exterior differentials of $\theta_0, ..., \theta_k$,

$$
(3.2.19)\quad
\begin{aligned}
d\theta_0 &= d\,\overset{\vee}{p}_{(k)i}\wedge\delta x^i,\\
d\theta_1 &= d\,\overset{\vee}{p}_{(k)i}\wedge\delta y^{(1)i} + \overset{\vee}{p}_{(k)i}\wedge d\delta y^{(1)i},\\
&\cdots\cdots\cdots\cdots\cdots\cdots\cdots\cdots\\
d\theta_k &= d\,\overset{\vee}{p}_{(k)i}\wedge\delta y^{(k)i} + \overset{\vee}{p}_{(k)i}\wedge d\delta y^{(k)i}
\end{aligned}
$$

have the same property of homogeneity.

The second terms of $d\theta_0, ..., d\theta_k$, the exterior differentials of 1-forms $\delta y^{(1)i}, ..., \delta y^{(k)i}$ are calculated by means of formulas:

$$
(3.2.20)\quad
\begin{aligned}
d\delta y^{(\alpha)i} = {} & \frac{1}{2}\underset{(0\alpha)}{R^i_{jm}}dx^m\wedge dx^j + \sum_{\gamma=1}^{k}\underset{(\gamma\alpha)}{B^i_{jm}}dy^{(\gamma)m}\wedge dx^j + \\
& + \sum_{\beta,\gamma=1}^{k}\underset{(\alpha\gamma)}{\overset{(\beta)}{C^i_{jm}}}\,dy^{(\gamma)m}\wedge dy^{(\alpha)j},
\end{aligned}
$$

where $\underset{(\alpha\alpha)}{\overset{(\alpha)}{C^i_{jm}}} = 0$, and the coefficients from the right hand side can be calculated using the following Lie brackets:

(3.2.21)
$$[\frac{\delta}{\delta x^j}, \frac{\delta}{\delta x^k}] = \underset{(01)}{R^i_{jh}} \frac{\delta}{\delta y^{(1)i}} + ... + \underset{(0k)}{R^i_{jh}} \frac{\delta}{\delta y^{(k)i}},$$
$$[\frac{\delta}{\delta x^j}, \frac{\delta}{\delta y^{(\alpha)h}}] = \underset{(\alpha 1)}{B^i_{jh}} \frac{\delta}{\delta y^{(1)i}} + ... + \underset{(\alpha k)}{B^i_{jh}} \frac{\delta}{\delta y^{(k)i}},$$
$$[\frac{\delta}{\delta y^{(\alpha)j}}, \frac{\delta}{\delta y^{(\beta)h}}] = \underset{(\alpha\beta)}{\overset{(1)}{C^i_{jh}}} \frac{\delta}{\delta y^{(1)i}} + ... + \underset{(\alpha\beta)}{\overset{(k)}{C^i_{jh}}} \frac{\delta}{\delta y^{(k)i}},$$
$$(\alpha, \beta = 1, ..., k).$$

The whole previous theory can be applied to the following Lagrangians associated to the Finsler space $F^{(k)n}$

(3.2.22)
$$F_1^2 = g_{ij} z^{(1)i} z^{(1)j}, ..., F_k^2 = g_{ij} z^{(k)i} z^{(k)j}$$

especially in the cases of the particular Finsler spaces of order k, *Prol* $^k\mathcal{R}^n$ or *Prol*$^k F^n$ (see ch.2). The Lagrangians $F_1^2, ..., F_k^2$ are positive functions and are $2, 4, ..., 2k$ -homogeneous, respectively.

The autoparallel curves of the Cartan nonlinear connection N are characterized by the system of differential equations

$$\frac{\delta y^{(1)i}}{dt} = ... = \frac{\delta y^{(k)i}}{dt} = 0,$$
$$y^{(1)i} = \frac{dx^i}{dt}, ..., y^{(k)i} = \frac{1}{k!} \frac{dx^{(k)i}}{dt^k}.$$

3.3 The Cartan Metrical N-Linear Connection

Let N be the Cartan nonlinear connection of the Finsler space of order k, $F^{(k)n} = (M, F)$ having the adapted basis (3.2.10) and its dual (3.2.12).

The lift of the fundamental tensor field g_{ij} is given by (2.6.1),

(3.3.1)
$$\mathbb{G} = g_{ij} dx^i \otimes dx^j + g_{ij} \delta y^{(1)i} \otimes \delta y^{(1)j} + ... + g_{ij} \delta y^{(k)i} \otimes \delta y^{(k)j}.$$

Theorem 3.3.1 *$\mathbb{G}$ from (3.3.1) is a Riemannian structure on $\widetilde{T^kM}$ which depend only on the fundamental function F of the space $F^{(k)n}$. The terms of $\mathbb{G}$ are $0, 2, ..., 2k$ homogeneous, respectively.*

Notice that $\mathbb{G}$ is not homogeneous. We can construct an homogeneous one using the Lagrangians (3.2.22).

Namely

(3.3.1')
$$\check{\mathbb{G}} = g_{ij} dx^i \otimes dx^j + \frac{1}{F_1^2} g_{ij} \delta y^{(1)i} \otimes \delta y^{(1)j} + ... + \frac{1}{F_k^2} g_{ij} \delta y^{(k)i} \otimes \delta y^{(k)j}.$$

$\overset{\vee}{\mathbb{G}}$ is a Riemannian structure on the manifold $\widetilde{T^kM}$ determined only by the fundamental function $\mathbb{F}$ and it is 0-homogeneous.

In the following we consider the Riemannian structure $\mathbb{G}$ from (3.3.1).

An N-linear connection D is compatible with $\mathbb{G}$ if

$$D_X\mathbb{G} = 0,\ \forall X \in \mathcal{X}(\widetilde{T^kM}).$$

Applying the Theorem 2.6.1, we have:

Theorem 3.3.2 *For a Finsler space of order k, $F^{(k)n} = (M, F)$, the following properties hold:*

1°. There exists an unique N-linear connection D on $\widetilde{T^kM}$ verifying the axioms:

A_1 *N is the Cartan nonlinear connection,*

A_2 $g_{ij|h} = 0,$

A_3 $g_{ij} \underset{(\alpha)}{|}_h = 0,$

A_4 $F^i_{jk} = F^i_{kj},$

A_5 $\underset{(\alpha)}{C}{}^i_{jk} = \underset{(\alpha)}{C}{}^i_{kj}$ $(\alpha = 1, ..., k)$.

2°. The coefficients $C\Gamma(N) = (F^i_{jk}, \underset{(1)}{C}{}^i_{jk}, ..., \underset{(k)}{C}{}^i_{jk})$ of D are given by the generalized Christoffel symbols (2.6.3), $(F^i_{jk} = L^i_{jk})$.

3°. D depends only on the fundamental function F of the space $F^{(k)n}$.

The metrical N linear -connection D from the previous theorem will be called the *Cartan metrical N-linear connection* of the space $F^{(k)n}$ and denoted by $C\Gamma(N)$.

Of course, the torsion d-tensor fields and the curvature d-tensor fields of $C\Gamma(N)$ can be written without difficulties. Such that we have

$$\underset{(0)}{T}{}^i_{jk} = 0, \quad \underset{(\alpha)}{S}{}^i_{jk} = 0, (\alpha = 1, ..., k). \tag{3.3.2}$$

Also we can calculate the deflection tensor of $C\Gamma(N)$:

$$\overset{(\alpha)}{D}{}^i_j = z^{(\alpha)i}{}_{|j}, \quad \overset{(\beta\alpha)}{d}{}^i_j = z^{(\beta)i} \overset{(\alpha)}{|}_j .$$

The coefficients $C\Gamma(N) = (F^i_{jh}, \underset{(1)}{C}{}^i_{jk}, ..., \underset{(k)}{C}{}^i_{jk})$ are $0, -1, ..., -k$ -homogeneous.

The d-tensors of curvature of $C\Gamma(N)$ satisfy the identities:

$$\begin{aligned} &g_{sj}R_i{}^s{}_{hm} + g_{is}R_j{}^s{}_{hm} = 0, \\ &g_{sj}\underset{(\alpha)}{P}{}_i{}^s{}_{hm} + g_{is}\underset{(\alpha)}{P}{}_j{}^s{}_{hm} = 0, \\ &g_{sj}\underset{(\alpha\beta)}{S}{}_i{}^s{}_{hm} + g_{is}\underset{(\alpha\beta)}{S}{}_j{}^s{}_{hm} = 0 \end{aligned} \tag{3.3.3}$$

Notice that the equations $\underset{(0\alpha)}{R}{}^{\ i}_{jh} = 0, (\alpha = 1, ..., k)$ characterize the integrability of the Cartan nonlinear connection.

The connection 1-forms ω^i_j of the Cartan metrical N-linear connection $C\Gamma(N)$ are given by:

$$\omega^i{}_j = F^i_{jh}dx^k + \underset{(1)}{C^i_{jh}}\delta y^{(1)h} + ... + \underset{(k)}{C^i_{jh}}\delta y^{(k)h}. \tag{3.3.4}$$

Theorem 3.3.3 *The structure equations of the Cartan metrical N-linear connection $C\Gamma(N)$ of the Finsler space $F^{(k)n}$ are given by:*

$$\begin{gathered} d(dx^i) - dx^m \wedge \omega^i{}_m = -\overset{(0)}{\Omega^i}, \\ d(\delta y^{(\alpha)i}) - \delta y^{(\alpha)m} \wedge \omega^i{}_m = -\overset{(\alpha)}{\Omega^i}, \ (\alpha = 1, ..., k), \\ d\omega^i{}_j - \omega^m_j \wedge \omega^i{}_m = -\Omega^i{}_j, \end{gathered} \tag{3.3.5}$$

where $\overset{(0)}{\Omega^i}, \overset{(\alpha)}{\Omega^i}$, *are the 2-forms of torsion:*

$$\overset{(0)}{\Omega^i} = \underset{(1)}{C^i_{jh}}dx^j \wedge \delta y^{(1)h} + ... + \underset{(k)}{C^i_{jh}}dx^j \wedge \delta y^{(k)h}$$

$$\overset{(\alpha)}{\Omega^i} = \frac{1}{2}\underset{(0\alpha)}{R^i_{jh}}dx^j \wedge dx^h + \sum_{\gamma=1}^{k} \underset{(\gamma\alpha)}{B^i_{jm}}dx^j \wedge \delta y^{(\gamma)m} + \tag{3.3.6}$$

$$+ \sum_{\gamma=1}^{k} \underset{(\alpha\gamma)}{\overset{(\beta)}{C^i_{jh}}} \delta y^{(\beta)j} \wedge \delta y^{(\gamma)h} - (F^i_{jh}dx^j + \sum_{\gamma=1}^{k} \underset{(\gamma)}{C^i_{jh}}\delta y^{(\gamma)j})\delta y^{(\alpha)h}$$

and $\Omega^i{}_j$ *are the 2-forms of curvature:*

$$\Omega^i{}_j = \frac{1}{2}R^i_{j\ pq}dx^p \wedge dx^q + \sum_{\gamma=1}^{k} \underset{(\gamma)}{P}{}^{\ i}_{j\ pq}dx^p \wedge \delta y^{(\gamma)q} + \sum_{\beta,\gamma=1}^{k} \underset{(\beta\gamma)}{S}{}^{\ i}_{j\ pq}\delta y^{(\beta)p} \wedge \delta y^{(\gamma)q}. \tag{3.3.7}$$

Now, we can obtain the Bianchi identities of the Cartan metrical N-linear connection $C\Gamma(N)$ if we apply the exterior differential to the system

of equations (3.3.5) and calculate $d\overset{(0)}{\Omega^i}, d\overset{(\alpha)}{\Omega^i}$ and $d\Omega^i{}_j$ from (3.3.6) and (3.3.7), modulo the system (3.3.5).

Finally, consider the tensor field $\mathbb{F}$ determined by the Cartan nonlinear connection N.

$$\mathbb{F} = -\frac{\delta}{\delta y^{(k)i}} \otimes dx^i + \frac{\delta}{\delta x^i} \otimes \delta y^{(k)i}. \tag{3.3.8}$$

It is not difficult to prove that $C\Gamma(N)$ has the property $D_X\mathbb{F}=0$. Indeed,

$$D_{\frac{\delta}{\delta x^h}}\mathbb{F} = -(D_{\frac{\delta}{\delta x^h}}\frac{\delta}{\delta y^{(k)i}})\otimes dx^i - \frac{\delta}{\delta y^{(k)i}}\otimes D_{\frac{\delta}{\delta x^h}} dx^i +$$

$$(D_{\frac{\delta}{\delta x^h}}\frac{\delta}{\delta x^i})\otimes \delta y^{(k)i} + \frac{\delta}{\delta x^i}\otimes D_{\frac{\delta}{\delta x^h}}\delta y^{(k)i} =$$

$$-F^m_{ih}\frac{\delta}{\delta y^{(k)m}}\otimes dx^i + F^i_{mh}\frac{\delta}{\delta y^{(k)i}}\wedge dx^m + F^m_{ih}\frac{\delta}{\delta x^m}\otimes\delta y^{(k)i} - F^i_{mh}\frac{\delta}{\delta x^i}\otimes\delta y^{(k)m} = 0.$$

We proceed analogously in the case $D_{\frac{\delta}{\delta y^{(\alpha)i}}}\mathbb{F}=0, (\alpha=1,...,k)$.

Since $C\Gamma(N)$ has the property $D_X\mathbb{G}=0$, it follows:

Theorem 3.3.4 *The structures* $\mathbb{G}$ *and* $\mathbb{F}$ *determine a Riemannian* $(k-1)n$*- almost contact structure on* T^kM*, which depends only on the fundamental Finsler function of Finsler space* $F^{(k)n}=(M,F)$.

In the case $k=1$, the triple $(\widetilde{TM},\mathbb{F},\mathbb{G})$ is an almost Kählerian space.

Chapter 4

The Geometry of the Dual of k-Tangent Bundle

In the book [115] one studies the geometry of the dual of the 2-tangent bundle, which has been used to investigate the Hamilton spaces of order 2. Now, we consider this problem for the general case, $k \geq 1$.

The dual bundle $T^{*k}M$ of the k-tangent bundle must have the same properties as the cotangent bundle T^*M with respect to tangent bundle TM. Such that the manifold $T^{*k}M$ should have the same dimension $(k+1)n$ as the manifold T^kM. $T^{*k}M$ should carry a natural presymplectic structure and at least one Poisson structure. The manifolds T^kM and $T^{*k}M$ should be locally diffeomorphic.

The dual bundle $T^{*k}M$ plays a main role in construction of the notion of Hamilton space of order k.

In the present chapter we introduce the bundle $T^{*k}M$ and point out the main geometrical natural object fields, that live on the differentiable manifold T^kM.

4.1 The Dual Bundle $(T^{*k}M, \pi^{*k}, M)$

Definition 4.1.1 *We call the dual bundle of k-tangent bundle (T^kM, π^k, M) the differentiable bundle $(T^{*k}M, \pi^{*k}, M)$ whose total space is the fibered product:*

$$T^{*k}M = T^{k-1}M \times_M T^*M \tag{4.1.1}$$

and for which the canonical projection π^{*k} is

$$\pi^{*k} = \pi^{k-1} \times_M \pi^*. \tag{4.1.1'}$$

The previous fibered product has a differentiable structure given by that of the $(k-1)$-tangent bundle $(T^{k-1}M, \pi^{k-1}, M)$ and the cotangent bundle (T^*M, π^*, M). For $k = 1$, we have $T^{*1}M = T^*M$ and $\pi^{*1} = \pi^*$. Sometimes we denote $(T^{*k}M, \pi^{*k}, M)$ by $T^{*k}M$.

A point $u \in T^{*k}M$ will be denoted by $u = (x, y^{(1)}, ..., y^{(k-1)}, p)$, $\pi^{*k}(u) = x$. The projections on the factors of the $T^{*k}M$ from (4.1.1) are

$$\pi^{*k}_{k-1} : T^{*k}M \to T^{k-1}M, \quad \pi^{*k}_{k-1}(x, y^{(1)}, ..., y^{(k-1)}, p) = (x, y^{(1)}, ..., y^{(k-1)}),$$

$$\overline{\pi}^* : T^{*k}M \to T^*M, \quad \overline{\pi}^*(x, y^{(1)}, ..., y^{(k-1)}, p) = (x, p)$$

and the canonical projection $\pi^{*k} : T^{*k}M \to M$.

Therefore, the following diagram is commutative:

$$\begin{array}{ccccc} & & T^{*k}M & & \\ & \overset{\pi^{*k}_{k-1}}{\swarrow} & & \overset{\overline{\pi}^*}{\searrow} & \\ T^{k-1}M & & \Big\downarrow \pi^{*k} & & T^*M \\ & \underset{\pi^{k-1}}{\searrow} & & \underset{\pi^*}{\swarrow} & \\ & & M & & \end{array}$$

Let $(x^i, y^{(1)i}, ..., y^{(k-1)i}, p_i)$, $(i = 1, 2, ..., n = \dim M)$ be the coordinates of a point $u = (x, y^{(1)}, ..., y^{(k-1)}, p) \in T^{*k}M$ in a local chart $\left((\pi^{*k})^{-1}(U), \Phi\right)$ on $T^{*k}M$.

The change of coordinates on the manifold $T^{*k}M$ is:

$$\begin{aligned} &\widetilde{x}^i = \widetilde{x}^i(x^1, ..., x^n), \ \det\left(\frac{\partial \widetilde{x}^i}{\partial x^j}\right) \neq 0, \\ &\widetilde{y}^{(1)i} = \frac{\partial \widetilde{x}^i}{\partial x^j} y^{(1)j}, \\ &\dots\dots\dots\dots\dots\dots\dots\dots\dots\dots\dots\dots\dots\dots \\ &(k-1)\widetilde{y}^{(k-1)i} = \frac{\partial \widetilde{y}^{(k-2)i}}{\partial x^j} y^{(1)j} + \dots + (k-1)\frac{\partial \widetilde{y}^{(k-2)i}}{\partial y^{(k-2)j}} y^{(k-1)j}, \\ &\widetilde{p}_i = \frac{\partial x^j}{\partial \widetilde{x}^i} p_j. \end{aligned} \tag{4.1.2}$$

where the following equalities hold:

$$\frac{\partial \widetilde{y}^{(\alpha)i}}{\partial x^j} = \frac{\partial \widetilde{y}^{(\alpha+1)i}}{\partial y^{(1)j}} = \dots = \frac{\partial \widetilde{y}^{(k-1)i}}{\partial y^{(k-1-\alpha)j}}; \ (\alpha = 0, ..., k-2; y^{(0)} = x). \tag{4.1.3}$$

Sometimes (cf. Ch. 1) $y^{(1)i}$, ..., $y^{(k-1)i}$ will be called *accelerations of order* 1, 2, ..., $k-1$, respectively and p_i will be called *momenta.* $T^{*k}M$ is a real manifold of dimension $(k+1)n$. So it has the same dimension as that of the manifold T^kM.

The natural basis of the vector space $T_u(T^{*k}M)$ at the point $u \in T^{*k}M$,

$\left\{\frac{\partial}{\partial x^i}|_u, \frac{\partial}{\partial y^{(1)i}}|_u, ..., \frac{\partial}{\partial y^{(k-1)i}}|_u, \frac{\partial}{\partial p_i}|_u\right\}$ is transformed under (4.1.2) as follows:

(4.1.4)

$$\frac{\partial}{\partial x^i} = \frac{\partial \widetilde{x}^j}{\partial x^i}\frac{\partial}{\partial \widetilde{x}^j} + \frac{\partial \widetilde{y}^{(1)j}}{\partial x^i}\frac{\partial}{\partial \widetilde{y}^{(1)j}} + \cdots + \frac{\partial \widetilde{y}^{(k-1)j}}{\partial x^i}\frac{\partial}{\partial \widetilde{y}^{(k-1)j}} + \frac{\partial \widetilde{p}_j}{\partial x^i}\frac{\partial}{\partial \widetilde{p}_j},$$

$$\frac{\partial}{\partial y^{(1)i}} = \frac{\partial \widetilde{y}^{(1)j}}{\partial y^{(1)i}}\frac{\partial}{\partial \widetilde{y}^{(1)j}} + \frac{\partial \widetilde{y}^{(2)j}}{\partial y^{(1)i}}\frac{\partial}{\partial \widetilde{y}^{(2)j}} + \cdots + \frac{\partial \widetilde{y}^{(k-1)j}}{\partial y^{(1)i}}\frac{\partial}{\partial \widetilde{y}^{(k-1)j}},$$

....

$$\frac{\partial}{\partial y^{(k-2)i}} = \frac{\partial \widetilde{y}^{(k-2)j}}{\partial y^{(k-2)i}}\frac{\partial}{\partial \widetilde{y}^{(k-2)j}} + \frac{\partial \widetilde{y}^{(k-1)j}}{\partial y^{(k-2)i}}\frac{\partial}{\partial \widetilde{y}^{(k-1)j}},$$

$$\frac{\partial}{\partial y^{(k-1)i}} = \frac{\partial \widetilde{y}^{(k-1)j}}{\partial y^{(k-1)i}}\frac{\partial}{\partial \widetilde{y}^{(k-1)j}},$$

$$\frac{\partial}{\partial p_i} = \frac{\partial x^i}{\partial \widetilde{x}^j}\frac{\partial}{\partial \widetilde{p}_j}$$

calculated at the point $u \in T^{*k}M$.

The Jacobian matrix of the transformation (4.1.2) at the point $u \in T^{*k}M$ is:

$$(4.1.5)\qquad J_k = \begin{pmatrix} \frac{\partial \widetilde{x}^j}{\partial x^i} & 0 & 0 & \cdots & 0 & 0 \\ \frac{\partial \widetilde{y}^{(1)j}}{\partial x^i} & \frac{\partial \widetilde{y}^{(1)j}}{\partial y^{(1)i}} & 0 & \cdots & 0 & 0 \\ \cdots & \cdots & \cdots & \cdots & \cdots & \cdots \\ \frac{\partial \widetilde{y}^{(k-1)j}}{\partial x^i} & \frac{\partial \widetilde{y}^{(k-1)j}}{\partial y^{(1)i}} & \frac{\partial \widetilde{y}^{(k-1)j}}{\partial y^{(2)i}} & \cdots & \frac{\partial \widetilde{y}^{(k-1)j}}{\partial y^{(k-1)i}} & 0 \\ \frac{\partial \widetilde{p}_j}{\partial x^i} & 0 & 0 & \cdots & 0 & \frac{\partial x^j}{\partial \widetilde{x}^i} \end{pmatrix}.$$

It follows

$$(4.1.5')\qquad \det J_k(u) = \left[\det\left(\frac{\partial \widetilde{x}^j}{\partial x^i}(u)\right)\right]^{k-1}.$$

Theorem 4.1.1 *1^0 If k is an odd number, then $T^{*k}M$ is an orientable manifold.*

2^0 If k is an even number, the manifold $T^{*k}M$ is orientable if and only if the base manifold M is orientable.

The form (4.1.5) of the Jacobian matrix implies the following transformation, with respect to (4.1.2), of the natural cobasis $\{dx^i, dy^{(1)i}, ..., dy^{(k-1)i}, dp_i\}$, at a point $u \in T^{*k}M$:

$$
\begin{aligned}
&d\widetilde{x}^i = \frac{\partial \widetilde{x}^i}{\partial x^j} dx^j, \\
&d\widetilde{y}^{(1)i} = \frac{\partial \widetilde{y}^{(1)i}}{\partial x^j} dx^j + \frac{\partial \widetilde{y}^{(1)i}}{\partial y^{(1)j}} dy^{(1)j}, \\
&\dots\dots\dots\dots\dots\dots\dots\dots\dots\dots\dots\dots\dots\dots\dots\dots\dots\dots \\
&d\widetilde{y}^{(k-1)i} = \frac{\partial \widetilde{y}^{(k-1)i}}{\partial x^j} dx^j + \frac{\partial \widetilde{y}^{(k-1)i}}{\partial y^{(1)j}} dy^{(1)j} + \dots + \frac{\partial \widetilde{y}^{(k-1)i}}{\partial y^{(k-1)j}} dy^{(k-1)j}, \\
&d\widetilde{p}_j = \frac{\partial \widetilde{p}_j}{\partial x^i} dx^i + \frac{\partial x^i}{\partial \widetilde{x}^j} dp_i.
\end{aligned}
\tag{4.1.6}
$$

Also, we can prove without difficulties the following theorem:

Theorem 4.1.2 *If the differentiable manifold M is paracompact, then the differentiable manifold $T^{*k}M$ is paracompact.*

Consider the category Man of differentiable manifolds.

There exists a covariant functor $T^{*k} : Man \to Man$ in which the differentiable mappings $f : M \to N$, analytical expressed by $x^{i'} = x^{i'}(x^1, ..., x^n)$, ($i'$, $j' = 1', 2', ..., n' = \dim N$) give the mappings $T^{*k}f : T^{*k}M \to T^{*k}N$, in the form

$$
\begin{aligned}
&x^{i'} = x^{i'}(x^1, ..., x^n), \\
&y^{(1)i'} = \frac{\partial x^{i'}}{\partial x^j} y^{(1)j}, \\
&\dots\dots\dots\dots\dots\dots\dots\dots\dots\dots\dots\dots\dots\dots\dots\dots\dots\dots \\
&(k-1) y^{(k-1)i'} = \frac{\partial y^{(k-2)i'}}{\partial x^j} y^{(1)j} + \dots + (k-1) \frac{\partial y^{(k-2)i'}}{\partial y^{(k-2)j}} y^{(k-1)j}, \\
&\frac{\partial x^{i'}}{\partial x^j} p_{i'} = p_j.
\end{aligned}
\tag{4.1.7}
$$

This fact will be used in the theory of subspaces in Hamilton spaces of order k.

4.2 Vertical Distributions. Liouville Vector Fields

The null section $\mathbf{0} : M \to T^{*k}M$ of the projection π^{*k} is defined by $\mathbf{0} : x \in M \to (x, 0, ..., 0) \in T^{*k}M$. As usual we denote $\widetilde{T^{*k}M} = T^{*k}M \setminus \{\mathbf{0}\}$.

The tangent bundle of the manifold $T^{*k}M$, $(TT^{*k}M, d\pi^{*k}, TM)$, allows to define the vertical subbundle $VT^{*k}M = \ker d\pi^{*k}$. We get the vertical distribution V, formed by the fibres of $VT^{*k}M$. V is locally generated by the set of vector fields $\left\{ \frac{\partial}{\partial y^{(1)i}}, \dots, \frac{\partial}{\partial y^{(k-1)i}}, \frac{\partial}{\partial p_i} \right\}$ at every point $u \in T^{*k}M$.

So, V is an integrable distribution and its dimension is kn.
It is convenient to adopt the notation:

$$\overset{\cdot}{\partial}{}^i = \frac{\partial}{\partial p_i}. \tag{4.2.1}$$

Taking into account the relations (4.1.4) we can consider the following sub-distributions of V:

V_{k-1}, locally generated by $\left\{\frac{\partial}{\partial y^{(k-1)i}}\right\}$. Its dimension is n and it is integrable.

V_{k-2}, locally generated by $\left\{\frac{\partial}{\partial y^{(k-2)i}}, \frac{\partial}{\partial y^{(k-1)i}}\right\}$. This has dimension $2n$ and it is integrable, too and so on.

V_1, locally generating by $\left\{\frac{\partial}{\partial y^{(1)i}}, \cdots, \frac{\partial}{\partial y^{(k-1)i}}\right\}$. Its dimension is $(k-1)n$ and it is also integrable. Of course, we have the sequence of inclusions:

$$V_{k-1} \subset V_{k-2} \subset \cdots \subset V_1 \subset V.$$

The transformation of vector field $\overset{\cdot}{\partial}{}^i$ from (4.1.4), $\overset{\cdot}{\partial}{}^i = \frac{\partial x^i}{\partial \widetilde{x}^j} \overset{\cdot}{\widetilde{\partial}}{}^j$, shows that we have one more vertical distribution W_k, locally generated by the vector fields $\left\{\overset{\cdot}{\partial}{}^i\right\}$ at the points $u \in \left(\pi^{*k}\right)^{-1}(U)$. Its dimension is n and it is an integrable distribution, too.

We conclude by

Proposition 4.2.1 *The following direct sum of vector spaces holds:*

$$V_u = V_{1,u} \oplus W_{1,u}\ , \quad \forall u \in T^{*k}M. \tag{4.2.2}$$

Using again (4.1.4) we obtain without difficulties

Theorem 4.2.1 *1⁰ The following operators in the algebra of functions $\mathcal{F}(T^{*k}M)$:*

$$\begin{aligned}
\overset{1}{\Gamma} &= y^{(1)i}\frac{\partial}{\partial y^{(k-1)i}}, \\
\overset{2}{\Gamma} &= y^{(1)i}\frac{\partial}{\partial y^{(k-2)i}} + 2y^{(2)i}\frac{\partial}{\partial y^{(k-1)i}}, \\
&\cdots\cdots\cdots\cdots\cdots\cdots\cdots\cdots\cdots\cdots \\
\overset{k-1}{\Gamma} &= y^{(1)i}\frac{\partial}{\partial y^{(1)i}} + \cdots + (k-1)y^{(k-1)i}\frac{\partial}{\partial y^{(k-1)i}}
\end{aligned} \tag{4.2.3}$$

and

$$C^* = p_i\, \overset{\cdot}{\partial}{}^i \tag{4.2.4}$$

are vector fields on $T^{*k}M$. *They are independent vector fields on the manifold* $\widetilde{T^{*k}M}$.

2⁰ The function

$$\varphi = p_i y^{(1)i} \tag{4.2.5}$$

is a scalar function on the manifold $T^{*k}M$.

Evidently, $\overset{1}{\Gamma}$ belongs to the distribution V_{k-1}, $\overset{2}{\Gamma}$ belongs to V_{k-2}, ..., $\overset{k-1}{\Gamma}$ belongs to the distribution V_1 and the vector field C^* belongs to the distribution W_k.

$\overset{1}{\Gamma}$, ..., $\overset{k-1}{\Gamma}$ are called *the Liouville vector fields* and C^* is *the Hamilton vector field* on $T^{*k}M$.

The Liouville vector fields $\overset{1}{\Gamma}$, ..., $\overset{k-1}{\Gamma}$ are linearly independent on $\widetilde{T^{*k}M}$ and exactly as in ch. 1 we can prove:

Theorem 4.2.2 *For any differentiable function* $H : \widetilde{T^{*k}M} \to \mathbf{R}$, d_0H, d_1H, ..., $d_{k-2}H$ *defined by*

$$\begin{aligned} d_0H &= \frac{\partial H}{\partial y^{(k-1)i}} dx^i, \\ d_1H &= \frac{\partial H}{\partial y^{(k-2)i}} dx^i + \frac{\partial H}{\partial y^{(k-1)i}} dy^{(1)i}, \\ &\cdots\cdots\cdots \\ d_{k-2}H &= \frac{\partial H}{\partial y^{(1)i}} dx^i + \frac{\partial H}{\partial y^{(2)i}} dy^{(1)i} + \cdots + \frac{\partial H}{\partial y^{(k-1)i}} dy^{(k-2)i} \end{aligned} \tag{4.2.6}$$

are fields of 1-form on $\widetilde{T^{*k}M}$.

Proposition 4.2.2 *1⁰* $d_{k-1}H$ *given by*

$$d_{k-1}H = \frac{\partial H}{\partial x^i} dx^i + \frac{\partial H}{\partial y^{(1)i}} dy^{(1)i} + \cdots + \frac{\partial H}{\partial y^{(k-1)i}} dy^{(k-1)i} \tag{4.2.7}$$

is not a field of 1-form.

2⁰ Under a transformation of coordinate on $T^{*k}M$, $d_{k-1}H$ *transforms as follows*

$$d_{k-1}H = d_{k-1}\widetilde{H} + \dot{\widetilde{\partial}}^j\, \widetilde{H} \frac{\partial \widetilde{p}_j}{\partial x^i} dx^i. \tag{4.2.8}$$

3⁰ If $\dot{\partial}^i\, H = 0$, *then* $d_{k-1}H$ *is a field of 1-form.*

Proposition 4.2.3 *The relation between the differential* dH *and* $d_{k-1}H$ *are given by*

$$dH = d_{k-1}H + \dot{\partial}^i\, H dp_i. \tag{4.2.9}$$

Indeed,

$$dH = \frac{\partial H}{\partial x^i}dx^i + \frac{\partial H}{\partial y^{(1)i}}dy^{(1)i} + \cdots + \frac{\partial H}{\partial y^{(k-1)i}}dy^{(k-1)i} + \dot{\partial}^i\, H dp_i =$$

$$= d_{k-1}H + \dot{\partial}^i\, H dp_i.$$

Remark The differential dH being invariant under the coordinate transformations (4.1.2), i.e. $dH = d\widetilde{H}$, from (4.2.9) it follows the rule of transformation (4.2.8) of $d_{k-1}H$.

If $H = \varphi = p_i y^{(1)i}$, then $d_0 H = \cdots = d_{k-3}H = 0$ and

$$\omega = d_{k-2}\varphi = p_i dx^i. \tag{4.2.10}$$

ω is called *the Liouville 1-form* on the manifold $\widetilde{T^{*k}M}$.

The exterior differential $d\omega$ of the Liouville 1-form ω is expressed by

$$\theta = d\omega = dp_i \wedge dx^i. \tag{4.2.11}$$

Using (4.1.6) we can prove the invariance of 2-form θ with respect to (4.1.2).

Now, based on the previous results we obtain:

Theorem 4.2.3 *1^0 The differential forms ω and θ are globally defined on the manifold $\widetilde{T^{*k}M}$.*

2^0 θ is a closed 2-form, i.e. $d\theta = 0$.

3^0 θ is a 2-form of rank $2n$. It is a presymplectic structure on $\widetilde{T^{*k}M}$.

Proof: 1^0 ω and θ are invariant with respect to a change of coordinates (4.1.2).

2^0 $\theta = d\omega$ implies $d\theta = 0$.

3^0 $\theta = dp_i \wedge dx^i$ is a 2-form of rank $2n < (k+1)n = \dim T^{*k}M$, for $k > 1$. Consequently θ is a presymplectic structure on $T^{*k}M$.

Remarks 1^0 For $k = 1$, ω and θ are the

Poincaré-Cartan forms on the cotangent bundle T^*M.

2^0 The previous theorem shows the existence of a natural presymplectic structure on $\widetilde{T^{*k}M}$.

4.3 The Structures J and J^*

There exists a tangent structure J on $T^{*k}M$ defined as usual by the endomorphism $J : \mathcal{X}(T^{*k}M) \to \mathcal{X}(T^{*k}M)$:

$$\begin{aligned} &J\left(\frac{\partial}{\partial x^i}\right) = \frac{\partial}{\partial y^{(1)i}}, J\left(\frac{\partial}{\partial y^{(1)i}}\right) = \frac{\partial}{\partial y^{(2)i}}, \ldots, \\ &J\left(\frac{\partial}{\partial y^{(k-2)i}}\right) = \frac{\partial}{\partial y^{(k-1)i}}, J\left(\frac{\partial}{\partial y^{(k-1)i}}\right) = 0, J(\dot{\partial}^i) = 0 \end{aligned} \tag{4.3.1}$$

at every point $u \in \widetilde{T^{*k}M}$.

By means of (4.1.4) one proves without difficulties

Theorem 4.3.1 *1⁰ J is globally defined on $T^{*k}M$.*

2⁰ J is a tensor field of type $(1,1)$, locally expressed by

$$J = \frac{\partial}{\partial y^{(1)i}} \otimes dx^i + \frac{\partial}{\partial y^{(2)i}} \otimes dy^{(1)i} + \cdots + \frac{\partial}{\partial y^{(k-1)i}} \otimes dy^{(k-2)i}. \tag{4.3.2}$$

3⁰ The structure J is integrable.

4⁰ $J \circ J \circ \cdots \circ J = J^k = 0$.

5⁰ $\ker J = V_{k-1} \oplus W_k$, $Im\ J = V_1$.

6⁰ rank $||J|| = (k-1)n$.

According to the above theorem we may call J *the $k-1$ -tangent structure.* The endomorphism J applied to the Liouville vector fields gives us:

$$J(\overset{1}{\Gamma}) = 0,\ J(\overset{2}{\Gamma}) = \overset{1}{\Gamma},\ ...,\ J(\overset{k-1}{\Gamma}) = \overset{k-2}{\Gamma}, \tag{4.3.3}$$

and

$$J(C^*) = 0. \tag{4.3.3'}$$

Let us consider a vector field $X \in \mathcal{X}(T^{*k}M)$, locally expressed by:

$$X = \overset{(0)i}{X} \frac{\partial}{\partial x^i} + \overset{(1)i}{X} \frac{\partial}{\partial y^{(1)i}} + \cdots + \overset{(k-1)i}{X} \frac{\partial}{\partial y^{(k-1)i}} + X_i\, \dot{\partial}^i\ . \tag{4.3.4}$$

Proposition 4.3.1 *1⁰ For any vector field $X \in \mathcal{X}(T^{*k}M)$, $\overset{1}{X}, ..., \overset{k-1}{X}$ given by*

$$\overset{1}{X} = JX,\ \overset{2}{X} = J^2X,\ ...,\ \overset{k-1}{X} = J^{k-1}X. \tag{4.3.4'}$$

are vector fields.

2⁰ If X is given by (4.3.4), then we have:

$$\overset{1}{X} = \overset{(0)i}{X} \frac{\partial}{\partial y^{(1)i}} + \cdots + \overset{(k-2)i}{X} \frac{\partial}{\partial y^{(k-1)i}},$$

$$\overset{2}{X} = \overset{(0)i}{X} \frac{\partial}{\partial y^{(2)i}} + \cdots + \overset{(k-3)i}{X} \frac{\partial}{\partial y^{(k-1)i}},$$

....

$$\overset{k-1}{X} = \overset{(0)i}{X} \frac{\partial}{\partial y^{(k-1)i}}.$$

3⁰ The vector field $\overset{1}{X}$ belongs to the vertical distribution V_1, $\overset{2}{X}$ belongs to the distribution V_2, ..., $\overset{k-1}{X}$ belongs to the distribution V_{k-1}.

Now consider the adjoint of $k-1$-tangent structure J. It is the endomorphism $J^*: \mathcal{X}^*(T^{*k}M) \to \mathcal{X}^*(T^{*k}M)$ defined by

$$\begin{aligned} &J^*(dy^{(k-1)i}) = dy^{(k-2)i}, \; ..., \; J^*(dy^{(1)i}) = dx^i, \\ &J^*(dx^i) = 0, \; J^*(dp_i) = 0. \end{aligned} \tag{4.3.5}$$

Using (4.3.5) and (4.1.6) we obtain:

Theorem 4.3.2 *1⁰ J^* is globally defined on $T^{*k}M$.*
2⁰ J^* is a tensor field of type $(1,1)$ on $T^{*k}M$, i.e.

$$J^* = dx^i \otimes \frac{\partial}{\partial y^{(1)i}} + dy^{(1)i} \otimes \frac{\partial}{\partial y^{(2)i}} + \cdots + dy^{(k-2)i} \otimes \frac{\partial}{\partial y^{(k-1)i}}. \tag{4.3.6}$$

3⁰ rank $||J^*|| = (k-1)n$.
4⁰ J^* is an integrable structure.

J^* is called *the $k-1$-adjoint tangent structure.*

J^* can be extended to an endomorphism of the exterior algebra $\Lambda(T^{*k}M)$ as follows:

$$\begin{aligned} &J^* f = f, \; \forall f \in \mathcal{F}(T^{*k}M), \\ &(J^*\omega)(X_1, ..., X_q) = \omega(JX_1, ..., JX_q), \; \forall \omega \in \Lambda^q(T^{*k}M). \end{aligned} \tag{4.3.7}$$

Let be $\omega \in \Lambda^1(T^{*k}M)$ and consider

$$\overset{1}{\omega} = J^*\omega, \; ..., \; \overset{k-1}{\omega} = J^{*(k-1)}\omega. \tag{4.3.7'}$$

Then $\overset{1}{\omega}, ..., \overset{k-1}{\omega}$ are 1-form fields.

In particular, we get

$$J^* dH = d_{k-2}H, \; ..., \; J^{*(k-1)}dH = d_0 H. \tag{4.3.8}$$

The $k-1$ adjoint structure J^* allows to introduce the vertical differential operator in the exterior algebra $\Lambda(T^{*k}M)$, [74].

Taking into account the operator of differentiation:

$$d = \frac{\partial}{\partial x^i}dx^i + \frac{\partial}{\partial y^{(1)i}}dy^{(1)i} + \cdots + \frac{\partial}{\partial y^{(k-1)i}}dy^{(k-1)i} + \dot{\partial}^i \, dp_i$$

we define the following operators:

$$(4.3.9)\quad \begin{aligned} d_{k-2} &= J^*d = \frac{\partial}{\partial y^{(1)i}}dx^i + \cdots + \frac{\partial}{\partial y^{(k-1)i}}dy^{(k-2)i},\\ d_{k-3} &= J^{*2}d = \frac{\partial}{\partial y^{(2)i}}dx^i + \cdots + \frac{\partial}{\partial y^{(k-1)i}}dy^{(k-3)i},\\ &\cdots\cdots\cdots\cdots\cdots\cdots\cdots\cdots\\ d_1 &= J^{*(k-2)}d = \frac{\partial}{\partial y^{(k-2)i}}dx^i + \frac{\partial}{\partial y^{(k-1)i}}dy^{(1)i},\\ d_0 &= J^{*(k-1)}d = \frac{\partial}{\partial y^{(k-1)i}}dx^i. \end{aligned}$$

Clearly, these operators d_0, ..., d_{k-2} and d do not depend on the transformation of coordinates on the manifold $T^{*k}M$.

For any $H \in \mathcal{F}(T^{*k}M)$, d_0H, ..., $d_{k-2}H$ are given by Theorem 4.2.2. d_0, ..., d_{k-2} are *the vertical operators of differentiation.*

They can be extended to the exterior algebra $\Lambda(T^{*k}M)$ if we give their restrictions to $\Lambda^0(T^{*k}M)$ and $\Lambda^1(T^{*k}M)$.

As we already have seen d_0H, ..., $d_{k-2}H$ are expressed in (4.2.6) and dH is the differential of H. The restrictions to $\Lambda^1(T^{*k}M)$ are defined by

$$(4.3.10)\quad \begin{cases} d_\alpha(dx^i) = 0,\ d_\alpha(dp_i) = 0,\ (\alpha = 0, ..., k-2),\\ d_\alpha(dy^{(\beta)i}) = 0,\ \ (\alpha = 1, ..., k-2;\ \beta = 1, ..., k-1),\\ d(dx^i) = 0,\ d(dy^{(\beta)i}) = 0,\ d(dp_i) = 0. \end{cases}$$

In this case d_0, ..., d_{k-2} and d are the antiderivations of degree 1.

Proposition 4.3.2 *The vertical differential operators d_0, ..., d_{k-2} and d have the property*

$$(4.3.11)\quad d_\alpha \circ d_\alpha = 0,\ (\alpha = 0, ..., k-2),\ \ d \circ d = 0.$$

For instance, applying the exterior differential d to the 1-forms d_0H , d_1H, ..., $d_{k-2}H$, written in the form

$$(4.3.12)\quad \begin{aligned} d_0H &= \overset{(0)}{p_i}\, dx^i,\\ d_1H &= \overset{(1)}{p_i}\, dx^i + \overset{(0)}{p_i}\, dy^{(1)i},\\ &\cdots\cdots\cdots\cdots\cdots\cdots\\ d_{k-2}H &= \overset{(k-2)}{p_i}\, dx^i + \overset{(k-3)}{p_i}\, dy^{(1)i} + \cdots + \overset{(0)}{p_i}\, dy^{(k-2)i}, \end{aligned}$$

with

$$(4.3.12')\quad \overset{(0)}{p_i} = \frac{\partial H}{\partial y^{(k-1)i}},\ \overset{(1)}{p_i} = \frac{\partial H}{\partial y^{(k-2)i}},\ ...,\ \overset{(k-2)}{p_i} = \frac{\partial H}{\partial y^{(1)i}},$$

we obtain

$$dd_0H = d\overset{(0)}{p_i} \wedge dx^i,$$

$$\text{(4.3.13)} \qquad dd_1H = d\overset{(1)}{p_i} \wedge dx^i + d\overset{(0)}{p_i} \wedge dy^{(1)i},$$

$$\cdots\cdots\cdots\cdots\cdots\cdots\cdots\cdots\cdots\cdots\cdots\cdots\cdots\cdots$$

$$dd_{k-2}H = d\overset{(k-2)}{p_i} \wedge dx^i + d\overset{(k-3)}{p_i} \wedge dy^{(1)i} + \cdots + d\overset{(0)}{p_i} \wedge dy^{(k-2)i}.$$

The previous formulae (4.3.12) are similar to the Jacobi-Ostrogradski formulae (2.4.2) from the Lagrange spaces of order k, $L^{(k)n}$. The equalities (4.3.13) are important in the geometrical theory of the Hamiltonians $H(x, y^{(1)}, ..., y^{(k-1)}, p)$.

4.4 Canonical Poisson Structures on $T^{*k}M$

In this section we state that on the manifold $T^{*k}M$ there exists at least a Poisson structure. Let us consider the brackets:

$$\text{(4.4.1)} \qquad \{f, g\}_0 = \frac{\partial f}{\partial x^i}\frac{\partial g}{\partial p_i} - \frac{\partial f}{\partial p_i}\frac{\partial g}{\partial x^i}, f, g \in \mathcal{F}(T^{*k}M),$$

$$\text{(4.4.1')} \qquad \{f, g\}_\alpha = \frac{\partial f}{\partial y^{(\alpha)i}}\frac{\partial g}{\partial p_i} - \frac{\partial f}{\partial p_i}\frac{\partial g}{\partial y^{(\alpha)i}}, (\alpha = 1, ..., k-1).$$

Theorem 4.4.1 *The bracket $\{\cdot,\cdot\}_\alpha$, $(\alpha = k-1)$ is a canonical Poisson structure on the manifold $T^{*k}M$.*

Proof: Indeed, remarking that with respect to a change of local coordinates on $T^{*k}M$, $\frac{\partial f}{\partial y^{(\alpha)i}}\frac{\partial g}{\partial p_i}$ has a geometrical meaning, it follows

1^0 For any $f, g \in \mathcal{F}(T^{*k}M)$, $\{f, g\}_{k-1}$ is a differentiable function on $T^{*k}M$.

2^0 $\{f, g\}_{k-1} = -\{g, f\}_{k-1}$.

3^0 $\{f, g\}_{k-1}$ is **R**-linear in every argument.

4^0 The Jacobi identities are verified:

$$\{\{f, g\}_{k-1}, h\}_{k-1} + \{\{g, h\}_{k-1}, f\}_{k-1} + \{\{h, f\}_{k-1}, g\}_{k-1} = 0.$$

Also we remark here

5^0 $\{\cdot, gh\}_{k-1} = \{\cdot, g\}_{k-1} h + \{\cdot, h\}_{k-1} g$.
We can see, without dificulties that every bracket $\{f, g\}_0$ and $\{f, g\}_\alpha$, $(\alpha = 1, ..., k-2)$ verifies 2^0, 3^0, 4^0 and 5^0 but they do not satisfy the property 1^0.

The restrictions of these brackets to the special submanifolds immersed in $T^{*k}M$ can induce Poisson structures. As we shall see in ch. 8, §3, the restriction of $\{f, g\}_0$, $f, g \in \mathcal{F}(\Sigma_0)$ to the submanifold Σ_0:

$$\Sigma_0 = \{(x, y^{(1)}, ..., y^{(k-1)}, p) \in T^{(*k)}M | y^{(1)i} = \cdots = y^{(k-1)i} = 0\}$$

has this property.

Using the notion of nonlinear connection N, studied in chapter 6 of the present monograph, we can take an adapted basis $\frac{\delta}{\delta x^i}, \frac{\delta}{\delta y^{(\alpha)i}}, (\alpha = 1, ..., k-2)$, $\frac{\partial}{\partial y^{(k-1)i}}$ and remarking that $\frac{\delta f}{\delta x^i}\frac{\partial g}{\partial p_i}$ and $\frac{\delta f}{\delta y^{(\alpha)i}}\frac{\partial g}{\partial p_i}$, $(\alpha = 1, ..., k-2)$, have geometrical meaning we can construct the following brackets:

$$\text{(4.4.2)}\qquad \begin{aligned} \{f,g\}_0^N &= \frac{\delta f}{\delta x^i}\frac{\partial g}{\partial p_i} - \frac{\delta g}{\delta x^i}\frac{\partial f}{\partial p^i}, f, g \in \mathcal{F}(T^{*k}M), \\ \{f,g\}_\alpha^N &= \frac{\delta f}{\delta y^{(\alpha)i}}\frac{\partial g}{\partial p_i} - \frac{\partial f}{\partial p_i}\frac{\delta g}{\delta y^{(\alpha)i}}, (\alpha = 1, ..., k-2). \end{aligned}$$

We can check, without difficulties

Proposition 4.4.1 *Every bracket from (4.4.2) has the properties 1^0, 2^0, 3^0, 5^0 from Theorem 4.4.1.*

We shall see later that for some particular nonlinear connections, the brackets (4.4.2) have also the property 4^0.

4.5 Homogeneity

The notion of homogeneity for the functions $H(x, y^{(1)}, ..., y^{(k-1)}, p)$ defined on the manifold $T^{*k}M$ can be considered with respect to the vertical variables $y^{(1)i}$, ..., $y^{(k-1)i}$, as well as with respect to the momenta p_i, respectively.

Indeed, any homothety $h_a : T^{*k}M \to T^{*k}M$

$$h_a(x, y^{(1)}, y^{(2)}, ..., y^{(k-1)}, p) = (x, ay^{(1)}, a^2y^{(2)}, ..., a^{k-1}y^{(k-1)}, p)$$

is preserved by the transformations of local coordinates (4.1.2) on $T^{*k}M$.

Let H_y be a group of transformation on $T^{*k}M$:

$$H_y = \{h_a | a \in \mathbf{R}^+\}.$$

The orbit of a point $u_0 = (x_0, y_0^{(1)}, ..., y_0^{(k-1)}, p^0)$ by H_y is given by

$$x^i = x_0^i,\ y^{(1)i} = ay_0^{(1)i},\ ...,\ y^{(k-1)i} = a^{k-1}y_0^{(k-1)i},\ p_i = p_i^0,\ a \in \mathbf{R}^+.$$

The tangent vector at the point $u_0 = h_1(u_0)$ is the Liouville vector field $\overset{k-1}{\Gamma}$ at a point u_0:

$$\overset{k-1}{\Gamma}(u_0) = y_0^{(1)i}\frac{\partial}{\partial y^{(1)i}}|_{u_0} + 2y_0^{(2)i}\frac{\partial}{\partial y^{(2)i}}|_{u_0} + \cdots + (k-1)y_0^{(k-1)i}\frac{\partial}{\partial y^{(k-1)i}}|_{u_0}.$$

Definition 4.5.1 *A function $H : T^{*k}M \to \mathbf{R}$ differentiable on $\widetilde{T^{*k}M}$ and continuous on the null section of the projection π^{*k} is called homogeneous of degree $r \in \mathbf{Z}$ with respect to $(y^{(1)}, ..., y^{(k-1)})$ if*

$$\text{(4.5.1)}\qquad H \circ h_a = a^r H, \quad \forall a \in \mathbf{R}^+.$$

Exactly as in the section 3.1, ch. 3, it follows:

Theorem 4.5.1 *A function $H \in \mathcal{F}(T^{*k}M)$, differentiable on $\widetilde{T^{*k}M}$ and continuous on the null section is r-homogeneous with respect to $(y^{(1)}, ..., y^{(k-1)})$ if and only if*

$$\mathcal{L}_{\overset{k-1}{\Gamma}} H = rH, \tag{4.5.2}$$

where $\mathcal{L}_{\overset{k-1}{\Gamma}}$ is the Lie derivative with respect to the Liouville vector field $\overset{k-1}{\Gamma}$.

Evidently, (4.5.2) can be written in the form

$$y^{(1)i}\frac{\partial H}{\partial y^{(1)i}} + 2y^{(2)i}\frac{\partial H}{\partial y^{(2)i}} + \cdots + (k-1)y^{(k-1)i}\frac{\partial H}{\partial y^{(k-1)i}} = rH. \tag{4.5.2'}$$

As is usually (see Ch.3, §3.1) the notion of homogeneity can be extended to the vector fields on $\widetilde{T^{*k}M}$.

A vector field X on $\widetilde{T^{*k}M}$ is r-homogeneous with respect to $(y^{(1)}, ..., y^{(k-1)})$ if

$$X \circ h_a = a^{r-1} h_a^* \circ X, \ \forall a \in \mathbf{R}^+. \tag{4.5.3}$$

One proves [115]:

Theorem 4.5.2 *A vector field $X \in \mathcal{X}(\widetilde{T^{*k}M})$ is r-homogeneous with respect to $(y^{(1)}, ..., y^{(k-1)})$ if and only if*

$$\mathcal{L}_{\overset{k-1}{\Gamma}} X = (r-1)X. \tag{4.5.4}$$

Of course, $\mathcal{L}_{\overset{k-1}{\Gamma}} X = [\overset{k-1}{\Gamma}, X]$.

Consequently, $\dfrac{\partial}{\partial x^i}, \dfrac{\partial}{\partial y^{(1)i}}, ..., \dfrac{\partial}{\partial y^{(k-1)i}}, \dfrac{\partial}{\partial p_i}$ are 0 , 1, ..., $2-k$, 0 homogeneous, respectively, with respect to $(y^{(1)}, ..., y^{(k-1)})$.

The following properties hold:

1^0 If H is s-homogeneous and $X \in \mathcal{X}(\widetilde{T^{*k}M})$ is r-homogeneous, then HX is $s+r$-homogeneous, with respect to $(y^{(1)}, ..., y^{(k-1)})$.

2^0 If H is s-homogeneous and $X \in \mathcal{X}(\widetilde{T^{*k}M})$ is r-homogeneous, then XH is $s+r-1$-homogeneous, with respect to $(y^{(1)}, ..., y^{(k-1)})$.

A q-form $\omega \in \Lambda^q(\widetilde{T^{*k}M})$ is s-homogeneous, with respect to $(y^{(1)}, ..., y^{(k-1)})$ if

$$\omega \circ h_a^* = a^s \omega, \ \forall a \in \mathbf{R}^+. \tag{4.5.5}$$

As we know, [115] the following theorem holds:

Theorem 4.5.3 *A q-form ω is s-homogeneous with respect to $(y^{(1)},...,y^{(k-1)})$ if and only if*

$$\mathcal{L}_{\overset{k-1}{\Gamma}}\omega = s\omega. \tag{4.5.6}$$

The 1-forms dx^i, $dy^{(1)i}$, ..., $dy^{(k-1)i}$, dp_i are homogeneous of degree 0, 1, ..., $k-1$, 0 respectively, with respect to $(y^{(1)},...,y^{(k-1)})$.

As we remarked above it is important to study the notion of homogeneity with respect to momenta.

Let H_p be the group of homothety:

$$H_p = \left\{h'_a : \widetilde{T^{*k}M} \to \widetilde{T^{*k}M} | a \in \mathbf{R}^+\right\}$$

where $h'_a(x,y^{(1)},...,y^{(k-1)},p) = (x,y^{(1)},...,y^{(k-1)},ap)$. A function $H \in T^{*k}M$, differentiable on $\widetilde{T^{*k}M}$ and continuous on the null section of π^{*k} is homogeneous of degree r, with respect to the momenta p_i, if

$$H \circ h'_a = a^r H, \quad \forall a \in \mathbf{R}^+. \tag{4.5.7}$$

In other words:

$$H(x,y^{(1)},...,y^{(k-1)},ap) = a^r H(x,y^{(1)},...,y^{(k-1)},p), \quad \forall a \in \mathbf{R}^+. \tag{4.5.7'}$$

We have, [115]:

Theorem 4.5.4 *A function $H \in \mathcal{F}(T^{*k}M)$, differentiable on $\widetilde{T^{*k}M}$ and continuous on the null section is r-homogeneous with respect to p_i if and only if*

$$\mathcal{L}_{C^*}H = rH. \tag{4.5.8}$$

But $\mathcal{L}_{C^*}H = C^*H = p_i\dfrac{\partial H}{\partial p_i}$.

Remark It is not difficult to see that if H is differentiable on $T^{*k}M$, then the r-homogeneous function H is a polinom of degree r in the variables p_i.

A vector field $X \in \mathcal{X}(\widetilde{T^{*k}M})$ is homogeneous of degree r if

$$X \circ h'_a = a^{r-1}h'^*_a \circ X, \; \forall a \in \mathbf{R}^+.$$

We have

Theorem 4.5.5 *A vector field $X \in \mathcal{X}(\widetilde{T^{*k}M})$ is r-homogeneous with respect to p_i if and only if:*

$$\mathcal{L}_{C^*}X = (r-1)X. \tag{4.5.9}$$

This result implies that the vector fields $\frac{\partial}{\partial x^i}, \frac{\partial}{\partial y^{(1)i}}, ..., \frac{\partial}{\partial y^{(k-1)i}}, \frac{\partial}{\partial p_i}$ are 1, 1, ..., 1, 0 homogeneous, respectively, with respect to p_i.

If H is s-homogeneous and X is r-homogeneous, then fX is $s+r$ -homogeneous and Xf is $s+r-1$-homogeneous, with respect to p_i.

For instance, if $H(x, y^{(1)}, ..., y^{(k-1)}, p)$ is a function r -homogeneous with respect to p_i, then

1^0 $\frac{\partial H}{\partial p_i}$ is $r-1$-homogeneous;

2^0 $\frac{\partial^2 H}{\partial p_i \partial p_j}$ is $r-2$ -homogeneous.

A q-form $\omega \in \Lambda^q(\widetilde{T^{*k}M})$ is r-homogeneous with respect to momenta p_i if

$$\omega \circ h_a^{\prime *} = a^r \omega, \ \forall a \in \mathbf{R}^+.$$

It follows

Theorem 4.5.6 *A q-form ω is r-homogeneous with respect to p_i if*

$$\mathcal{L}_{C^*}\omega = r\omega. \tag{4.5.10}$$

Consequently, dx^i, $dy^{(1)i}$, ..., $dy^{(k-1)i}$, dp_i are 0, 0, ..., 0, 1-homogeneous with respect to p_i, respectively.

Finally, we determine the degree of homogeneity of the function $\{f, g\}_{k=1}$.

Proposition 4.5.1 *If f and g are r and s-homogeneous functions with respect to p_i, then the function $\{f, g\}_{k-1}$ is homogeneous of degree $s+r-1$.*

For applications to the geometry of Cartan spaces of order k is important to have a special notion of homogeneity on the fibres of the bundle $T^{*k}M$.

More precisely, the homothety

$$\overline{h}_a : (x, y^{(1)}, ..., y^{(k-1)}, p) \in T^{*k}M \to (x, ay^{(1)}, ..., a^{k-1}y^{(k-1)}, a^k p) \in T^{*k}M$$

is preserved by the transformation of local coordinates (4.1.2) on $T^{*k}M$.

Let $H_{y,p}$ be the group of transformation on $T^{*k}M$:

$$H_{y,p} = \left\{ \overline{h}_a : (x, y^{(1)}, ..., y^{(k-1)}, p) \to (x, ay^{(1)}, ..., a^{k-1}y^{(k-1)}, a^k p) | a \in \mathbf{R}^+ \right\}.$$

The orbit of a point $u_0 = (x_0, y_0^{(1)}, ..., y_0^{(k-1)}, p^0)$ by $H_{y,p}$ is given by

$$x^i = x_0^i,\ y^{(1)i} = ay_0^{(1)i},\ ...,\ y^{(k-1)} = a^{k-1}y_0^{(k-1)i},\ p_i = a^k p_i^0,\ \forall a \in \mathbf{R}^+.$$

The tangent vector at the point $u_0 = \overline{h}_1(u_0)$ is the vector field $\overset{k-1}{\Gamma} + kC^*$ at the point u_0:

$$\overset{k-1}{\Gamma}(u_0) + kC^*(u_0) = y_0^{(1)i}\frac{\partial}{\partial y^{(1)i}}|_{u_0} + \cdots + (k-1)y_0^{(k-1)i}\frac{\partial}{\partial y^{(k-1)i}}|_{u_0} + kp_i^0\,\dot{\partial}^i\,|_{u_0}.$$

Definition 4.5.2 *A function* $H : T^{*k}M \to \mathbf{R}$ *differentiable on* $\widetilde{T^{*k}M}$ *and continuous on the null section of the projection* π^{*k} *is called homogeneous of degree* $r \in \mathbf{Z}$ *on the fibres of the bundle* $T^{*k}M$ *if*

$$H \circ \overline{h}_a = a^r H, \ \forall a \in \mathbf{R}^+. \tag{4.5.11}$$

Applying the usual methods it follows:

Theorem 4.5.7 *A function* H *on* $T^{*k}M$, *differentiable on* $\widetilde{T^{*k}M}$ *and continuous on the null section is* r*-homogeneous on the fibres of* $T^{*k}M$ *if and only if*

$$\mathcal{L}_{\overset{k-1}{\Gamma}+kC^*} H = rH. \tag{4.5.12}$$

If we expand (4.5.12), we can write it in the form

$$y^{(1)i}\frac{\partial H}{\partial y^{(1)i}} + \cdots + (k-1)y^{(k-1)i}\frac{\partial H}{\partial y^{(k-1)i}} + kp_i \dot{\partial}^i H = rH. \tag{4.5.12'}$$

A vector field X on $\widetilde{T^{*k}M}$ is r-homogeneous on the fibres of $T^{*k}M$ if

$$X \circ \overline{h}_a = a^{r-1}\overline{h}^*_a \circ X, \ \forall a \in \mathbf{R}^+. \tag{4.5.13}$$

It follows

Theorem 4.5.8 *A vector field* $X \in \mathcal{X}(\widetilde{T^{*k}M})$ *is* r*-homogeneous on the fibres of* $T^{*k}M$ *if and only if*

$$\mathcal{L}_{\overset{k-1}{\Gamma}+kC^*} X = (r-1)X. \tag{4.5.14}$$

Of course, (4.5.14) can be given in the form

$$\left[\overset{k-1}{\Gamma}, X\right] + k\,[C^*, X] = (r-1)X. \tag{4.5.14'}$$

Corollary 4.5.1 *1⁰ The vector fields* $\dfrac{\partial}{\partial x^i}, \dfrac{\partial}{\partial y^{(1)i}}, \ldots, \dfrac{\partial}{\partial y^{(k-1)i}}$ *and* $\dot{\partial}^i = \dfrac{\partial}{\partial p_i}$ *are* $1, 0, \ldots, 2-k, 1-k$ *homogeneous on the fibres of* $\widetilde{T^{*k}M}$, *respectively.*

2⁰ If $H \in \mathcal{F}(\widetilde{T^{*k}M})$ *is* s*-homogeneous and* $X \in \mathcal{X}(\widetilde{T^{*k}M})$ *is* r*-homogeneous on the fibres of* $T^{*k}M$ *then*

- *a.* HX *is* $r+s$*-homogeneous;*
- *b.* XH *is* $r+s-1$*-homogeneous.*

A q-form $\omega \in \Lambda^q(\widetilde{T^{*k}M})$ is s-homogeneous on the fibres of $T^{*k}M$ if

$$\omega \circ \overline{h}^*_a = a^s\omega, \ \forall a \in \mathbf{R}^+.$$

The following theorem holds:

Theorem 4.5.9 *A q-form* $\omega \in \Lambda^q(\widetilde{T^{*k}M})$ *is s-homogeneous on the fibres of* $T^{*k}M$ *if and only if*

$$\mathcal{L}_{\overset{k-1}{\Gamma}+kC^*}\,\omega = s\omega. \tag{4.5.15}$$

Corollary 4.5.2 *The 1-forms* dx^i, $dy^{(1)i}$, ..., $dy^{(k-1)i}$, dp_i *are* 0, 1, ..., $k-1$, k*-homogeneous on the fibres of* $T^{*k}M$.

We will apply these results in the study of the homogeneity on the fibres of $T^{*k}M$ of 1-forms d_0H, ..., $d_{k-2}H$ and, of course of the functions $\{f, g\}_{k=1}$.

Chapter 5

The Variational Problem for the Hamiltonians of Order k

The theory of higher order Hamiltonian systems and its applications in Analytical Mechanics are consistent only if we study the variational problem for the differentiable Hamiltonians of order k, $H(x, y^{(1)}, ..., y^{(k-1)}, p)$, [96, 98].

In this case the integral of action of H must be defined along curve c on the cotangent manifold T^*M by

$$I(c) = \int_0^1 [p_i \frac{dx^i}{dt} - \frac{1}{2}H(x, \frac{dx}{dt}, ..., \frac{1}{(k-1)!}\frac{d^{k-1}x}{dt^{k-1}}, p)]dt.$$

A local variation of c is a curve $\widetilde{c}(\varepsilon_1, \varepsilon_2)$ which depend on a vector field V^i and a covector field η_i. The integral of action $I(\widetilde{c}(\varepsilon_1, \varepsilon_2))$ depends on two parameters $\varepsilon_1, \varepsilon_2$. In order for the functional $I(c)$ to be an extremal value of the functionals $I(\widetilde{c}(\varepsilon_1, \varepsilon_2))$ it is necessary that

$$\left.\frac{\partial I(\widetilde{c}(\varepsilon_1, \varepsilon_2))}{\partial \varepsilon_\alpha}\right|_{\varepsilon_1=\varepsilon_2=0} = 0, \quad (\alpha = 1, 2)$$

These conditions allow to determine the Hamilton-Jacoby equations (5.1.17).

Introducing the higher order energies of H, $\mathcal{E}^{k-1}(H), ..., \mathcal{E}^1(H)$, a law of conservation of the energy $\mathcal{E}^{k-1}(H)$ is proved and a Nöther type theorem is formulated. This theory is valid in the case when the order k is greater then 1.

5.1 The Hamilton-Jacobi Equations

A function $H : T^{*k}M \to \mathbf{R}$ differentiable on $\widetilde{T^{*k}M}$ and continue on the nul section is called a differentiable Hamiltonian of order k. It depends on the variables

$(x^i, y^{(1)i}, ..., y^{(k-1)i}, p_i)$. So, it will be denoted by $H(x, y^{(1)}, ..., y^{(k-1)}, p)$.

Let us consider a curve

$$c : t \in [0,1] \to (x^i(t), p_i(t)) \in \widetilde{T^*M},$$

having the image in a local chart of the manifold $\widetilde{T^*M}$. The curve c can be analytically given by the equations:

$$x^i = x^i(t),\ p_i = p_i(t),\ t \in [0,1]. \tag{5.1.1}$$

The extension $\overset{\vee}{c}$ to the dual bundle $T^{*k}M$ is well determined. The extension $\overset{\vee}{c}$ is given by the equations

$$\begin{aligned} &x^i = x^i(t), \\ &y^{(\alpha)i}(t) = \frac{1}{\alpha!}\frac{d^\alpha x^i}{dt^\alpha}(t), (\alpha = 1, ..., k-1), \\ &p_i = p_i(t),\ t \in [0,1]. \end{aligned} \tag{5.1.2}$$

Also, we consider a vector field $V^i(t)$ and a covector field $\eta_i(t)$ along curve c, having the properties:

$$\begin{aligned} &V^i(0) = V^i(1) = 0, \eta_i(0) = \eta_i(1) = 0, \\ &\frac{d^\alpha V^i}{dt^\alpha}(0) = \frac{d^\alpha V^i}{dt^\alpha}(1) = 0, (\alpha = 1, ..., k-2). \end{aligned} \tag{5.1.3}$$

The variation $\overline{c}(\varepsilon_1, \varepsilon_2)$ of a curve c determined by the pair $(V^i(t), \eta_i(t))$ is defined by

$$\begin{aligned} &\overline{x}^i = x^i(t) + \varepsilon_1 V^i(t), \\ &\overline{p}_i = p_i(t) + \varepsilon_2 \eta_i(t), t \in [0,1], \end{aligned} \tag{5.1.4}$$

where ε_1 and ε_2 are constants, small in the absolute values, such that the image of the curve $\widetilde{c}(\varepsilon_1, \varepsilon_2)$ belongs to the same domain of chart on $\widetilde{T^*M}$ as the image of curve c. The extension of $\overline{c}(\varepsilon_1, \varepsilon_2)$ is the curve $\overset{\vee}{\overline{c}}(\varepsilon_1, \varepsilon_2)$ given by the equations:

$$\begin{aligned} &\overline{x}^i = x^i(t) + \varepsilon_1 V^i(t), \\ &\overline{y}^{(\alpha)i} = \frac{1}{\alpha!}\left(\frac{d^\alpha x^i}{dt^\alpha} + \varepsilon_1 \frac{d^\alpha V^i}{dt^\alpha}\right),\ (\alpha = 1, ..., k-1), \\ &\overline{p}_i = p_i(t) + \varepsilon_2 \eta_i(t),\ t \in [0,1]. \end{aligned} \tag{5.1.5}$$

The integral of action for the Hamiltonian $H(x, y^{(1)}, ..., y^{(k-1)}, p)$ along curve c is defined, like an extension of the classical form, by

$$(5.1.6)\quad I(c) = \int_0^1 [p_i(t)\frac{dx^i}{dt}(t) - \frac{1}{2}H(x(t), \frac{dx}{dt}(t), ..., \frac{1}{(k-1)!}\frac{d^{k-1}x}{dt^{k-1}}(t), p(t))]dt.$$

Evidently, $p_i\frac{dx^i}{dt} - \frac{1}{2}H(x, \frac{dx}{dt}, ..., \frac{1}{(k-1)!}\frac{d^{k-1}x}{dt^{k-1}}, p)$ is a differentiable Hamiltonian on the curve c.

The integral of action $I(\overline{c}(\varepsilon_1, \varepsilon_2))$ is:

(5.1.7)

$$I(\overline{c}(\varepsilon_1, \varepsilon_2)) = \int_0^1 [(p + \varepsilon_2\eta)(\frac{dx}{dt} + \varepsilon_1\frac{dV}{dt}) - \frac{1}{2}H(x + \varepsilon_1 V, \frac{dx}{dt} + \varepsilon_{11}\frac{dV}{dt}), ...$$

$$... \frac{1}{(k-1)!}(\frac{d^{k-1}x}{dt^{k-1}} + \varepsilon_1\frac{d^{k-1}V}{dt^{k-1}}), p + \varepsilon_2\eta]dt.$$

The necessary conditions in order that $I(c)$ is an extremal value of $I(\overline{c}(\varepsilon_1, \varepsilon_2))$ are:

$$(5.1.8)\qquad \frac{\partial I(\overline{c}(\varepsilon_1, \varepsilon_2))}{\partial \varepsilon_1}|_{\varepsilon_1=\varepsilon_2=0} = 0, \frac{\partial I(\overline{c}(\varepsilon_1, \varepsilon_2))}{\partial \varepsilon_2}|_{\varepsilon_1=\varepsilon_2=0} = 0.$$

In our conditions of differentiability, using the equality (5.1.7), we get the equations:

(5.1.9)

$$\int_0^1 [p_i(t)\frac{dV^i}{dt}(t) - \frac{1}{2}(\frac{\partial H}{\partial x^i}V^i + \frac{\partial H}{\partial y^{(1)i}}\frac{dV^i}{dt} + ... + \frac{1}{(k-1)!}\frac{\partial H}{\partial y^{(k-1)i}}\frac{d^{k-1}V^i}{dt^{k-1}})]dt = 0$$

and

$$(5.1.10)\qquad \int_0^1 [\frac{dx^i}{dt} - \frac{1}{2}\frac{\partial H}{\partial p_i}]\eta_i dt = 0.$$

So, we obtain:

Theorem 5.1.1 *The necessary conditions for $I(c)$ to be an extremal value of the functional $I(\overline{c}(\varepsilon_1, \varepsilon_2)$ are given by the equations (5.1.9) and (5.1.10).*

The previous equations imply the Hamilton-Jacobi equations of the Hamiltonian H. To prove this we need to introduce some new notions.

Consider the following main invariants, [96]:

$$(5.1.11)\qquad I^1(H) = \mathcal{L}_{\overset{1}{\Gamma}}H,\ I^2(H) = \mathcal{L}_{\overset{2}{\Gamma}}H, ..., I^{k-1}(H) = \mathcal{L}_{\overset{k-1}{\Gamma}}(H),$$

in which $\mathcal{L}$ is the Lie operator of derivation and

$$\overset{1}{\Gamma}= y^{(1)i}\frac{\partial}{\partial y^{(k-1)i}},$$

$$\overset{2}{\Gamma}= y^{(1)i}\frac{\partial}{\partial y^{(k-2)i}}+2y^{(2)i}\frac{\partial}{\partial y^{(k-1)i}}, \tag{5.1.12}$$

$$\overset{k-1}{\Gamma} = y^{(1)i}\frac{\partial}{\partial y^{(1)i}}+2y^{(2)i}\frac{\partial}{\partial y^{(2)i}}+...+(k-1)y^{(k-1)i}\frac{\partial}{\partial y^{(k-1)i}}$$

are the Liouville vector fields on the manifold $T^{*k}M$. Also, we define the invariants:

(5.1.13)

$$I^1_V(H)=V^i\frac{\partial H}{\partial y^{(k-1)i}},$$

$$I^2_V(H)=V^i\frac{\partial H}{\partial y^{(k-2)i}}+\frac{dV^i}{dt}\frac{\partial H}{\partial y^{(k-1)i}},$$

....

$$I^{k-1}_V(H)=V^i\frac{\partial H}{\partial y^{(1)i}}+\frac{dV^i}{dt}\frac{\partial H}{\partial y^{(2)i}}+...+\frac{1}{(k-2)!}\frac{d^{k-2}V^i}{dt^{k-2}}\frac{\partial H}{\partial y^{(k-1)i}}).$$

For $V^i=\frac{dx^i}{dt}$, the invariants (5.1.13) are the same with the invariants $I^1(H),...,I^{(k-1)}(H)$ along curve c. An important notation is as follows:

(5.1.14)

$$\overset{\circ}{E}_i(H)=\frac{dp_i}{dt}+\frac{1}{2}[\frac{\partial H}{\partial x^i}-\frac{d}{dt}\frac{\partial H}{\partial y^{(1)i}}+...+(-1)^{k-1}\frac{1}{(k-1)!}\frac{d^{k-1}}{dt^{k-1}}\frac{\partial H}{\partial y^{(k-1)i}}]$$

Later we prove that $\overset{\circ}{E}_i(H)$ is a d-covector field along curve c.
By a straighforward calculus we can prove:

Lemma 5.1.1 *The following identity holds:*

$$p_i\frac{dV^i}{dt}-\frac{1}{2}[\frac{\partial H}{\partial x^i}V^i+\frac{\partial H}{\partial y^{(1)i}}\frac{dV^i}{dt}+...+\frac{1}{(k-1)!}\frac{\partial H}{\partial y^{(k-1)i}}\frac{d^{k-i}V^i}{dt^{k-1}}]=$$

$$=-\overset{\circ}{E}_i(H)V^i+\frac{d}{dt}(p_iV^i)-\frac{1}{2}\frac{d}{dt}[I^{k-1}_V(H)-\frac{1}{2!}\frac{d}{dt}I^{k-2}_V(H)+... \tag{5.1.15}$$

$$+(-1)^{k-2}\frac{1}{(k-1)!}\frac{d^{k-2}}{dt^{k-2}}I^1_V(H)].$$

Using the previous Lemma, we can prove:

Theorem 5.1.2 *The equations (5.1.9) and (5.1.10) are equivalent to the equations*

$$(5.1.16)\qquad \int_0^1 \overset{\circ}{E}_i(H)V^i dt = 0;\quad \int_0^1 [\frac{dx^i}{dt} - \frac{1}{2}\frac{\partial H}{\partial p_i}]\eta_i dt = 0.$$

Proof. By means of the identity (5.1.15) the equations (5.1.9) can be written:

$$\int_0^1 \{-\overset{\circ}{E}_i(H)V^i + \frac{d}{dt}[p_i V^i - \frac{1}{2}(I_V^{k-1}(H) - \frac{1}{2!}\frac{d}{dt}I_V^{k-2}(H) + ...+$$
$$+(-1)^{k-2}\frac{1}{(k-1)!}\frac{d^{k-2}}{dt^{k-2}}I_V^1)]\}dt = 0.$$

Integrating and taking into account the conditions (5.1.3) and the expression of the invariants (5.1.13) we obtain the announced result. q.e.d.

Now, the equations (5.1.16) in which V^i and η_i are arbitrary lead to the following Hamilton-Jacoby equations:

Theorem 5.1.3 *In order for the integral of action $I(c)$, (5.1.6) to be an extremal value for the functionals $I(\overline{c}(\varepsilon_1, \varepsilon_2))$, (5.1.7) it is necessary that the curve c to satisfy the following Hamilton-Jacobi equations*

(5.1.17)

$$\frac{dx^i}{dt} = \frac{1}{2}\frac{\partial H}{\partial p_i},$$

$$\frac{dp_i}{dt} = -\frac{1}{2}[\frac{\partial H}{\partial x^i} - \frac{d}{dt}\frac{\partial H}{\partial y^{(1)i}} + ... + (-1)^{(k-1)}\frac{1}{(k-1)!}\frac{d^{k-1}}{dt^{k-1}}\frac{\partial H}{\partial y^{(k-1)i}}],$$

where

$$(5.1.17')\qquad y^{(1)i} = \frac{dx^i}{dt}, ..., y^{(k-1)i} = \frac{1}{(k-1)!}\frac{d^{k-1}x^i}{dt^{k-1}}.$$

Evidently, the second equation (5.1.17) is equivalent to $\overset{\circ}{E}_i(H) = 0$.

Another important property is expressed in the following theorem:

Theorem 5.1.4 *$\overset{\circ}{E}_i(H)$ is a covector field.*

This result can be proved by a direct calculation. Another way is as follows. With respect to a change of local coordinates on $\widetilde{T^{*k}M}$ we have

$$\int_0^1 \overset{\circ}{\widetilde{E}}_i(\widetilde{H})\widetilde{V}^i dt - \int_0^1 \overset{\circ}{E}_i(H)V^i dt = \int_0^1 [\overset{\circ}{\widetilde{E}}_i(\widetilde{H})\frac{\partial \widetilde{x}^i}{\partial x^j} - \overset{\circ}{E}_j(H)]V^j dt = 0$$

Since V^i is an arbitrary vector field we obtain

$$\overset{\circ}{\widetilde{E}}_i(\widetilde{H})\frac{\partial \widetilde{x}^i}{\partial x^j} = \overset{\circ}{E}_j(H).$$

The previous property shows that the equation $\overset{\circ}{E}_i(H) = 0$ has a geometrical meaning.

5.2 Zermelo Conditions

The integral of action (5.1.6) is defined for the parametrized curves $c : t \in [0,1] \to (x^i(t), p_i(t)) \in \widetilde{T^*M}$. The problem is when it does not depend on the parametrization of curve c.

Consider a differentiable diffeomorphism $\widetilde{t} = \widetilde{t}(t), t \in [0,1]$ which defines a new parametrization of the curve c. If $a = \widetilde{t}(0), b = \widetilde{t}(1)$, then c will be represented by

$$c : \widetilde{t} \in [a,b] \to (\widetilde{x}^i(t), \widetilde{p}_i(\widetilde{t})) \in \widetilde{T^*M}.$$

In order that the integral of action $I(c)$ does not depend on the parametrization of curve c is necessary that:

$$\{\widetilde{p}_i\frac{d\widetilde{x}^i}{d\widetilde{t}} - \frac{1}{2}\widetilde{H}(\widetilde{x}, \frac{d\widetilde{x}}{d\widetilde{t}}, \ldots, \frac{1}{(k-1)!}\frac{d^{k-1}\widetilde{x}}{d\widetilde{t}^{k-1}}, \widetilde{p})\}\frac{d\widetilde{t}}{dt} =$$

$$(5.2.1) \qquad = p_i\frac{dx^i}{d\widetilde{t}}\frac{d\widetilde{t}}{dt} - \frac{1}{2}H[x, \frac{dx}{d\widetilde{t}}\frac{d\widetilde{t}}{dt}, \frac{1}{2!}\frac{d}{dt}(\frac{dx}{d\widetilde{t}}\frac{d\widetilde{t}}{dt}), \ldots$$

$$\ldots, \frac{1}{(k-1)!}\frac{d^{k-2}}{dt^{k-2}}(\frac{dx}{d\widetilde{t}}\frac{d\widetilde{t}}{dt}), p].$$

The previous equality holds for any diffeomorpfism $\widetilde{t} = \widetilde{t}(t)$.

If we take the derivatives of (5.2.1) with respect to $\frac{d\widetilde{t}}{dt}$ and take $\widetilde{t} = t$ we obtain

$$(5.2.2) \qquad H = y^{(1)i}\frac{\partial H}{\partial y^{(1)i}} + \ldots + (k-1)y^{(k-1)i}\frac{\partial H}{\partial y^{(k-1)i}}.$$

If we take again the derivative of (5.2.1) with respect to $\frac{d^2\widetilde{t}}{dt^2}$ and consider $\widetilde{t} = t$ we have

$$(5.2.2') \qquad 0 = y^{(1)i}\frac{\partial H}{\partial y^{(2)i}} + \ldots + (k-2)y^{(k-2)i}\frac{\partial H}{\partial y^{(k-2)i}}.$$

And so on.

Therefore we have:

Theorem 5.2.1 *The necessary conditions that the integral of action $I(c)$, (5.1.6) does not depend on the parametrization of the curve c are the following ones:*

$$I^{k-1}(H) = H, I^{k-2}(H) = 0, ..., I^1(H) = 0. \tag{5.2.3}$$

Indeed, the equation (5.2.2) can be written as $H = \underset{\Gamma}{\mathcal{L}_1} H = I^{k-1}(\dot{H})$. The equation (5.2.2') is expressed by $0 = \underset{\Gamma}{\mathcal{L}_2} H = I^{(k-2)}(H)$, etc. Q.E.D.

The equations (5.2.3) will be called the Zermelo conditions. They were introduced for the higher order Lagrangians by Kazuo Kondo [70] and are fundamental for the definition of the Kawaguchi spaces [63].

These conditions are very restrictive for the Hamiltonians H. Indeed, let us consider the Hessian of H, with respect to momenta p_i. It has the elements:

$$g^{ij} = \frac{1}{2} \frac{\partial^2 H}{\partial p_i \partial p_j}. \tag{5.2.4}$$

In next chapter we shall prove that g^{ij} is a tensor field and its is called the fundamental tensor field of the Hamiltonian H.

Now is not difficult to prove the following result:

Theorem 5.2.2 *If the Hamiltonian H satisfies the Zermelo conditions (5.2.3), then its fundamental tensor g^{ij} has the properties:*

$$\underset{\Gamma}{\mathcal{L}_1} g^{ij} = ... = \underset{\Gamma}{\mathcal{L}_{k-2}} g^{ij} = 0, \underset{\Gamma}{\mathcal{L}_{k-1}} g^{ij} = g^{ij}. \tag{5.2.5}$$

Proof. Remarking that the invariants $I^1(H) = \underset{\Gamma}{\mathcal{L}_1} H, ..., I^{(k-1)} = \underset{\Gamma}{\mathcal{L}_{k-1}} H$ have the properties:

$$\dot{\partial}^i \dot{\partial}^i \, I^{(k-1)}(H) = I^{(k-1)}(\dot{\partial}^i \dot{\partial}^i \, H), ..., \dot{\partial}^i \dot{\partial}^i \, I^1(H) = I^1(\dot{\partial}^i \dot{\partial}^i \, H)$$

and using the Zermelo conditions (5.2.3), it follows the equations (5.2.5).

The first conditions (5.2.3), $I^{k-1}(H) = H$ and theorem 4.5.1 allow to prove:

Theorem 5.2.3 *A necessary condition that the Zermelo conditions (5.2.3) be verified is that the Hamiltonian H is 1-homogeneous with respect to the variables $y^{(1)i}, ..., y^{(k-1)i}$.*

Corollary 5.2.1 *If the differentiable Hamiltonian H is not 1-homogeneous with respect to $y^{(1)i}, ..., y^{(k-1)i}$ then the Zermelo conditions (5.2.3) are not verified.*

Corollary 5.2.2 *If the differentiable Hamiltonian H is 1-homogeneous with respect to $y^{(1)i}, ..., y^{(k-1)i}$ and 2-homogeneous with respect to p_i, then H is $2k+1$-homogeneous on the fibres of $T^{*k}M$.*

Indeed, $\mathcal{L}_{\underset{\Gamma}{k-1} + kC^*} H = (1 + 2k)H$.

Corollary 5.2.3 *If the differentiable Hamiltonian H has the properties:*
a. It satisfies the Zermelo conditions
b. H is $mk+1$-homogeneous on the fibres of $T^{*k}M$, ($m \in Z$), then H is m-homogeneous with respect to momenta p_i.

Remark 5.2.1 *The Hamiltonian $H = p_i y^{(1)i}$ satisfies the Zermelo conditions*

5.3 Higher Order Energies. Conservation of Energy $\mathcal{E}^{k-1}(H)$

For a differentiable Hamiltonian H the following invariants are important, [98]:

(5.3.1)
$$\mathcal{E}^{k-1}(H) = $$
$$= I^{k-1}(H) - \frac{1}{2!}\frac{d}{dt}I^{k-2}(H) + ... + (-1)^{k-2}\frac{1}{(k-1)!}\frac{d^{k-2}}{dt^{k-2}}I^1(H) - H,$$

$$\mathcal{E}^{k-2}(H) = $$
$$= -\frac{1}{2!}\frac{d}{dt}I^{k-2}(H) + \frac{1}{3!}\frac{d}{dt}I^{k-3}(H) + ... + (-1)^{k-3}\frac{1}{(k-1)!}\frac{d^{k-3}}{dt^{k-3}}I^1(H),$$

....

$$\mathcal{E}^1(H) = (-1)^{k-2}\frac{1}{(k-1)!}I^1(H).$$

The invariants $\mathcal{E}^{k-1}(H), \mathcal{E}^{k-2}(H), ..., \mathcal{E}^1(H)$ are called *the energies of order* $k-1, k-2, ..., 1$ of the Hamiltonian H, respectively.

Their expressions justify the invariant character for each. And these energies are essential for studying the Nother symmetries of H.

In order to prove the law of conservation for the energy $\mathcal{E}^{k-1}(H)$ we need some preliminary considerations.

Lemma 5.3.1 *If the differentiable Hamiltonian H has the property*

$$\frac{dx^i}{dt} = \frac{1}{2}\frac{\partial H}{\partial p_i}, \tag{5.3.2}$$

then we have

$$\frac{dH}{dt} = 2\overset{\circ}{E}_i(H)\frac{dx^i}{dt} + \frac{d}{dt}[I^{k-1}(H) - \frac{1}{2!}\frac{d}{dt}I^{k-2}(H) + ... \tag{5.3.3}$$
$$... + (-1)^{k-2}\frac{1}{(k-1)!}\frac{d^{k-2}}{dt^{k-2}}I^1(H)].$$

Proof. The condition (5.3.2) implies:
$$\frac{dH}{dt} = \frac{\partial H}{\partial x^i}\frac{dx^i}{dt} + \frac{\partial H}{\partial p_i}\frac{dp_i}{dt} + (\frac{\partial H}{\partial y^{(1)i}}\frac{dy^{(1)i}}{dt} + ... + \frac{\partial H}{\partial y^{(k-1)i}}\frac{dy^{(k-1)i}}{dt}) =$$

$$= 2\frac{dp_i}{dt}\frac{dx^i}{dt} + \{\frac{\partial H}{\partial x^i}\frac{dx^i}{dt} + \frac{\partial H}{\partial y^{(1)i}}\frac{dy^{(1)i}}{dt} + ... + \frac{\partial H}{\partial y^{(k-1)i}}\frac{dy^{(k-1)i}}{dt}\} =$$

$$= -2p_i\frac{d^2x^i}{dt^2} + \{\frac{\partial H}{\partial x^i}\frac{dx^i}{dt} + \frac{\partial H}{\partial y^{(1)i}}\frac{dy^{(1)i}}{dt} + ... + \frac{\partial H}{\partial y^{(k-1)i}}\frac{dy^{(k-1)i}}{dt}\} + 2\frac{d}{dt}(p_i\frac{dx^i}{dt}).$$

Using Lemma 4.1.1 for $V^i = \frac{dx^i}{dt}, \frac{dV^i}{dt} = \frac{d^2x^i}{dt^2}, ...$ the previous equality leads to the formula (5.3.3). Q.E.D.

Now, we can prove without difficulties

Theorem 5.3.1 *If the differentiable Hamiltonian H satisfies the first Hamilton-Jacobi equations (5.3.2) then the variation of the energy of order $k-1$, $\mathcal{E}^{k-1}(H)$ is given by:*

$$\frac{d\mathcal{E}^{k-1}(H)}{dt} = -\overset{\circ}{E_i}(H)\frac{dx^i}{dt}. \tag{5.3.4}$$

Clearly, this formula leads to an interesting result

Theorem 5.3.2 *Along every solution curve c of the Hamilton-Jacobi equations (5.1.17) the energy of order $k-1$, $\mathcal{E}^{k-1}(H)$, is conserved.*

Finally, we notice the following theorems:

Theorem 5.3.3 *If the differentiable Hamiltonian H verifies the Zermelo conditions (5.2.3), then the energies of H, $\mathcal{E}^{k-1}(H), ..., \mathcal{E}^1(H)$ vanish.*

Theorem 5.3.4 *If the differentiable Hamiltonian H verifies the Zermelo conditions (5.2.3), then along any curve c which satisfies the first Hamilton-Jacobi equations (5.1.17) we have*

$$\overset{\circ}{E_i}(H)\frac{dx^i}{dt} = 0.$$

5.4 The Jacobi-Ostrogradski Momenta

The theory of Jacoby-Ostrogradski momenta of the Lagrangians of order k, briefly presented in the Chapter 2, can be extended to the Hamiltonians of order k.

Indeed, from (5.3.1) we see that the energy of order $k-1$, $\mathcal{E}^{k-1}(H)$ is a polynomial function of degree one in the higher order accelerations $\frac{dx^i}{dt}, ..., \frac{d^{k-1}x^i}{dt^{k-1}}$. So, we have

$$(5.4.1) \qquad \mathcal{E}^{k-1}(H) = p_{(1)i}\frac{dx^i}{dt} + ... + p_{(k-1)i}\frac{d^{k-1}x^i}{dt^{k-1}} - H$$

A straighfoward calculus shows that $p_{(1)i}, ..., p_{(k-1)i}$ are given by

(5.4.2)

$$p_{(1)i} = \frac{\partial H}{\partial y^{(1)i}} - \frac{1}{2!}\frac{d}{dt}\frac{\partial H}{\partial y^{(2)i}} + ... + (-1)^{k-2}\frac{1}{(k-1)!}\frac{d^{k-2}}{dt^{k-2}}\frac{\partial H}{\partial y^{(k-1)i}},$$

$$p_{(2)i} = \frac{1}{2!}\frac{\partial H}{\partial y^{(2)i}} - \frac{1}{3!}\frac{d}{dt}\frac{\partial H}{\partial y^{(3)i}} + ... + (-1)^{k-3}\frac{1}{(k-1)!}\frac{d^{k-3}}{dt^{k-3}}\frac{\partial H}{\partial y^{(k-1)i}},$$

....

$$p_{(k-1)i} = \frac{1}{(k-1)!}\frac{\partial H}{\partial y^{(k-1)i}}.$$

The entries $p_{(1)i}, ...p_{(k-1)i}$ are called the Jacobi-Ostrogradski momenta of the Hamiltonian H.

Using the rule of transformation (4.1.4) of the natural basis of the module $\mathcal{X}(T^{*k}M)$ and the definition (5.4.2) of the Jacobi-Ostrogradski momenta $p_{(1)i}, ..., p_{(k-1)i}$ we obtain

Proposition 5.4.1 *With respect to the transformations of local coordinates (4.1.2) on the manifold $T^{*k}M$ the Jacobi-Ostrogradski momenta $p_{(1)i}, ..., p_{(k-1)i}$ are transformed as follows:*

$$p_{(1)i} = \frac{\partial \widetilde{y}^{(1)m}}{\partial y^{(1)i}}\widetilde{p}_{(1)m} + ... + \frac{\partial \widetilde{y}^{(k-1)m}}{\partial y^{(1)i}}\widetilde{p}_{(k-1)m},$$

$$(5.4.3) \qquad p_{(2)i} = \frac{\partial \widetilde{y}^{(2)m}}{\partial y^{(2)i}}\widetilde{p}_{(2)m} + ... + \frac{\partial \widetilde{y}^{(k-1)m}}{\partial y^{(2)i}}\widetilde{p}_{(k-1)m},$$

....

$$p_{(k-1)i} = \frac{\partial \widetilde{y}^{(k-1)m}}{\partial y^{(k-1)i}}\widetilde{p}_{(k-1)m}.$$

Let us consider the following 1-form fields on $T^{*k}M$:

$$p_{(1)} = p_{(1)i}dx^i + p_{(2)i}dy^{(1)i} + ... + p_{(k-1)i}dy^{(k-2)i},$$

$$(5.4.4) \qquad p_{(2)} = p_{(2)i}dx^i + p_{(3)i}dy^{(1)i} + ... + p_{(k-1)i}dy^{(k-3)i},$$

....

$$p_{(k-1)} = p_{(k-1)i}dx^i.$$

Proposition 5.4.2 *With respect the transformations of coordinates on the manifold $T^{*k}M$ we have*

$$(5.4.5) \qquad p_{(\alpha)} = \widetilde{p}_{(\alpha)}, \quad (\alpha = 1, ..., k-1).$$

Indeed, (5.4.3), (5.4.4) and (4.1.6), have as consequence the equalities (5.4.5).

The relations between the momenta p_i and the Jacobi-Ostrogradski momenta $p_{(1)i}$ are given by:

Lemma 5.4.1 *The following identity holds:*

$$\frac{1}{2}\frac{dp_{(1)i}}{dt} = -\overset{\circ}{E}_i(H) + \left(\frac{dp_i}{dt} + \frac{1}{2}\frac{\partial H}{\partial x^i}\right). \tag{5.4.6}$$

Indeed, (5.1.14) and (5.4.2) imply the last equality. Now, we can formulate

Theorem 5.4.1 *Along the solution curves of the Hamilton-Jacobi equations (5.1.17), the following Hamilton-Jacobi-Ostrogradski equations hold:*

$$\begin{aligned}
&\frac{\partial \mathcal{E}^{k-1}(H)}{\partial p_{(\alpha)i}} = \frac{d^\alpha x^i}{dt^\alpha}, \quad (\alpha = 1, ..., k-1),\\
&\frac{\partial \mathcal{E}^{k-1}(H)}{\partial x^i} = -\frac{dp_{(1)i}}{dt} + 2\frac{dp_i}{dt},\\
&\frac{\partial \mathcal{E}^{k-1}(H)}{\partial y^{(\alpha)i}} = -\alpha!\frac{dp_{(\alpha+1)i}}{dt}, \quad \alpha = (1, ..., k-1).
\end{aligned} \tag{5.4.7}$$

Proof. The energy $\mathcal{E}^{k-1}(H)$, (5.4.1) can be written in the form

$$\mathcal{E}^{k-1}(H) = p_{(1)i}y^{(1)i} + 2!p_{(2)i}y^{(2)i} + ... + (k-1)!p_{(k-1)i}y^{(k-1)i} - H, \tag{5.4.8}$$

where

$$y^{(\alpha)i} = \frac{1}{\alpha!}\frac{d^\alpha x^i}{dt^\alpha}, (\alpha = 1, ..., k-1).$$

Therefore the equations $(5.4.7)_1$ and $(5.4.7)_3$ hold. In order to prove (5.4.7) $_2$ we remark that (5.4.8) implies $\dfrac{\partial \mathcal{E}^{k-1}(H)}{\partial x^i} = -\dfrac{\partial H}{\partial x^i}$ and using Lemma 5.4.1 we obtain the equations $(5.4.7)_2$ .Q.E.D.

¿From Lemma 5.4.1 we deduce also

Theorem 5.4.2 *Along the solution curves of the Hamilton-Jacobi equations (5.1.17) we have*

$$\frac{dx^i}{dt} = \frac{1}{2}\frac{\partial H}{\partial p_i}; \frac{dp_i}{dt} = -\frac{1}{2}\frac{\partial H}{\partial x^i} + \frac{1}{2}\frac{\partial p_{(1)i}}{dt}. \tag{5.4.9}$$

5.5 Nöther Type Theorems

In this section we will define the notion of symmetry of a differentiable Hamiltonian H and will prove two Nöther type theorems, using the model of symmetries from the Lagrangian theory, (Ch.2).

Consider a differentiable Hamiltonian H_0 with the properties:

$$\frac{\partial H_0}{\partial y^{(k-1)i}} = 0, \frac{\partial H_0}{\partial p_i} = 0 \tag{5.5.1}$$

Proposition 5.5.1 *If H_0 is a differentiable Hamiltonian, having the properties (5.5.1), then the following identities hold:*

$$\begin{aligned}
&\frac{\partial}{\partial p_i}\frac{dH_0}{dt} = \frac{d}{dt}\frac{\partial H_0}{\partial p_i} = 0,\\
&\frac{\partial}{\partial x^i}\frac{dH_0}{dt} = \frac{d}{dt}\frac{\partial H_0}{\partial x^i},\\
&\frac{\partial}{\partial y^{(1)i}}\frac{dH_0}{dt} = \frac{d}{dt}\frac{\partial H_0}{\partial y^{(1)i}} + \frac{\partial H_0}{\partial x^i},\\
&\frac{\partial}{\partial y^{(2)i}}\frac{dH_0}{dt} = \frac{d}{dt}\frac{\partial H_0}{\partial y^{(2)i}} + 2\frac{\partial H_0}{\partial y^{(1)i}},\\
&\dots\dots\dots\dots\dots\dots\dots\dots\dots\dots\dots\dots\\
&\frac{\partial}{\partial y^{(k-2)i}}\frac{dH_0}{dt} = \frac{d}{dt}\frac{\partial H_0}{\partial y^{(k-2)i}} + (k-2)\frac{\partial H_0}{\partial y^{(k-3)i}},\\
&\frac{\partial}{\partial y^{(k-1)i}}\frac{dH_0}{dt} = (k-1)\frac{\partial H_0}{\partial y^{(k-2)i}}.
\end{aligned} \tag{5.5.2}$$

Indeed, along curve $c : t \in [0,1] \to (x^i(t), p_i(t)) \in \widetilde{T^*M}$ we have

$$\frac{dH_0}{dt} = \frac{\partial H_0}{\partial x^i}y^{(1)i} + 2\frac{\partial H_0}{\partial y^{(1)i}}y^{(2)i} + ... + (k-1)\frac{\partial H_0}{\partial y^{(k-2)i}}y^{(k-1)i}.$$

By a straightforward calculus we deduce the previous identities.

Proposition 5.5.2 *If H_0 is a differentiable Hamiltonian which satisfies the equations (5.5.1), then the following identities hold:*

$$\begin{aligned}
&\frac{1}{2}\frac{\partial}{\partial p_i}(H + \frac{dH_0}{dt}) = \frac{1}{2}\frac{\partial H}{\partial p_i},\\
&\overset{\circ}{E}_i (H + \frac{dH_0}{dt}) = \overset{\circ}{E}_i (H).
\end{aligned} \tag{5.5.3}$$

Indeed, taking into account the proposition 5.5.1 and the expression (5.1.14) of the covector $\overset{\circ}{E}_i(H)$ we have $(5.5.3)_1$ and

$$\overset{\circ}{E}_i\left(H+\frac{dH_0}{dt}\right)=\overset{\circ}{E}_i(H)+\frac{1}{2}\left[\frac{\partial}{\partial x^i}\frac{\partial H_0}{dt}-\frac{d}{dt}\frac{\partial}{\partial y^{(1)i}}\frac{dH_0}{dt}+\right.$$
$$\left.+\frac{1}{2!}\frac{d^2}{dt^2}\frac{\partial}{\partial y^{(2)i}}\frac{dH_0}{dt}+\ldots+(-1)^{k-1}\frac{1}{(k-1)!}\frac{d^{k-1}}{dt^{k-1}}\frac{\partial}{\partial y^{(k-1)i}}\frac{dH_0}{dt}\right]=\overset{\circ}{E}_i(H).$$

Now, it is easy to prove

Theorem 5.5.1 *H_0 being an arbitrary differentiable Hamiltonian with the properties (5.5.1), then the Hamiltonians H and $H+\frac{dH_0}{dt}$ have the same Hamilton-Jacobi equations:*

$$\frac{dx^i}{dt}=\frac{1}{2}\frac{\partial H}{\partial p_i},\quad \overset{\circ}{E}_i(H)=0 \tag{5.5.4}$$

Indeed, (5.5.4) are the consequence of the equations (5.5.3) Q.E.D.

Corollary 5.5.1 *If H_0 is an arbitrary differentiable Hamiltonian with the properties $\frac{\partial H_0}{\partial y^{(k-1)i}}=0$, $\frac{\partial H_0}{\partial p_i}=0$, then the integrals of action:*

$$I(c)=\int_0^1\left[p_i\frac{dx^i}{dt}-H\left(x,\frac{dx}{dt},\ldots,\frac{1}{(k-1)!}\frac{d^{k-1}x}{dt^{k-2}},p\right)\right]dt \tag{5.5.5}$$

and

$$I'(c)=\int_0^1\left[p_i\frac{dx^i}{dt}-\frac{1}{2}H\left(x,\frac{dx}{dt},\ldots,\frac{1}{(k-1)!}\frac{d^{k-1}x}{dt^{k-2}},p\right)-\right.$$
$$\left.-\frac{1}{2}\frac{dH_0}{dt}\left(x,\frac{dx}{dt},\ldots,\frac{1}{(k-2)!}\frac{d^{k-2}x}{dt^{k-2}}\right)\right]dt \tag{5.5.6}$$

determine the same Hamilton-Jacobi equations (5.5.4).

Be means of the previous result we can formulate

Definition 5.5.1 *A symmetry of the differentiable Hamiltonian $H(x,y^{(1)},\ldots,y^{(k-1)},p)$ is a C^∞-diffeomorphism $\varphi:\widetilde{T^*M}\times R\to\widetilde{T^*M}\times R$ which preserves the variational principle of the integral of action, expressed in the Corollary 5.5.1.*

We consider the local symmetries only. Therefore we study the infinitesimal symmetries defined on an open set $\pi^{*-1}(U)\times(a,b)$ in the infinitesimal form:

$$
\begin{aligned}
& x'^i(t') = x^i(t) + \varepsilon_1 V^i(t), \\
(5.5.7) \qquad & p'_i(t') = p_i(t) + \varepsilon_2 \eta_i(t), \\
& t' = t + \varepsilon_1 \tau_1(t) + \varepsilon_2 \tau_2(t),
\end{aligned}
$$

where $\varepsilon_1, \varepsilon_2$ are real numbers, sufficiently small in absolute values so that the points (x, p, t) and (x', p', t') belong to the same set $\pi^{*-1}(U) \times (a, b)$, where the curve

$$c : t \in [0, 1] \to (x^i(t), p_i(t), t) \in \pi^{*-1}(U) \times (a, b)$$

is given. By $V^i(t)$ we mean a vector field $V^i(x(t))$ and $\eta_i(t)$ denotes a covector field $\eta_i(x(t))$, along c. The pair $(V^i(t), \eta_i(t))$ satisfies (5.1.3) and the differentiable functions $\tau_1(t), \tau_2(t)$ satisfy the conditions $\tau_\alpha(0) = \tau_\alpha(1) = 0, (\alpha = 1, 2)$.

In the following considerations the terms of higher order in $\varepsilon_{1,}\varepsilon_2$, will be neglected.

The infinitesimal transformation (5.5.7) is a symmetry of the differentiable Hamiltonian $H(x, y^{(1)}, ..., y^{(k-1)}, p)$ if for any differentiable Hamiltonian $H_0(x, y^{(1)}, ..., y^{(k-2)})$ the following equations hold:

$$
\begin{aligned}
& [p'_i \frac{dx'^i}{dt'} - \frac{1}{2} H(x', \frac{dx'}{dt'}, ..., \frac{1}{(k-1)!} \frac{d^{k-1}x'}{dt'^{k-1}}, p')]dt' = \\
(5.5.8) \qquad & = p_i \frac{dx^i}{dt} - \frac{1}{2} H(x, \frac{dx}{dt}, ..., \frac{1}{(k-1)!} \frac{d^{k-1}}{dt^{k-1}}, p) - \\
& - \frac{1}{2} \frac{dH_0}{dt}(x, \frac{dx}{dt}, ..., \frac{1}{(k-2)!} \frac{d^{k-2}x}{dt^{k-2}})]dt.
\end{aligned}
$$

From (5.5.7) we deduce

$$
\begin{aligned}
& \frac{dt'}{dt} = 1 + \varepsilon_1 \frac{d\tau_1}{dt} + \varepsilon_2 \frac{d\tau_2}{dt}, \\
(5.5.9) \qquad & \frac{dx'^i}{dt'} = \frac{dx^i}{dt} + \varepsilon_1 (\frac{dV^i}{dt} - \varphi^i_1) + \varepsilon_2 \Psi^i_1, \\
& \dots\dots\dots\dots\dots\dots\dots\dots\dots\dots\dots\dots, \\
& \frac{d^{k-1}x'^i}{dt'^{k-1}} = \frac{d^{k-1}x^i}{dt^{k-1}} + \varepsilon_1 (\frac{d^{k-1}V^i}{dt^{k-1}} - \varphi^i_{k-1}) + \varepsilon_2 \Psi^i_{k-1},
\end{aligned}
$$

where

$$\text{(5.5.10)}\quad \begin{aligned} \varphi_1^i &= \frac{dx^i}{dt}\frac{d\tau_1}{dt}, \\ \varphi_2^i &= \binom{2}{1}\frac{d^2x^i}{dt^2}\frac{d\tau_1}{dt} + \binom{2}{2}\frac{dx^i}{dt}\frac{d^2\tau_1}{dt^2}, \\ &\dots\dots\dots\dots, \\ \varphi_{k-1}^i &= \binom{k-1}{1}\frac{d^{k-1}x^i}{dt^{k-1}}\frac{d\tau_1}{dt} + \binom{k-1}{2}\frac{d^{k-2}x^i}{dt^{k-2}}\frac{d^2\tau_1}{dt^2} + \dots \\ &+ \binom{k-1}{k-1}\frac{dx^i}{dt}\frac{d^{k-1}\tau_1}{dt^{k-1}} \end{aligned}$$

and similarly, for $\Psi_1^i, ..., \Psi_{k-1}^i$:

$$\text{(5.5.10')}\quad \begin{aligned} \Psi_1^i &= \frac{dx^i}{dt}\frac{d\tau_2}{dt}, \\ \Psi_2^i &= \binom{2}{1}\frac{d^2x^i}{dt^2}\frac{d\tau_2}{dt} + \binom{2}{2}\frac{dx^i}{dt}\frac{d^2\tau_2}{dt^2}, \\ &\dots\dots\dots\dots, \\ \Psi_{k-1}^i &= \binom{k-1}{1}\frac{d^{k-1}x^i}{dt^{k-1}}\frac{d\tau_2}{dt} + \binom{k-1}{2}\frac{d^{k-2}x^i}{dt^{k-2}}\frac{d^2\tau_2}{dt^2} + \dots \\ &+ \binom{k-1}{k-1}\frac{dx^i}{dt}\frac{d^{k-1}\tau_2}{dt^{k-1}}. \end{aligned}$$

The previous formulas are proved starting from (5.5.7) and taking into account the following expressions:

$$\frac{dx'^i}{dt'} = \left(\frac{dx^i}{dt} + \varepsilon_1\frac{dV^i}{dt}\right)\frac{dt}{dt'} = \left(\frac{dx^i}{dt} + \varepsilon_1\frac{dV^i}{dt}\right)\left(1 - \varepsilon_1\frac{d\tau_1}{dt} - \varepsilon_2\frac{d\tau_2}{dt}\right) =$$

$$= \frac{dx^i}{dt} + \varepsilon_1\left(\frac{dV^i}{dt} - \varphi_1^i\right) - \varepsilon_2\Psi_1^i, \text{etc.}$$

By means of the formulas (5.5.9) the equality (5.5.8) becomes:

$$\{(p_i + \varepsilon_2\eta_i)[\frac{dx^i}{dt} + \varepsilon_1\left(\frac{dV^i}{dt} - \varphi_1^i\right) - \varepsilon_2\Psi_1^i] - \frac{1}{2}H[x + \varepsilon_1 V, \frac{dx}{dt} + \varepsilon_1\left(\frac{dV}{dt} - \varphi_1\right) -$$

$$\varepsilon_2\Psi_1, ..., \frac{1}{(k-1)!}\left(\frac{d^{k-1}x}{dt^{k-1}} + \varepsilon_1\left(\frac{d^{k-1}V}{dt^{k-1}} - \varphi_{k-1}\right) - \varepsilon_2\Psi_{k-1}\right), p + \varepsilon_2\eta]\}$$

$$\left(1 + \varepsilon_1\frac{d\tau_1}{dt} + \varepsilon_2\frac{d\tau_2}{dt}\right) = p_i\frac{dx^i}{dt} - \frac{1}{2}H(x, \frac{dx}{dt}, ..., \frac{1}{(k-1)!}\frac{d^{k-1}x}{dt^{k-1}}, p) -$$

$$\frac{1}{2}\frac{dH_0}{dt}(x, \frac{dx}{dt}, ..., \frac{1}{(k-2)!}\frac{d^{k-2}x}{dt^{k-2}}).$$

Using the Taylor expansion with respect to $\varepsilon_1, \varepsilon_2$ and neglecting the terms of higher order in $\varepsilon_1, \varepsilon_2$ we obtain

$$p_i\frac{dV}{dt} - \frac{1}{2}[\frac{\partial H}{\partial x^i}V^i + \frac{\partial H}{\partial y^{(1)i}}\frac{dV^i}{dt}, ..., \frac{1}{(k-1)!}\frac{\partial H}{\partial y^{(k-1)i}}\frac{d^{k-1}V^i}{dt^{k-1}}]+$$

$$(5.5.11)\qquad +\frac{1}{2}[\frac{\partial H}{\partial y^{(1)i}}\varphi_1^i + \frac{1}{2!}\frac{\partial H}{\partial y^{(2)i}}\varphi_2^i + ... + \frac{1}{(k-1)!}\frac{\partial H}{\partial y^{(k-1)i}}\varphi_{k-1}^i]-$$

$$-\frac{1}{2}H\frac{d\tau_1}{dt} = \frac{1}{2}\frac{d\Phi_1}{dt}$$

and

$$(5.5.11')\qquad \left(\frac{dx^i}{dt} - \frac{1}{2}\frac{\partial H}{\partial p_i}\right)\eta_i + \frac{1}{2}[\frac{\partial H}{\partial y^{(1)i}}\Psi_1^i + \frac{1}{2!}\frac{\partial H}{\partial y^{(2)i}}\Psi_2^i + ...$$

$$... + \frac{1}{(k-1)!}\frac{\partial H}{\partial y^{(k-1)i}}\Psi_{k-1}^i] - \frac{1}{2}H\frac{d\tau_2}{dt} = \frac{d\Phi_2}{dt}$$

where

$$(5.5.12)\qquad \varepsilon_1\Phi_1 + \varepsilon_2\Phi_2 = H_0$$

As a consequence we have that the functions Φ_1 and Φ_2 are depending only on the variables $(x^i, y^{(1)i}, ..., y^{(k-2)i})$ and, of course, these can be arbitrarly chosen.

Taking into account of Lemma 5.1.1 and remarking the identity

$$\frac{\partial H}{\partial y^{(1)i}}\varphi_1^i + \frac{1}{2!}\frac{\partial H}{\partial y^{(2)i}}\varphi_2^i + ... + \frac{1}{(k-1)!}\frac{\partial H}{\partial y^{(k-1)i}}\varphi_{k-1}^i =$$

$$= \frac{d\tau_1}{dt}I^{k-1}(H) + \frac{1}{2!}\frac{d^2\tau_1}{dt^2}I^{k-2}(H) + ... + \frac{1}{(k-1)!}\frac{d^{k-1}\tau_1}{dt^{k-1}}I^1(H)$$

and similar identities for Ψ_α and τ_2, the equations (5.5.11) and (5.5.11') can be expressed more simple.

Proposition 5.5.3 *The equations (5.5.11) and (5.5.11') are echivalent to*

$$-\overset{\circ}{E}_i(H)V^i + \frac{d}{dt}(p_iV^i) - \frac{1}{2}\frac{d}{dt}[I_V^{k-1}(H) - \frac{1}{2!}\frac{d}{dt}I_V^{k-2}(H) + ...$$

$$(5.5.13)\qquad ... + (-1)^{k-2}\frac{1}{(k-1)!}\frac{d^{k-2}}{dt^{k-1}}I_V^1(H)]+$$

$$\frac{1}{2}[\frac{d\tau_1}{dt}I^{k-1}(H) + \frac{1}{2!}\frac{d^2\tau_1}{dt^2}I^{k-2}(H) + ...$$

$$... + \frac{1}{(k-1)!}\frac{d^{k-1}\tau_1}{dt^{k-1}}I^1(H)] - \frac{1}{2}H\frac{d\tau_1}{dt} = \frac{1}{2}\frac{d\Phi_1}{dt}$$

and

(5.5.13')
$$(\frac{dx^i}{dt}-\frac{1}{2}\frac{\partial H}{\partial p_i})\eta_i+\frac{1}{2}[\frac{d\tau_2}{dt}I^{k-1}(H)+\frac{1}{2!}\frac{d^2\tau_2}{dt^2}I^{k-2}(H)+...+$$
$$...+\frac{1}{(k-1)!}\frac{d^{k-1}\tau_1}{dt^{k-1}}I^1(H)]-\frac{1}{2}H\frac{d\tau_2}{dt}=\frac{1}{2}\frac{d\Phi_2}{dt}.$$

Now, we can prove by a straighforward calculus

Lemma 5.5.1 *The following identities hold*
(5.5.14)
$$\frac{d\tau_\alpha}{dt}I^{k-1}(H)+\frac{1}{2!}\frac{d^2\tau_\alpha}{dt^2}I^{k-2}(H)+...+\frac{1}{(k-1)!}\frac{d^{k-1}\tau_\alpha}{dt^{k-1}}I^1(H)-H\frac{d\tau_\alpha}{dt}=$$
$$=-\tau_\alpha\frac{d}{dt}\mathcal{E}^{k-1}(H)+\frac{d}{dt}[\tau_\alpha\mathcal{E}^{k-1}(H)-\frac{d\tau_\alpha}{dt}\mathcal{E}^{k-2}(H)+...+$$
$$+(-1)^{k-2}\frac{d^{k-2}\tau_\alpha}{dt^{k-2}}\mathcal{E}^1(H)],\quad (\alpha=1,2).$$

In the previous equality $\mathcal{E}^{k-1}(H),...\mathcal{E}^1(H)$ are the energies of order $(k-1,...1)$ respectively, expressed in (5.3.1).

Theorem 5.5.2 *For any infinitesimal symmetry (5.5.7) the left hand sides of the following equalities do not depend on the variables $y^{(k-1)i}$ and momenta p_i.*
(5.5.15)
$$-\overset{\circ}{E}_i(H)V^i+\frac{d}{dt}(p_iV^i)-\frac{1}{2}\frac{d}{dt}[I_V^{k-1}(H)-\frac{1}{2!}\frac{dI_V^{k-2}(V)}{dt}+...$$
$$...+(-1)^{k-2}\frac{1}{(k-1)!}\frac{d^{k-2}I_V^1(H)}{dt^{k-2}}]-\frac{1}{2}\tau_1\frac{d\mathcal{E}^{k-1}(H)}{dt}+\frac{1}{2}\frac{d}{dt}[\tau_1\mathcal{E}^{k-1}(H)-$$
$$\frac{d\tau_1}{dt}\mathcal{E}^{k-2}(H)+...+(-1)^{k-2}\frac{d^{k-2}\tau_1}{dt^{k-2}}\mathcal{E}^1(H)]=\frac{1}{2}\frac{d\Phi_1}{dt}.$$
$$(\frac{dx^i}{dt}-\frac{1}{2}\frac{\partial H}{\partial p_i})\eta_i-\frac{1}{2}\tau_2\frac{d\mathcal{E}^{k-1}(H)}{dt}+\frac{1}{2}\frac{d}{dt}[\tau_2\mathcal{E}^{k-1}(H)-\frac{d\tau_2}{dt}\mathcal{E}^{k-2}(H)+...$$
$$+(-1)^{k-2}\frac{d^{k-2}\mathcal{E}^1(H)}{dt^{k-2}}]=\frac{1}{2}\frac{d\Phi_2}{dt}.$$

Proof. Indeed, the equalities (5.5.15) result from proposition 5.5.3 and Lemma 5.5.1. But the functions Φ_1 and Φ_2 satisfies the conditions

$$\frac{\partial\Phi_\alpha}{\partial y^{(k-1)i}}=0,\ \frac{\partial\Phi_\alpha}{\partial p_i}=0,\ (\alpha=1,2)\text{ Q.E.D.}$$

Now, we can prove a Nöther type theorem:

Theorem 5.5.3 *For any infinitesimal symmetry (5.5.7) of a differentiable Hamiltonian $H(x,y^{(1)},...,y^{(k-1)},p)$ and for any differentiable functions Φ_1,Φ_2*

with the properties $\dfrac{\partial\Phi_\alpha}{\partial y^{(k-1)i}} = \dfrac{\partial\Phi_\alpha}{\partial p_i} = 0, (\alpha = 1,2)$, *the following functions* $\mathcal{F}_1^k(H,\Phi_1), \mathcal{F}_2^k(H,\Phi_2)$ *are conserved along the solutions curves of the Hamilton-Jacobi equations*

$$\frac{dx^i}{dt} = \frac{1}{2}\frac{\partial H}{\partial p_i}, \overset{\circ}{E_i}(H) = 0:$$

(5.5.16)

$$\mathcal{F}_1^k(H,\Phi_1) =$$
$$= p_iV^i - \frac{1}{2}[I_V^{k-1}(H) - \frac{1}{2!}\frac{dI_V^{k-2}(H)}{dt} + ... + (-1)^{k-2}\frac{1}{(k-1)!}\frac{dI_V^{k-2}(H)}{dt^{k-2}}] +$$
$$+\frac{1}{2}[\tau_1\mathcal{E}^{k-1}(H) - \frac{d\tau_1}{dt}\mathcal{E}^{k-2}(H) + ... + (-1)^{k-2}\frac{d^{k-2}\tau_1}{dt^{k-2}}\mathcal{E}^1(H) - \Phi_1],$$

$$\mathcal{F}_2^k(H,\Phi_2) = \tau_2\mathcal{E}^{k-1}(H) - \frac{d\tau_2}{dt}\mathcal{E}^{k-2}(H) + ... + (-1)^{k-2}\frac{d^{k-2}\tau_2}{dt^{k-2}}\mathcal{E}^1(H) - \Phi_2.$$

Proof. Taking into account the Theorem 5.5.2 and the expression of the function $\mathcal{F}_1^k(H,\Phi_1)$ and $\mathcal{F}_2^k(H,\Phi_2)$ it follows that the following equations hold:

$$\begin{aligned} &\overset{\circ}{E_i}(H)V^i = \frac{d\mathcal{F}_1^k(H,\Phi_1)}{dt} - \frac{1}{2}\tau_1\frac{d\mathcal{E}^{k-1}(H)}{dt}, \\ &(\frac{dx^i}{dt} - \frac{1}{2}\frac{\partial H}{\partial p_i})\eta_i = \frac{1}{2}\frac{d\mathcal{F}_2^k(H,\Phi_2)}{dt} - \frac{1}{2}\tau_2\frac{d\mathcal{E}^{k-1}(H)}{dt}. \end{aligned} \tag{5.5.17}$$

According to the Hamilton-Jacobi equations (5.1.17) and the law of conservation of energy $\mathcal{E}^{k-1}(H)$, the last equations (5.5.17) implies

$$\frac{d\mathcal{F}_1(H,\Phi_1)}{dt} = 0, \frac{d\mathcal{F}_2(H,\Phi_2)}{dt} = 0.$$

The functions $\mathcal{F}_1^k(H,\Phi_1)$ and $\mathcal{F}_2^k(H,\Phi_2)$ depend on the invariants $I_V^1(H)$, ..., $I_V^{k-1}(H)$, the energies of order $1,...,k-1$, $\mathcal{E}^1(H),...,\mathcal{E}^{k-1}(H)$ and the arbitrary functions $\Phi_\alpha(x,y^{(1)},...,y^{(k-2)})$, $(\alpha = 1,2)$.

In particular, if the Zermelo conditions (5.2.3) are verified then the energies $\mathcal{E}^1(H),...,\mathcal{E}^{k-1}(H) = 0$. Assuming $\Phi_2 = 0$, the previous Nöther Theorem, becomes:

Theorem 5.5.4 *For any infinitesimal symmetry (5.5.7) of a differentiable Hamiltonian* $H(x,y^{(1)},...,y^{(k-1)},p)$ *, which satisfies the Zermelo conditions (5.2.3), and for any differentiable function* $\Phi(x,y^{(1)},...,y^{(k-1)})$ *, along the solution curves of the Hamilton-Jacobi equations* $\dfrac{dx^i}{dt} = \dfrac{1}{2}\dfrac{\partial H}{p_i}, \overset{\circ}{E_i}(H) = 0$, *the following function is constant:*

$$\mathcal{F}^k(H,\Phi) = p_i V^i - \frac{1}{2}[I_V^{k-1}(H) - \frac{1}{2!}\frac{dI_V^{k-2}(H)}{dt} + \ldots + \tag{5.5.18}$$
$$+(-1)^{k-2}\frac{1}{(k-1)!}\frac{d^{k-2}I_V^1(H)}{dt^{k-2}} + \Phi].$$

The theory from this chapter will be applied to the geometrical study of the Hamilton space of order k.

Chapter 6

Dual Semispray. Nonlinear Connections

The notion of semispray from the geometry of higher order Lagrange spaces has a dual correspondent in the geometrical theory of the Hamilton spaces of order k. The same remark is true concerning the concept of nonlinear connection which is canonical related with that of semispray.

6.1 Dual Semispray

Definition 6.1.1 *A dual k-semispray on $T^{*k}M$ is a vector field $S \in \mathcal{X}(\widetilde{T^{*k}M})$ with the property*

$$JS = \overset{k-1}{\Gamma}, \tag{6.1.1}$$

where $\overset{k-1}{\Gamma}$ is the Liouville vector field

$$\overset{k-1}{\Gamma} = y^{(1)i}\frac{\partial}{\partial y^{(1)i}} + \cdots + (k-1)y^{(k-1)i}\frac{\partial}{\partial y^{(k-1)i}}. \tag{6.1.1'}$$

We have:

Proposition 6.1.1 *A dual k-semispray S can be locally represented by*
(6.1.2)

$$\begin{aligned} S &= y^{(1)i}\frac{\partial}{\partial x^i} + 2y^{(2)i}\frac{\partial}{\partial y^{(1)i}} + \cdots + (k-1)y^{(k-1)i}\frac{\partial}{\partial y^{(k-2)i}} + \\ &+ k\xi^i(x, y^{(1)}, ..., y^{(k-1)}, p)\frac{\partial}{\partial y^{(k-1)i}} + \eta_i(x, y^{(1)}, ..., y^{(k-1)}, p)\frac{\partial}{\partial p_i}. \end{aligned}$$

Indeed, S from (6.1.2) satisfies the equation (6.1.1) for an arbitrary system of functions $\{\xi^i\}$ and $\{\eta_i\}$ ($i = 1, ..., n$), J being the $k-1$-tangent endomorphism (§4.3). The systems of functions $\{\xi^i(x, y^{(1)}, ..., y^{(k-1)}, p)\}$ and $\{\eta_i(x, y^{(1)}, ..., y^{(k-1)}, p)\}$ are called *the coefficients of the dual k-semispray S*.

Since S is a vector field on $T^{*k}M$ it follows that the functions ξ^i and η_i, defined on every local chart of the manifold $T^{*k}M$ are important geometrical object fields on the manifold $T^{*k}M$.

So, we have

Theorem 6.1.1 *With respect to the transformation (4.1.2) of the local coordinates on $T^{*k}M$, the systems of functions $\{\xi^i\}$ and $\{\eta_i\}$ transform as follows:*

(6.1.3)
$$k\widetilde{\xi}^i = k\frac{\partial \widetilde{y}^{(k-1)i}}{\partial y^{(k-1)j}}\xi^j + (k-1)\frac{\partial \widetilde{y}^{(k-1)i}}{\partial y^{(k-2)j}}y^{(k-1)j} + \cdots + 2\frac{\partial \widetilde{y}^{(k-1)i}}{\partial y^{(1)j}}y^{(2)j} + \frac{\partial \widetilde{y}^{(k-1)i}}{\partial x^j}y^{(1)j},$$

$$\widetilde{\eta}_i = \frac{\partial x^j}{\partial \widetilde{x}^i}\eta_j + \frac{\partial \widetilde{p}_i}{\partial x^j}y^{(1)j}. \tag{6.1.4}$$

Proof: S being a vector field, from (6.1.2), (4.1.4) we have:

$$S = y^{(1)i}\left\{\frac{\partial \widetilde{x}^m}{\partial x^i}\frac{\partial}{\partial \widetilde{x}^m} + \frac{\partial \widetilde{y}^{(1)m}}{\partial x^i}\frac{\partial}{\partial \widetilde{y}^{(1)m}} + \cdots + \frac{\partial \widetilde{y}^{(k-1)m}}{\partial x^i}\frac{\partial}{\partial \widetilde{y}^{(k-1)m}} + \frac{\partial \widetilde{p}_m}{\partial x^i}\frac{\partial}{\partial \widetilde{p}_m}\right\} +$$

$$+2y^{(2)i}\left\{\frac{\partial \widetilde{y}^{(1)m}}{\partial y^{(1)i}}\frac{\partial}{\partial \widetilde{y}^{(1)m}} + \frac{\partial \widetilde{y}^{(2)m}}{\partial y^{(1)i}}\frac{\partial}{\partial \widetilde{y}^{(2)m}} + \cdots + \frac{\partial \widetilde{y}^{(k-1)m}}{\partial y^{(1)i}}\frac{\partial}{\partial \widetilde{y}^{(k-1)m}}\right\} +$$

....

$$+(k-1)y^{(k-1)i}\left\{\frac{\partial \widetilde{y}^{(k-2)m}}{\partial y^{(k-2)i}}\frac{\partial}{\partial \widetilde{y}^{(k-2)m}} + \frac{\partial \widetilde{y}^{(k-1)m}}{\partial y^{(k-2)i}}\frac{\partial}{\partial \widetilde{y}^{(k-1)m}}\right\} +$$

$$+k\xi^i\frac{\partial \widetilde{y}^{(k-1)m}}{\partial y^{(k-1)i}}\frac{\partial}{\partial \widetilde{y}^{(k-1)m}} + \eta_i\frac{\partial x^i}{\partial \widetilde{x}^m}\frac{\partial}{\partial \widetilde{p}_m}.$$

Taking into account (6.1.4) one obtains:

$$\begin{aligned} S \;=\;& \widetilde{y}^{(1)m}\frac{\partial}{\partial \widetilde{x}^m} + 2\widetilde{y}^{(2)m}\frac{\partial}{\partial \widetilde{y}^{(1)m}} + \cdots + (k-1)\widetilde{y}^{(k-1)m}\frac{\partial}{\partial \widetilde{y}^{(k-2)m}} + \\ +& \left(y^{(1)i}\frac{\partial \widetilde{y}^{(k-1)m}}{\partial x^i} + 2y^{(2)i}\frac{\partial \widetilde{y}^{(k-1)m}}{\partial y^{(1)i}} + \cdots + (k-1)y^{(k-1)i}\frac{\partial \widetilde{y}^{(k-1)m}}{\partial y^{(k-1)i}} + \right. \\ +& \left. k\xi^i\frac{\partial \widetilde{y}^{(k-1)m}}{\partial y^{(k-1)i}}\right)\frac{\partial}{\partial \widetilde{y}^{(k-1)m}} + \left(y^{(1)i}\frac{\partial \widetilde{p}_m}{\partial x^i} + \eta_i\frac{\partial x^i}{\partial \widetilde{x}^m}\right)\frac{\partial}{\partial \widetilde{p}_m}. \end{aligned}$$

Now, writting S in the form

$$\begin{aligned} &S = \widetilde{y}^{(1)m}\frac{\partial}{\partial \widetilde{x}^m} + 2\widetilde{y}^{(2)m}\frac{\partial}{\partial \widetilde{y}^{(1)m}} + \cdots + \\ &+(k-1)\widetilde{y}^{(k-1)m}\frac{\partial}{\partial \widetilde{y}^{(k-2)m}} + k\widetilde{\xi}^m\frac{\partial}{\partial y^{(k-1)m}} + \widetilde{\eta}_m\frac{\partial}{\partial \widetilde{p}_m} \end{aligned}$$

and identifying with the previous expression of S we obtain the relations (6.1.3) and (6.1.4). q.e.d.

Theorem 6.1.2 *If on every domain of local chart of the manifold $T^{*k}M$ the systems of functions $\{\xi^i\}$ and $\{\eta_i\}$, $(i = 1, ..., n)$ are given, such that, with respect to (4.1.2) the formulae (6.1.3) and (6.1.4) hold, then S from (6.1.2) is a dual k-semispray on $T^{*k}M$.*

The proof follow the usual way, [115].

Proposition 6.1.2 *The integral curves of the dual k-semispray S, from (6.1.2) are the solution curves of the system of differential equations:*

(6.1.5)

$$\begin{cases} \dfrac{dx^i}{dt} = y^{(1)}, \ \dfrac{dy^{(1)i}}{dt} = 2y^{(2)}, \ ..., \ \dfrac{dy^{(k-2)i}}{dt} = (k-1)y^{(k-1)i}, \\ \\ \dfrac{dy^{(k-1)i}}{dt} = k\xi^i(x, y^{(1)}, ..., y^{(k-1)}, p), \ \dfrac{dp_i}{dt} = \eta_i(x, y^{(1)}, ..., y^{(k-1)}, p). \end{cases}$$

Notice that this system is equivalent to the following:

$$\begin{cases} \dfrac{dx^i}{dt} = y^{(1)}, \ \dfrac{1}{2!}\dfrac{d^2x^i}{dt^2} = y^{(2)}, \ ..., \ \dfrac{1}{(k-1)!}\dfrac{d^{k-1}x^i}{dt^{k-1}} = y^{(k-1)i}, \\ \\ \dfrac{1}{k!}\dfrac{d^kx^i}{dt^k} = \xi^i(x, \dfrac{dx}{dt}, ..., \dfrac{1}{(k-1)!}\dfrac{d^{k-1}x}{dt^{k-1}}, p), \\ \dfrac{dp_i}{dt} = \eta_i(x, \dfrac{dx}{dt}, ..., \dfrac{1}{(k-1)!}\dfrac{d^{k-1}x}{dt^{k-1}}, p). \end{cases} \qquad (6.1.5')$$

The problem of integration of this system of differential equations is solved by usual methods. The solutions $\{(x^i(t), p_i(t)), \ t \in (a, b)\}$ are curves on the cotangent manifold T^*M.

Concerning the homogeneity of S, we see that every term which does not include ξ^i or η_i has the degree of homogeneity 2 on the fibres of $T^{*k}M$. Therefore we have:

Proposition 6.1.3 *The dual k-semispray S, from (6.1.2) is 2-homogeneous on the fibres of $T^{*k}M$ if and only if the coefficients ξ^i are k-homogeneous and η_i are $k+1$-homogeneous on the fibres of $T^{*k}M$.*

A dual k-semispray S, 2-homogeneous on the fibres of $T^{*k}M$ is called *a k-spray.*

Remarking that the rule (6.1.3) of transformation of the coefficients ξ^i of a k-semispray S is same rule with that of the coordinates $y^{(k)i}$ on the total space of bundle of accelerations T^kM we can give a remarkable geometrical meaning of the coefficients ξ^i:

Proposition 6.1.4 *Every dual k-semispray S on the manifold $T^{*k}M$ with the coefficients (ξ^i, η_i) determine a bundle morphism*

$$\xi : (x, y^{(1)}, ..., y^{(k-1)}, p) \in T^{*k}M \to (x, y^{(1)}, ..., y^{(k-1)}, y^{(k)}) \in T^kM$$

defined by

(6.1.6)

$$x^i = x^i,\ y^{(1)i} = y^{(1)i},\ ...,\ y^{(k-1)i} = y^{(k-1)i},\ y^{(k)i} = \xi^i(x, y^{(1)}, ..., y^{(k-1)}, p).$$

Moreover, the bundle morphism ξ *is a local diffeomorphism if and only if*

$$rank \left\| \dot{\partial}^i\ \xi^j \right\| = n.$$

Indeed, the mapping $\xi : T^{*k}M \to T^kM$ does not depend on the transformation of local coordinates. It is local invertible if and only if

$rank \left\| \dot{\partial}^i\ \xi^j \right\| = n.$ Q.E.D.

We shall see in Chapter 8, that the bundle morphism ξ defined in (6.1.6) is uniquely determined by the Legendre transformation between a Lagrange spaces of order k, $L^{(k)n} = (M, L)$ and a Hamilton spaces of order k, $H^{(k)n} = (M, H)$.

Consequently, if the bundle morphism ξ from (6.1.6) is apriori given we can consider the dual k-semispray

(6.1.7)

$$S_\xi = y^{(1)i}\frac{\partial}{\partial x^i} + 2y^{(2)i}\frac{\partial}{\partial y^{(1)i}} + \cdots + (k-1)y^{(k-1)i}\frac{\partial}{\partial y^{(k-2)i}} + k\xi^i\frac{\partial}{\partial y^{(k-1)i}} + \eta_i\ \dot{\partial}^i\ .$$

In this case S_ξ is characterized only by the coefficients $\eta_i(x, y^{(1)}, ..., y^{(k-1)}, p)$.

The values of 1-forms d_0H, ..., $d_{k-2}H$ from (4.3.14), (4.3.5) on the vector fields S_ξ are as follows:

$$d_0H(S_\xi) = \overset{(0)}{p_i}\ y^{(1)i},$$

(6.1.8)
$$d_1H(S_\xi) = \overset{(1)}{p_i}\ y^{(1)i} + 2\ \overset{(0)}{p_i}\ y^{(2)i},$$

....

$$d_{k-2}H(S_\xi) = \overset{(k-2)}{p_i}\ y^{(1)i} + 2\ \overset{(k-3)}{p_i}\ y^{(2)i} + \cdots + (k-1)\ \overset{(0)}{p_i}\ y^{(k-1)i}.$$

These scalar fields does not depend on the coefficients ξ^i and η_i of S_ξ. Also the 1-form dH leads to the formula

(6.1.8')

$$dH(S_\xi) = \frac{\partial H}{\partial x^i}y^{(1)i} + 2\frac{\partial H}{\partial y^{(1)i}}y^{(2)i} + \cdots + (k-1)\frac{\partial H}{\partial y^{(k-2)i}}y^{(k-1)i} +$$
$$+ k\frac{\partial H}{\partial y^{(k-1)i}}\xi^i + \dot{\partial}^i\ H\eta_i = S_\xi(H).$$

The existence of a k-semispray on the manifold $T^{*k}M$ is assured by the following theorem

Theorem 6.1.3 *If the base manifold* M *is paracompact, then on the manifold* $T^{*k}M$ *there exists the dual* k*-semispray* S_ξ*, with apriori given bundle morphism* ξ.

Proof: Assuming that the manifold M is paracompact by the Theorem 4.1.2, the manifold $T^{*k}M$ is paracompact, too. We shall see (Ch. 8) that a bundle morphism ξ, defined in (6.1.6) exists. Now, let $\gamma_{ij}(x)$, $x \in M$, be a Riemann metric on M and $\gamma^i_{jk}(x)$ its Christoffel symbols.

Setting

$$\eta_j = \gamma^i_{jh}(x) p_i y^{(1)h} \tag{6.1.9}$$

we can prove that the rule of transformation of the system of functions $\{\eta_i\}$, with respect to (4.1.2) is just (6.1.4). Applying Theorem 6.1.2 we obtain a k-semispray S_ξ, with the coefficients η_i.

Another properties of S are given in §4.3. We have

$$\overset{k-1}{\Gamma} = JS, \ \overset{k-2}{\Gamma} = J^2 S, \ ..., \ \overset{1}{\Gamma} = J^{(k-1)} S.$$

6.2 Nonlinear Connections

The notion of nonlinear connection on the total space of the dual bundle $(T^{*k}M, \pi^{*k}, M)$ can be introduced by the classical method, [115].

Definition 6.2.1 *A nonlinear connection on the manifold $T^{*k}M$ is a regular distribution N on the $T^{*k}M$ supplementary to the vertical distribution V, i.e*

$$T_u(T^{*k}M) = N_u \oplus V_u, \ \forall u \in T^{*k}M. \tag{6.2.1}$$

Taking into account the Proposition 4.2.1 it follows that the distribution N has the property

$$T_u(T^{*k}M) = N_u \oplus V_{1,u} \oplus W_{k,u}. \tag{6.2.1'}$$

Locally V_1 is generated by the system of vector fields $(\frac{\partial}{\partial y^{(1)i}}, ..., \frac{\partial}{\partial y^{(k-1)i}})$ and W_k is generated locally by $(\frac{\partial}{\partial p_i})$.

As usual, we shall write these systems of tangent vectors $\left(\frac{\partial}{\partial y^{(1)i}}, ..., \frac{\partial}{\partial y^{(k-1)i}}\right)$ as $(\overset{(1)}{\partial_i}, ..., \overset{(k-1)}{\partial_i})$ and $(\frac{\partial}{\partial p_i})$ as $(\dot{\partial}_i)$, respectively.

It follows that the local dimension of the distribution N is n, local dimension of the distribution V_1 is $(k-1)n$ and that of distribution W_k is n.

Consider a nonlinear connection N on $T^{*k}M$ and denote by h and v the projectors determined by direct decomposition (6.2.1). Then we have:

$$h + v = Id, h^2 = h, v^2 = v, hv = vh = 0.$$

As usual we denote

$$X^H = hX,\ X^V = vX,\ \forall X \in \mathcal{X}(T^{*k}M). \tag{6.2.2}$$

A horizontal lift, with respect to N is a $\mathcal{F}(M)$-linear mapping $l_h : \mathcal{X}(M) \to \mathcal{X}(T^{*k}M)$ which has the properties

$$v \circ l_h = 0,\ d\pi^{*k} \circ l_h = I_d.$$

There exists an unique local basis adapted to the *horizontal* distribution N. It is given by

$$\frac{\delta}{\delta x^i} = l_h\left(\frac{\partial}{\partial x^i}\right), (i = 1, ..., n). \tag{6.2.3}$$

The linearly independent vector fields $\frac{\delta}{\delta x^i}$, $(i = 1, ..., n)$ can be uniquely written in the form:

$$\frac{\delta}{\delta x^i} = \frac{\partial}{\partial x^i} - \underset{(1)}{N_i^j}\frac{\partial}{\partial y^{(1)j}} - \dots - \underset{(k-1)}{N_i^j}\frac{\partial}{\partial y^{(k-1)j}} + N_{ij}\frac{\partial}{\partial p_j}. \tag{6.2.4}$$

The systems of functions $\underset{(1)}{N_i^j}, ..., \underset{(k-1)}{N_i^j}, N_{ij}$ are called the *coefficients* of the nonlinear connection N.

They determine an important object fields on the manifold $T^{*k}M$. Indeed, a change of local coordinates (4.1.2) and (6.2.3) imply:

$$\frac{\delta}{\delta x^i} = \frac{\partial \tilde{x}^j}{\partial x^i}\frac{\delta}{\delta \tilde{x}^j}, \tag{6.2.5}$$

since

$$\frac{\delta}{\delta x^i}|_u = l_h\left(\frac{\delta}{\delta x^i}|_x\right) = l_h\left(\frac{\partial \tilde{x}^j}{\partial x^i}\frac{\partial}{\partial \tilde{x}^j}\right)_x = \left(\frac{\partial \tilde{x}^j}{\partial x^i}\frac{\delta}{\delta \tilde{x}^j}\right)_u, x = \pi^{*k}(u).$$

It is not difficult to prove that the formula (6.2.5) has the following consequence:

Theorem 6.2.1 *1°. With respect to changes of coordinate on $T^{*k}M$, (4.1.2),*

the coefficients of a nonlinear connection N are transformed by the rule
(6.2.6)

$$\underset{(1)}{\widetilde{N}}{}^i_m \frac{\partial \widetilde{x}^m}{\partial x^j} = \underset{(1)}{N}{}^m_j \frac{\partial \widetilde{x}^i}{\partial x^m} - \frac{\partial \widetilde{y}^{(1)i}}{\partial x^j},$$

$$\underset{(2)}{\widetilde{N}}{}^i_m \frac{\partial \widetilde{x}^m}{\partial x^j} = \underset{(2)}{N}{}^m_j \frac{\partial \widetilde{x}^i}{\partial x^m} + \underset{(1)}{N}{}^m_j \frac{\partial \widetilde{y}^{(1)i}}{\partial x^m} - \frac{\partial \widetilde{y}^{(2)i}}{\partial x^j},$$

....

$$\underset{(k-1)}{\widetilde{N}}{}^i_m \frac{\partial \widetilde{x}^m}{\partial x^j} = \underset{(k-1)}{N}{}^m_j \frac{\partial \widetilde{x}^i}{\partial x^m} + \underset{(k-2)}{N}{}^m_j \frac{\partial \widetilde{y}^{(1)i}}{\partial x^m} + \ldots + \underset{(1)}{N}{}^m_j \frac{\partial \widetilde{y}^{(k-2)i}}{\partial x^m} - \frac{\partial \widetilde{y}^{(k-1)i}}{\partial x^j},$$

$$\widetilde{N}_{ij} = \frac{\partial x^r}{\partial \widetilde{x}^i} \frac{\partial x^s}{\partial \widetilde{x}^j} N_{rs} + p_r \frac{\partial^2 x^r}{\partial \widetilde{x}^i \partial \widetilde{x}^j}.$$

2°. Conversely, if the system of functions ($\underset{(1)}{N}{}^j_i, \ldots, \underset{(k-1)}{N}{}^j_i, N_{ij}$) are given on every domain of local chart of the manifold $T^{*k}M$ such that the equations (6.2.6) are verified, then ($\underset{(1)}{N}{}^j_i, \ldots, \underset{(k-1)}{N}{}^j_i, N_{ij}$) are the coefficients of a nonlinear connection N on $T^{*k}M$.

The $(k-1)$-tangent structure J, defined in the section 3, ch.4, maps the horizontal distribution N into a vertical distribution $N_1 = J(N_0)$, $N_0 = N$ which is supplementary to the distribution V_2 in V_1. If we continue this process, we obtain:

(6.2.7)
$$N_0 = N, N_1 = J(N_0), N_2 = J^2(N_0), \ldots, N_{k-2} = J^{k-2}(N_0),$$
$$V_k = J^{k-1}(N_0).$$

Therefore, we have the direct sum of vector spaces:

(6.2.8)
$$V_{1,u} = N_{1,u} \oplus N_{2,u} \oplus \ldots \oplus N_{k-2,u} \oplus V_{k-1,u}, \ \forall u \in T^{*k}M$$

and

(6.2.8')
$$V_u = N_{1,u} \oplus N_{2,u} \oplus \ldots \oplus N_{k-2,u} \oplus V_{k-1,u} \oplus W_{k,u}, \forall u \in T^{*k}M.$$

Taking into account the formula (6.1.1), we get:

Theorem 6.2.2 *1°. A nonlinear connection N on the manifold $T^{*k}M$ gives rise to the direct sum of linear spaces for the tangent space $T_u(T^{*k}M)$:*

(6.2.9)
$$T_u(T^{*k}M) = N_{0,u} \oplus N_{1,u} \oplus \ldots \oplus N_{k-2,u} \oplus V_{k-1,u} \oplus W_{k,u}, \forall u \in T^{*k}M$$

2°. The adapted basis of every term of the previous direct sum, respectively is:

$$
(6.2.10)\qquad
\begin{aligned}
&\frac{\delta}{\delta x^i} = \frac{\partial}{\partial x^i} - \underset{(1)}{N_i^j}\frac{\partial}{\partial y^{(1)j}} - \cdots - \underset{(k-1)}{N_i^j}\frac{\partial}{\partial y^{(k-1)j}} + N_{ij}\frac{\partial}{\partial p_j},\\
&\frac{\delta}{\delta y^{(1)i}} = \frac{\partial}{\partial y^{(1)i}} - \underset{(1)}{N_i^j}\frac{\partial}{\partial y^{(2)j}} - \cdots - \underset{(k-2)}{N_i^j}\frac{\partial}{\partial y^{(k-1)j}},\\
&\cdots\cdots\cdots\cdots\cdots\cdots\cdots\cdots\cdots\cdots\cdots\cdots\\
&\frac{\delta}{\delta y^{(k-1)i}} = \frac{\partial}{\partial y^{(k-1)i}},\\
&\frac{\delta}{\delta p_i} = \frac{\partial}{\partial p_i}.
\end{aligned}
$$

Proof: 1°. The direct sum (6.2.9) is obtained from (6.2.1) and (6.2.8').
2°. $J(\frac{\delta}{\delta x^i}) = \frac{\delta}{\delta y^{(1)i}}$ is obtained by means of the definition of the $k-1$-tangent structure J. Also, $\frac{\delta}{\delta y^{(2)i}} = J(\frac{\delta}{\delta y^{(1)i}})$, ... leads to the formulae (6.2.10).
Remarking that (6.2.5) holds, it follows:

Proposition 6.2.1 *Under a change of local coordinates on $T^{*k}M$, the vector fields of the adapted basis:*

$$
(6.2.11)\qquad \{\frac{\delta}{\delta x^i}, \frac{\delta}{\delta y^{(1)i}}, ..., \frac{\delta}{\delta y^{(k-1)i}}, \frac{\delta}{\delta p_i}\}
$$

transform by the rule:

$$
(6.2.11')\qquad
\begin{aligned}
&\frac{\delta}{\delta x^i} = \frac{\partial \widetilde{x}^j}{\partial x^i}\frac{\delta}{\delta \widetilde{x}^j}, \frac{\delta}{\delta y^{(1)i}} = \frac{\partial \widetilde{x}^j}{\partial x^i}\frac{\delta}{\delta \widetilde{y}^{(1)j}}, ...,\\
&\frac{\delta}{\delta y^{(k-1)i}} = \frac{\partial \widetilde{x}^j}{\partial x^i}\frac{\delta}{\delta \widetilde{y}^{(k-1)j}}, \frac{\delta}{-\delta p_i} = \frac{\partial x^i}{\partial \widetilde{x}^j}\frac{\delta}{\delta \widetilde{p}_j}.
\end{aligned}
$$

According to (6.2.8') the vertical projector v is expressed by means of the projectors $v_1, ..., v_{k-1}, w_k$ determined by the distributions $N_1, N_2, ..., N_{k-2}, V_k$ and W_k as follows

$$v = v_1 + v_2 + ... + v_{k-1} + w_k.$$

If we consider also the horizontal projector h, determined by the distribution $N = N_0$ we have, for $\alpha = 1, ..., k-1$

$$
(6.2.12)\qquad h + v_1 + ... + v_{k-1} + w_k = Id,
$$

$$h^2 = h,\ v_\alpha^2 = v_\alpha,\ w_k^2 = w_k,$$

$$h \circ v_\alpha = v_\alpha \circ h = 0,\ h \circ w_\alpha = w_\alpha \circ h = 0,\ v_\beta \circ v_\alpha = v_\alpha \circ v_\beta = 0, \alpha \neq \beta.$$

As usual we put:

$$X^H = hX, X^{V_\alpha} = v_\alpha X,\ X^{W_k} = w_k X,\ \forall X \in \mathcal{X}(T^{*k}M). \tag{6.2.13}$$

In adapted basis we get

$$X^H = \overset{(0)}{X^i} \frac{\delta}{\delta x^i},\ X^{V_\alpha} = \overset{(\alpha)}{X^i} \frac{\delta}{\delta y^{(\alpha)i}},\ X^{W_k} = X_i \frac{\partial}{\partial p_i}, (\alpha = 1, ..., k-1). \tag{6.2.13'}$$

Therefore, the formulae (6.2.11') show that one has

Proposition 6.2.2 *With respect to (4.1.2), the coordinates of the vectors X^H, X^{V_α}, X^{W_k} are changed by the rule:*

$$\overset{(0)}{\widetilde{X}^i} = \frac{\partial \widetilde{x}^i}{\partial x^j} \overset{(0)}{X^j},\ \overset{(\alpha)}{\widetilde{X}^i} = \frac{\partial \widetilde{x}^i}{\partial x^j} \overset{(\alpha)}{X^j},\ \widetilde{X}_i = \frac{\partial x^j}{\partial \widetilde{x}^i} X_j. \tag{6.2.13''}$$

The following result is important:

Proposition 6.2.3 *1°. The distribution N_0 is integrable if and only if for any $X, Y \in \mathcal{X}(T^{*k}M)$:*

$$[X^H, Y^H]^{V_\alpha} = 0,\ [X^H, Y^H]^{W_k} = 0,\ (\alpha = 1, ..., k-1).$$

2°. The distributions $N_1, ..., N_{k-2}$ are integrable if and only if

$$[X^{V_\alpha}, Y^{V_\alpha}]^H = [X^{V_\alpha}, Y^{V_\alpha}]^{V_\beta} = [X^{V_\alpha}, Y^{V_\alpha}]^{W_k} = 0,$$

for $\alpha = 1, ..., k-2$, $\beta \neq \alpha$, $\beta = 1, ..., k-1$, respectively.

Indeed, the previous equations appear if we express that $[X^H, Y^H]$ is a vector field which belongs to the distribution N_0, $[X^{V_\alpha}, Y^{V_\alpha}]$ is a vector field which belongs to the distribution N_α, for $\alpha = 1, .., k-2$.

Notices that the distributions $V_{k-1}$and W_k are integrable. q.e.d

The notions of h-lift, v_α-lift or w_k-lift of a vector field $X = X^i(x)\frac{\partial}{\partial x^i} \in \mathcal{X}(M)$ are not difficult to define. If l_h is the horizontal lift to N_0, l_{v_α} are the lifts to N_α and $l_{v_{k-1}}$ is the lift to V_{k-1}, then we have

$$X^H = l_h X = X^i(x)\frac{\delta}{\delta x^i}, X^{V_\alpha} = l_{v_\alpha} X = X^i\,(x)\,\frac{\delta}{\delta y^{(\alpha)i}} \tag{6.2.14}$$

at the point $u \in T^{*k}M$, $\pi^{*k}(u) = x$.

In the case of an 1-form $\omega = \omega_i(x)dx^i$ from $\mathcal{X}^*(M)$ the lift $l_{w_k}(\omega)$ to the distribution W_k can be defined, at every point $u \in T^{*k}M$,with $\pi^{*k}(u) = x$,by

$$\omega^{W_k} = l_{w^\alpha}\omega = \omega_i\, \dot{\partial}^i\ . \tag{6.2.14'}$$

6.3 The Dual Coefficients of the Nonlinear Connection N

Throughout this chapter, we consider the coefficients $(\underset{(1)}{N^i_j}, ..., \underset{(k-1)}{N^i_j}, N_{ij})$ of the nonlinear connection N. By means of these coefficients, the vector fields from the adapted basis (6.2.11), are expressed . The dual basis of (6.2.11) will be denoted by

$$(\delta x^i, \delta y^{(1)i}, ..., \delta y^{(k-1)i}, \delta p_i). \tag{6.3.1}$$

Remark that the coefficients of the basis (6.3.1), called the dual coefficients of N, are expressed by the coefficients of the nonlinear connection but are not coincident with them.

First of all, the conditions of duality between the adapted basis (6.2.11) and its dual basis (6.3.1) impose the following form of the 1-form fields (6.3.1)

$$\begin{aligned} &\delta x^i = dx^i, \\ &\delta y^{(1)i} = dy^{(1)i} + \underset{(1)}{M^i_j} dx^j, \\ &\dots\dots\dots\dots\dots\dots\dots\dots \\ &\delta y^{(k-1)i} = dy^{(k-1)i} + \underset{(1)}{M^i_j} dy^{(k-2)i} + ... + \underset{(k-2)}{M^i_j} dy^{(1)j} + \underset{(k-1)}{M^i_j} dx^j, \\ &\delta p_i = dp_i - N_{ji} dx^j. \end{aligned} \tag{6.3.2}$$

We obtain, without difficulties, [115]

Proposition 6.3.1 *The dual coefficients* $(\underset{(1)}{M^i_j}, ..., \underset{(k-1)}{M^i_j}, N_{ji})$ *are uniquely determined by the coefficients* $(\underset{(1)}{N^i_j}, ..., \underset{(k-1)}{N^i_j}, N_{ij})$ *through the following formulae:*

$$\begin{aligned} &\underset{(1)}{M^i_j} = \underset{(1)}{N^i_j},\ \underset{(2)}{M^i_j} = \underset{(2)}{N^i_j} + \underset{(1)}{N^i_m}\underset{(1)}{M^m_j}, \\ &\dots\dots\dots\dots\dots\dots\dots\dots \\ &\underset{(k-1)}{M^i_j} = \underset{(k-1)}{N^i_j} + \underset{(k-2)}{N^i_m}\underset{(1)}{M^m_j} + ... + \underset{(1)}{N^i_m}\underset{(k-2)}{M^m_j}. \end{aligned} \tag{6.3.3}$$

The same formulae determine the coefficients $(\underset{(1)}{N^i_j}, ..., \underset{(k-1)}{N^i_j}, N_{ij})$ as functions by the dual coefficients.

These two dual basis (6.2.11) and (6.3.1) and the Proposition 6.1.1 lead to:

Proposition 6.3.2 *If we change the local coordinates on* $T^{*k}M$*, then the 1-form of dual basis*

$$(\delta x^i, \delta y^{(1)i}, ..., \delta y^{(k-1)i}, \delta p_i)$$

are transformed as follows:

$$
(6.3.4)\quad
\begin{aligned}
&\delta\widetilde{x}^i = \frac{\partial\widetilde{x}^i}{\partial x^j}\delta x^j, \delta\widetilde{y}^{(1)i} = \frac{\partial\widetilde{x}^i}{\partial x^j}\delta y^{(1)j}, ..., \delta\widetilde{y}^{(k-1)i} = \frac{\partial\widetilde{x}^i}{\partial x^j}\delta y^{(k-1)j},\\
&\delta\widetilde{p}_i = \frac{\partial x^i}{\partial\widetilde{x}^i}\delta p_j.
\end{aligned}
$$

Evidently, the properties (6.3.4) imply some special rules of transformation of dual coefficients. In this respect we remark the formulae, which can be easily deduced

$$
(6.3.5)\quad
\begin{aligned}
&\frac{\partial}{\partial x^i} = \frac{\delta}{\delta x^i} + \underset{(1)}{M}{}^j_i\frac{\delta}{\delta y^{(1)j}} + ... + \underset{(k-1)}{M}{}^j_i\frac{\delta}{\delta y^{(k-1)j}} + N_{ij}\frac{\delta}{\delta p_j},\\
&\frac{\partial}{\partial y^{(1)i}} = \frac{\delta}{\delta y^{(1)i}} + \underset{(1)}{M}{}^j_i\frac{\delta}{\delta y^{(2)j}} + ... + \underset{(k-2)}{M}{}^j_i\frac{\delta}{\delta y^{(k-1)j}},\\
&\cdots\cdots\cdots\cdots\cdots\cdots\\
&\frac{\partial}{\partial y^{(k-1)i}} = \frac{\delta}{\delta y^{(k-1)i}},\\
&\frac{\partial}{\partial p_i} = \frac{\delta}{\delta p_i},
\end{aligned}
$$

and

$$
(6.3.5')\quad
\begin{aligned}
&dx^i = \delta x^i,\\
&dy^{(1)i} = \delta y^{(1)i} - \underset{(1)}{N}{}^i_j\delta x^j,\\
&dy^{(2)i} = \delta y^{(2)i} - \underset{(1)}{N}{}^i_j\delta y^{(1)i} - \underset{(2)}{N}{}^i_j\delta x^j,\\
&\cdots\cdots\cdots\cdots\cdots\cdots\\
&dy^{(k-1)i} = \delta y^{(k-1)i} - \underset{(1)}{N}{}^i_j\delta y^{(k-2)i} - ... - \underset{(k-2)}{N}{}^i_j\,\delta y^{(1)j} - \underset{(k-1)}{N}{}^i_j\,\delta x^j,\\
&dp_i = \delta p_i + N_{ji}\delta x^j.
\end{aligned}
$$

Therefore, we obtain

Theorem 6.3.1 *1°. A change of local coordinates on $T^{*k}M$ implies for the dual* *coefficients of the nonlinear connection N, the following rule of transformations* (6.3.6)

$$\underset{(1)}{M}{}^m_j \frac{\partial \widetilde{x}^i}{\partial x^m} = \underset{(1)}{\widetilde{M}}{}^i_m \frac{\partial \widetilde{x}^m}{\partial x^j} + \frac{\partial \widetilde{y}^{(1)i}}{\partial x^j},$$

$$\underset{(2)}{M}{}^m_j \frac{\partial \widetilde{x}^i}{\partial x^m} = \underset{(2)}{\widetilde{M}}{}^i_m \frac{\partial \widetilde{x}^m}{\partial x^j} + \underset{(1)}{\widetilde{M}}{}^i_m \frac{\partial \widetilde{y}^{(1)m}}{\partial x^j} + \frac{\partial \widetilde{y}^{(2)i}}{\partial x^j},$$

....

$$\underset{(k-1)}{M}{}^m_j \frac{\partial \widetilde{x}^i}{\partial x^m} = \underset{(k-1)}{\widetilde{M}}{}^i_m \frac{\partial \widetilde{x}^m}{\partial x^j} + \underset{(k-2)}{\widetilde{M}}{}^i_m \frac{\partial \widetilde{y}^{(1)m}}{\partial x^j} + \ldots + \underset{(1)}{\widetilde{M}}{}^i_m \frac{\partial \widetilde{y}^{(k-2)m}}{\partial x^j} + \frac{\partial \widetilde{y}^{(k-1)i}}{\partial x^j},$$

$$\widetilde{N}_{ij} = \frac{\partial x^r}{\partial \widetilde{x}^i} \frac{\partial x^r}{\partial \widetilde{x}^i} N_{rs} + p_r \frac{\partial^2 x^r}{\partial \widetilde{x}^i \partial \widetilde{x}^j}.$$

2°. Conversely, if the system of function $(\underset{(1)}{M}{}^i_j, \ldots, \underset{(k-1)}{M}{}^i_j, N_{ij})$ *are given on every domain a local chart of the manifold* $T^{*k}M$ *such that the equations (6.3.6) are verified, then* $(\underset{(1)}{M}{}^i_j, \ldots, \underset{(k-1)}{M}{}^i_j, N_{ji})$ *are the dual coefficients of a nonlinear connection* N *on* $T^{*k}M$*.*

The proof can be done as in the usual cases, [115].

Some applications:

1) The vector fields $\overset{1}{\Gamma}, \overset{2}{\Gamma}, \ldots, \overset{k-1}{\Gamma}$ can be expressed in the adapted basis, in the form:

$$\overset{1}{\Gamma} = z^{(1)i} \frac{\delta}{\delta y^{(k-1)i}},$$

$$\overset{2}{\Gamma} = z^{(1)i} \frac{\delta}{\delta y^{(k-2)i}} + 2z^{(2)i} \frac{\delta}{\delta y^{(k-1)i}}, \tag{6.3.7}$$

....

$$\overset{k-1}{\Gamma} = z^{(1)i} \frac{\delta}{\delta y^{(1)i}} + 2z^{(2)i} \frac{\delta}{\delta y^{(2)i}} + \ldots + (k-1)\, z^{(k-1)} \frac{\delta}{\delta y^{(k-1)i}},$$

where

$$z^{(1)i} = y^{(1)i},$$

$$2z^{(2)i} = 2y^{(2)i} + \underset{(1)}{M}{}^i_m y^{(1)m}$$

(6.3.7')

....

$$(k-1)\, z^{(k-1)i} = (k-1)\, y^{(k-1)i} + (k-2) \underset{(1)}{M}{}^i_m y^{(k-2)m} + \ldots$$

$$\ldots + \underset{(k-2)}{M}{}^i_m\, y^{(1)m}.$$

With respect to (4.1.2), we have

$$\widetilde{z}^{(\alpha)i} = \frac{\partial \widetilde{x}^i}{\partial x^j} z^{(\alpha)j}, \ (\alpha = 1, \ldots, k-1) \tag{6.3.7''}$$

¿From (6.3.7") it follows that $z^{(1)i}, ..., z^{(k-1)i}$ have a geometrical meaning. They will be called the *Liouville d-vector fields.*

2) The operators $d_0, ..., d_{k-2}$ and d are represented in the adapted basis by the formulae

$$
\begin{aligned}
d_0 &= \frac{\delta}{\delta y^{(k-1)i}}\delta x^i, \\
d_1 &= \frac{\delta}{\delta y^{(k-2)i}}\delta x^i + \frac{\delta}{\delta y^{(k-1)i}}\delta y^{(1)i}, \\
&\dots\dots\dots\dots\dots\dots, \\
d_{k-2} &= \frac{\delta}{\delta y^{(1)i}}\delta x^i + \frac{\delta}{\delta y^{(2)i}}\delta y^{(1)i} + ... + \frac{\delta}{\delta y^{(k-1)i}}\delta y^{(k-2)i}
\end{aligned}
\tag{6.3.8}
$$

and

$$
d = \frac{\delta}{\delta x^i}\delta x^i + \frac{\delta}{\delta y^{(1)i}}\delta y^{(1)i} + ... + \frac{\delta}{\delta y^{(k-1)i}}\delta y^{(k-1)i} + \frac{\delta}{\delta p_i}\delta p_i. \tag{6.3.9}
$$

If $H \in \mathcal{F}(T^{*k}M)$, then we have the 1-forms

$$
\begin{aligned}
d_0 H &= \frac{\delta H}{\delta y^{(k-1)i}}\delta x^i, \\
d_1 H &= \frac{\delta H}{\delta y^{(k-2)i}}\delta x^i + \frac{\delta H}{\delta y^{(k-1)i}}\delta y^{(1)i}, \\
&\dots\dots\dots\dots\dots\dots, \\
d_{k-2} H &= \frac{\delta H}{\delta y^{(1)i}}\delta x^i + \frac{\delta H}{\delta y^{(2)i}}\delta y^{(1)i} + ... + \frac{\delta H}{\delta y^{(k-1)i}}\delta y^{(k-2)i}
\end{aligned}
\tag{6.3.10}
$$

and

$$
dH = \frac{\delta H}{\delta x^i}\delta x^i + \frac{\delta H}{\delta y^{(1)i}}\delta y^{(1)i} + ... + \frac{\delta H}{\delta y^{(k-1)i}}\delta y^{(k-1)i} + \frac{\delta H}{\delta p_i}\delta p_i. \tag{6.3.11}
$$

In the previous formulas

$$
\frac{\delta H}{\delta x^i}, ..., \frac{\delta H}{\delta y^{(k-1)i}} \tag{6.3.12}
$$

are d-covector and $\dfrac{\delta H}{\delta p_i}$ is a d-vector.

3) In the adapted basis (6.2.11) a dual k-semispray S_ξ can be written as follows

$$
\begin{aligned}
S_\xi = {} & z^{(1)i}\frac{\delta}{\delta x^i} + 2z^{(2)i}\frac{\delta}{\delta y^{(1)i}} + ... + (k-1)\, z^{(k-1)i}\frac{\delta}{\delta y^{(k-2)i}} + \\
& + k \overset{\vee}{\xi}{}^i \frac{\delta}{\delta y^{(k-1)i}} + \overset{\vee}{\eta}_i \frac{\delta}{\delta p_i},
\end{aligned}
\tag{6.3.13}
$$

where

$$(6.3.14)\qquad \begin{aligned} k \overset{\vee}{\xi}{}^{i} &= k\xi^i + (k-1)y^{(k-1)s}\underset{(1)}{M^i_s} + ... + y^{(1)s}\underset{(k-1)}{M^i_s},\\ \overset{\vee}{\eta}_i &= \eta_i + y^{(1)s}N_{si}. \end{aligned}$$

Evidently, $\overset{\vee}{\xi}{}^{i}$ is a d-vector field, and $\overset{\vee}{\eta}_i$ is a d-covector field. It is remarkable that the equation $\overset{\vee}{\eta}_i = 0$ has a geometrical meaning. Especially the sprays S_ξ with this property will be considered.

The interior products of the 1-forms $d_0H, ..., d_{k-2}H$ and dH with the vector field S_ξ are given by

(6.3.15)

$$d_0H(S_\xi) = z^{(1)i}\frac{\delta H}{\delta y^{(k-1)i}},$$

$$d_1H(S_\xi) = z^{(1)i}\frac{\delta H}{\delta y^{(k-2)i}} + 2z^{(2)i}\frac{\delta H}{\delta y^{(k-1)i}},$$

....,

$$d_{k-2}H(S_\xi) = z^{(1)i}\frac{\delta H}{\delta y^{(1)i}} + 2z^{(2)i}\frac{\delta H}{\delta y^{(2)i}} + ... + (k-1)z^{(k-1)i}\frac{\delta H}{\delta y^{(k-1)i}}$$

and

$$(6.3.16)\qquad \begin{aligned} dH(S_\xi) = {} & z^{(1)i}\frac{\delta H}{\delta x^i} + 2z^{(2)i}\frac{\delta H}{\delta y^{(1)i}} + ... + (k-1)z^{(k-1)i}\frac{\delta H}{\delta y^{(k-2)i}} + \\ & + k\overset{\vee}{\xi}{}^{i}\frac{\delta}{\delta y^{(k-1)i}} + \overset{\vee}{\eta}_i\frac{\delta H}{\delta p_i}. \end{aligned}$$

Consequently, we have

$$(6.3.16')\qquad dH(S_\xi) = S_\xi(H).$$

We shall use all these formulae in the theory of higher order Hamilton spaces.

An 1-form fields $\omega \in \mathcal{X}^*(T^{*k}M)$ can be uniquely written as

$$(6.3.17)\qquad \omega = \omega^H + \omega^{V_1} + ... + \omega^{V_{k-1}} + \omega^{W_k},$$

where

$$(6.3.18)\qquad \omega^H = \omega\circ h,\ \omega^{V_\alpha} = \omega\circ v_\alpha,\ (\alpha = 1,...,k-1),\ \omega^{W_k} = \omega\circ w_k.$$

These components can be easily written in the adapted basis.

For a function $H \in F(T^{*k}M)$ we deduce

$$(6.3.19)\qquad dH = (dH)^H + (dH)^{V_1} + ... + (dH)^{V_{k-1}} + (dH)^{W_k},$$

with

$$(6.3.19')\qquad (dH)^H = \frac{\delta H}{\delta x^i}dx^i, (dH)^{V_1} = \frac{\delta H}{\delta y^{(1)i}}\delta y^{(1)i}, ..., (dH)^{W_k} = \frac{\delta H}{\delta p_i}\delta p_i.$$

In the case of 1-forms $d_0H, ..., d_{k-2}H$ we have:

$$(d_0H)^{V_\alpha} = 0, (d_0H)^{W_k} = 0, (\alpha = 1, ..., k-1),$$

$$(d_1H)^{V_\alpha} = 0, (d_1H)^{W_k} = 0, (\alpha = 2, ..., k-1),$$

$$\dots\dots\dots\dots,$$

$$(d_{k-2}H)^{V_{k-1}} = 0, (d_{k-2}H)^{W_k} = 0.$$

4) The horizontal curves with respect to a nonlinear connection N, can be studied exactly as in Ch. 10, from the book [115].

Let $\gamma : I \to T^{*k}M$ be a parametrized curve, locally expressed by

$$(6.3.20)\qquad x^i = x^i(t), y^{(1)i} = y^{(1)i}(t), ..., y^{(k-1)i} = y^{(k-1)i}(t), p_i = p_i(t), t \in I.$$

For the tangent vector field $\frac{d\gamma}{dt}, t \in I$ one can write:

$$(6.3.20')\qquad \frac{d\gamma}{dt} = (\frac{d\gamma}{dt})^H + (\frac{d\gamma}{dt})^{V_1} + ... + (\frac{d\gamma}{dt})^{V_{k-1}} + (\frac{d\gamma}{dt})^{W_k}.$$

Using the basis (6.2.11), adapted to the direct decomposition (6.2.9), we obtain

$$(6.3.20'')\qquad \frac{d\gamma}{dt} = \frac{\delta x^i}{dt}\frac{\delta}{\delta x^i} + \frac{\delta y^{(1)i}}{dt}\frac{\delta}{\delta y^{(1)i}} + ... + \frac{\delta y^{(k-1)i}}{dt}\frac{\delta}{\delta y^{(k-1)i}} + \frac{\delta p_i}{dt}\frac{\delta}{\delta p_i}.$$

Taking into account the formulas (6.3.2) the coefficients of the tangent vector field $\frac{d\gamma}{dt}$ from (6.3.20") have the following expressions

$$(6.3.21)\qquad \begin{aligned} &\frac{\delta x^i}{dt} = \frac{dx^i}{dt},\\ &\frac{\delta y^{(1)i}}{dt} = \frac{dy^{(1)i}}{dt} + \underset{(1)}{M^i_m}\frac{dx^m}{dt},\\ &\dots\dots\dots\dots\\ &\frac{\delta y^{(k-1)i}}{dt} = \frac{dy^{(k-1)i}}{dt} + \underset{(1)}{M^i_m}\frac{dy^{(k-2)i}}{dt} + ... + \underset{(k-2)}{M^i_m}\frac{dy^{(1)i}}{dt} + \underset{(k-1)}{M^i_m}\frac{dx^m}{dt},\\ &\frac{\delta p_i}{dt} = \frac{dp_i}{dt} - N_{mi}\frac{dx^m}{dt}. \end{aligned}$$

An *horizontal curve* $\gamma : I \to T^{*k}M$, with respect to the nonlinear connection N is defined by the conditions

$$(\frac{d\gamma}{dt})^{V_1} = ... = (\frac{d\gamma}{dt})^{V_{k-1}} = 0, (\frac{d\gamma}{dt})^{W_k} = 0$$

It follows:

Theorem 6.3.2 *A parametrized curve $\gamma : I \to T^{*k}M$ is horizontal if and only if the following system of ordinary differential equations is verified:*

$$\frac{\delta y^{(1)i}}{dt} = ... = \frac{\delta y^{(k-1)i}}{dt} = 0, \frac{\delta p_i}{dt} = 0. \tag{6.3.22}$$

The horizontal curves with the property

$$y^{(1)i} = \frac{dx^i}{dt}, ..., y^{(k-1)i} = \frac{1}{(k-1)!}\frac{d^{k-1}x^i}{dt^{k-1}} \tag{*}$$

are called *the autoparalel curves* of the nonlinear connection N. These curves are characterized by (6.3.22) in the conditions (*) .

6.4 The Determination of the Nonlinear Connection by a Dual k-Semispray

The main problem concerning the notion of nonlinear connection N is wether a dual k-semispray S_ξ determines a nonlinear connection N or not. The answer is yes. We have the following important result.

Theorem 6.4.1 *Any dual k-semispray S_ξ with the coefficients (ξ^i, η_i) determines:*

1°. The dual coefficients $\underset{(1)}{M^i_j}, ..., \underset{(k-1)}{M^i_j}$, of a nonlinear connection N, by the formulas:

$$\underset{(1)}{M^i_j} = -\frac{\partial \xi^i}{\partial y^{(k-1)j}}, \underset{(2)}{M^i_j} = -\frac{\partial \xi^i}{\partial y^{(k-2)j}}, ..., \underset{(k-1)}{M^i_j} = -\frac{\partial \xi^i}{\partial y^{(1)j}}. \tag{6.4.1}$$

2°. The coefficients N_{ij} by the formula

$$N_{ij} = \frac{\delta \eta_i}{\delta y^{(1)j}}, \tag{6.4.2}$$

where the operators

$$\frac{\delta}{\delta y^{(1)j}} = \frac{\partial}{\partial y^{(1)j}} - \underset{(1)}{N^i_j}\frac{\partial}{\partial y^{(2)j}} - ... - \underset{(k-2)}{N^i_j}\frac{\partial}{\partial y^{(k-1)j}}, \underset{(\alpha)}{N^i_j},$$

$$(\alpha = 1, ..., k-2)$$

are determined by $\underset{(1)}{M}{}^i_j, ..., \underset{(k-2)}{M}{}^i_j$ *from (6.4.1).*

Proof: We prove that with respect to a change of local coordinates on the manifold $T^{*k}M$, the system of functions $\underset{(1)}{M}{}^i_j, ..., \underset{(k-1)}{M}{}^i_j$ from (6.4.1) obey the transformation (6.3.6).

By means of (6.1.3), we have

$$k\widetilde{\xi}^i = k\frac{\partial \widetilde{x}^i}{\partial x^j}\xi^j + (k-1)\frac{\partial \widetilde{y}^{(1)i}}{\partial x^j}y^{(k-1)j} + ... + \frac{\partial \widetilde{y}^{(k-1)i}}{\partial x^j}y^{(1)j}.$$

Applying the formula $\dfrac{\partial}{\partial y^{(k-1)m}} = \dfrac{\partial \widetilde{x}^s}{\partial x^m}\dfrac{\partial}{\partial \widetilde{y}^{(k-1)}}$, we deduce the first formula (6.3.6). By the same method, we establish inductively the other formula (6.3.6), for the coefficients (6.4.1).

In order to prove (6.4.2), we remark that the vector fields $\dfrac{\delta}{\delta y^{(1)i}}$, constructs by means of coefficients $\underset{(1)}{M}{}^i_j, ..., \underset{(k-1)}{M}{}^i_j$, from (6.4.1) has the law of transformation:

$$\frac{\delta}{\delta \widetilde{y}^{(1)i}} = \frac{\partial x^j}{\partial \widetilde{x}^i}\frac{\delta}{\delta y^{(1)j}}.$$

The rule of transformation of the coefficients η_i of S_ξ is

$$\widetilde{\eta}_i = \frac{\partial x^r}{\partial \widetilde{x}^i}\eta_r + \frac{\partial \widetilde{p}_i}{\partial x^r}y^{(1)r}.$$

These formula have as a consequence:

$$\frac{\delta \widetilde{\eta}_i}{\delta \widetilde{y}^{(1)j}} = \frac{\partial x^r}{\partial \widetilde{x}^i}\frac{\partial x^m}{\partial \widetilde{x}^j}\frac{\delta \eta_r}{\delta y^{(1)m}} + p_r\frac{\partial^2 x^r}{\partial \widetilde{x}^r \partial \widetilde{x}^j}.$$

This is the rule of transformation of the coefficients N_{ij} of a nonlinear connection N on the manifold $T^{*k}M$.

6.5 Lie Brackets. Exterior Differential

In the following it is important to determine the Lie brackets of the vector fields of the adapted basis (6.2.11) and the exterior differentials of the covector fields of adapted cobasis (6.3.1).

By a direct calculus we obtain

Proposition 6.5.1 *The following expressions of the Lie brackets hold true:*

$$[\frac{\delta}{\delta x^j}, \frac{\delta}{\delta x^h}] = \underset{(01)}{R}{}^i_{jh}\frac{\delta}{\delta y^{(1)i}} + ... + \underset{(0,k-1)}{R}{}^i_{jh}\frac{\delta}{\delta y^{(k-1)i}} + \underset{(0)}{R}{}_{ijh}\frac{\delta}{\delta p_i},$$

$$[\frac{\delta}{\delta x^j},\frac{\delta}{\delta y^{(\alpha)h}}]=\underset{(\alpha 1)}{B^i_{jh}}\frac{\delta}{\delta y^{(1)i}}+...+\underset{(\alpha,k-1)}{B^i_{jh}}\frac{\delta}{\delta y^{(k-1)i}}+\underset{(\alpha)}{B_{ijh}}\frac{\delta}{\delta p_i},$$

$$(6.5.1)\qquad [\frac{\delta}{\delta y^{(\alpha)j}},\frac{\delta}{\delta y^{(\beta)h}}]=\underset{(\alpha\beta)}{\overset{(1)}{C^i_{jh}}}\frac{\delta}{\delta y^{(1)i}}+...+\underset{(\alpha\beta)}{\overset{(k-1)}{C^i_{jh}}}\frac{\delta}{\delta y^{(k-1)i}}+\underset{(\alpha\beta)}{B_{ijh}}\frac{\delta}{\delta p_i},$$

$$[\frac{\delta}{\delta y^{(\alpha)j}},\frac{\delta}{\delta p_h}]=\underset{(\alpha 1)}{C^{ih}_j}\frac{\delta}{\delta y^{(1)i}}+...+\underset{(\alpha,k-1)}{C^{ih}_j}\frac{\delta}{\delta y^{(k-1)i}}+\underset{(\alpha)}{C^h_{ij}}\frac{\delta}{\delta p_i},$$

$$[\frac{\delta}{\delta x^j},\frac{\delta}{\delta p_h}]=\underset{(1)}{B^{ih}_j}\frac{\delta}{\delta y^{(1)i}}+...+\underset{(k-1)}{B^{ih}_j}\frac{\delta}{\delta y^{(k-1)i}}+\underset{(0)}{B^h_{ij}}\frac{\delta}{\delta p_i},$$

$\alpha,\beta=1,...k;\beta\leq\alpha$ *in which:*

$$\underset{(\alpha\beta)}{\overset{(1)}{C^i_{jh}}}=0,\quad 1\leq\alpha\leq\beta,$$

$$(6.5.2)\qquad \underset{(\alpha\beta)}{\overset{(2)}{C^i_{jh}}}=0,\quad 2\leq\alpha\leq\beta,$$

....

$$\underset{(\alpha\beta)}{\overset{(\gamma)}{C^i_{jh}}}=0,\quad \gamma\leq\alpha\leq\beta$$

and the coefficients R,B,C can be obtained by a straiforward calculus, $\underset{(0\alpha)}{R^i_{jh}}$ *being expressed by:*

$$\underset{(01)}{R^i_{jh}}=\underset{(01)}{r^i_{jh}},$$

$$(6.5.3)\qquad \underset{(02)}{R^i_{jh}}=\underset{(02)}{r^i_{jh}}+\underset{(1)}{M^i_s}\underset{(01)}{r^s_{jh}},$$

....

$$\underset{(0,k-1)}{R^i_{jh}}=\underset{(0,k-1)}{r^i_{jh}}+\underset{(1)}{M^i_s}\underset{(0,k-2)}{r^s_{jh}}+...+\underset{(k-2)}{M^i_s}\underset{(0,k-2)}{r^s_{jh}}$$

and

$$(6.5.3')\qquad R_{ijh}=\frac{\delta N_{ji}}{\delta x^h}-\frac{\delta N_{ki}}{\delta x^j},$$

with

$$(6.5.3")\qquad \underset{(0\alpha)}{r^i_{jh}}=\frac{\delta \underset{(\alpha)}{N^i_j}}{\delta x^h}-\frac{\delta \underset{(\alpha)}{N^i_h}}{\delta x^j},(\alpha=1,...k-1).$$

Taking into account the conditions of integrability of the nonlinear connection N, (cf Prop. 6.2.3) we have:

Theorem 6.5.1 *The nonlinear connection N is integrable if and only if the following equations hold:*

$$\underset{(0\alpha)}{R^i_{jh}} = 0, (\alpha = 1, ..., k-1), \underset{(0)}{R_{ijh}} = 0. \tag{6.5.4}$$

Indeed, the distribution N is integrable if and only if the vector fields $[\frac{\delta}{\delta x^j}, \frac{\delta}{\delta x^h}]$ belong to N. So, the equations (6.5.4) (which have a geometrical meaning) express the necessary and sufficient conditions for N to be integrable.

Similar characterizations hold for the case when each distribution $N_1, ..., N_{k-2}$ is integrable. (cf. Prop.6.2.3, 2°)

In order to determine the exterior differentials of 1-forms $(\delta x^i, \delta y^{(1)i}, ..., \delta y^{(k-1)i}, \delta p_i)$, we start from the formulae (6.3.2) and express the mentioned differentials with respect to the base of $\wedge^2(T^{*k}M)$ determined by $(\delta x^i, \delta y^{(1)i}, ..., \delta y^{(k-1)i}, \delta p_i)$.

¿From (6.3.2) one obtains:

$$\begin{aligned} & d\delta x^i = 0, \\ & d\delta y^{(1)i} = d\underset{(1)}{M^i_j} \wedge dx^j, \\ & d\delta y^{(2)i} = d\underset{(1)}{M^i_j} \wedge dy^{(1)j} + d\underset{(2)}{M^i_j} \wedge dx^j, \\ & \cdots\cdots\cdots \\ & d\delta y^{(k-1)i} = d\underset{(1)}{M^i_j} \wedge dy^{(k-2)j} + d\underset{(2)}{M^i_j} \wedge dy^{(k-3)j} + ... + d\underset{(k-1)}{M^i_j} \wedge dx^j \end{aligned} \tag{6.5.5}$$

and

$$d\delta p_i = -dN_{ji} \wedge dx^j. \tag{6.5.5'}$$

Substituting $(dx^i, dy^{(1)i}, ..., dy^{(k-1)i})$ from (6.3.5') we can write (6.5.5) in the following form

$$\begin{aligned} & d\delta y^{(\alpha)i} = \underset{(\alpha 0)}{P^i_j} \wedge dx^j + \underset{(\alpha 1)}{P^i_j} \wedge \delta y^{(1)j} + ... + \underset{(\alpha,\alpha-1)}{P^i_j} \wedge \delta y^{(\alpha-1)j}, \\ & (\alpha = 1, ..., k-1), \\ & d\delta p_i = P_{ij} \wedge dx^j, \end{aligned} \tag{6.5.6}$$

where $\underset{(\alpha\beta)}{P^i_j}$ are 1-forms, which should be calculated by means of formula

$$d\omega(X,Y) = X\omega(Y) - Y\omega(X) - \omega([X,Y]), \forall \omega \in \wedge^1(T^{*k}M)$$

and using the Lie brackets (6.5.1).

Theorem 6.5.2 *The exterior differentials of the 1-forms* $(\delta x^i, \delta y^{(1)i}, ..., \delta y^{(k-1)i}, \delta p_i)$ *are given by*

(6.5.7)

$$d\delta x^i = 0,$$

$$d\delta y^{(1)i} = \underset{(10)}{P^i_j} \wedge dx^j,$$

$$d\delta y^{(2)i} = \underset{(20)}{P^i_j} \wedge dx^j + \underset{(21)}{P^i_j} \wedge \delta y^{(1)j},$$

....

$$d\delta y^{(k-1)i} = \underset{(k-1,0)}{P^i_j} \wedge dx^j + \underset{(k-1,1)}{P^i_j} \wedge \delta y^{(1)j} + ... + \underset{(k-1,k-2)}{P^i_j} \wedge \delta y^{(k-2)j}$$

and

$$d\delta p_i = P_{ij} \wedge dx^j, \qquad (6.5.7')$$

where the 1-forms $\underset{(\alpha\beta)}{P^i_j}$ *and* P_{ij} *are given by*

$$\underset{(10)}{P^i_j} = \frac{1}{2}\underset{(01)}{R^i_{jm}} dx^m + \sum_{\gamma=1}^{k-1} \underset{(\gamma 1)}{B^i_{jm}} \delta y^{(\gamma)m} + \underset{(1)}{B^{im}_j} \delta p_m, \qquad (6.5.8_1)$$

$$(6.5.8_2) \quad \begin{cases} \underset{(20)}{P^i_j} = \frac{1}{2}\underset{(02)}{R^i_{jm}} dx^m + \sum_{\gamma=1}^{k-1} \underset{(\gamma 2)}{B^i_{jm}} \delta y^{(\gamma)m} + \underset{(2)}{B^{im}_j} \delta p_m, \\ \underset{(21)}{P^i_j} = -\underset{(12)}{B^i_{mj}} dx^m - \sum_{\gamma=1}^{k-1} \overset{(2)}{\underset{(\gamma 1)}{C^i_{mj}}} \delta y^{(\gamma)m} + \underset{(12)}{C^{im}_j} \delta p_m, \end{cases}$$

....

$$\underset{(\alpha 0)}{P^i_j} = \frac{1}{2}\underset{(0\alpha)}{R^i_{jm}} dx^m + \sum_{\gamma=1}^{k-1} \underset{(\gamma\alpha)}{B^i_{jm}} \delta y^{(\gamma)m} + \underset{(\alpha)}{B^{im}_j} \delta p_m,$$

$$(6.5.8_\alpha) \quad \underset{(\alpha 1)}{P^i_j} = -\underset{(1\alpha)}{B^i_{mj}} dx^m - \sum_{\gamma=1}^{k-1} \overset{(\alpha)}{\underset{(\gamma 1)}{C^i_{mj}}} \delta y^{(\gamma)m} + \underset{(1\alpha)}{C^{im}_j} \delta p_m,$$

....

$$\underset{(\alpha,\alpha-1)}{P^i_j} = -\underset{(\alpha-1,\alpha)}{B^i_{jm}} dx^m - \sum_{\gamma=1}^{k-1} \overset{(\alpha)}{\underset{(\gamma,\alpha-1)}{C^i_{mj}}} \delta y^{(\gamma)m} + \underset{(\alpha-1,\alpha)}{C^{im}_j} \delta p_m,$$

$$(\alpha = 3, ..., k-1)$$

and

$$P_{ij} = \frac{1}{2}\underset{(0)}{R}_{ijm}dx^m + \sum_{\gamma=1}^{k-1} \underset{(\gamma)}{B}_{ijm}\delta y^{(\gamma)m} + \underset{(0)}{B}{}^m_{ij}\delta p_m. \tag{6.5.9}$$

Remark 6.5.1 *In order to give a proof of the equations (6.5.2) we remark that*

$$d\delta y^{(\gamma)i}\left(\frac{\delta}{\delta y^{(\alpha)j}}, \frac{\delta}{\delta y^{(\beta)h}}\right) = -\overset{(\gamma)}{\underset{(\alpha\beta)}{C}}{}^i_{jh}$$

¿From (6.5.6) it follows that the left hand side of the previous formula vanishes for $\gamma < \alpha \leq \beta$.

Remark 6.5.2 *By this method we can calculate the coefficients for the expression of the Lie brackets (6.5.1).*

Indeed, $d\delta^{(\alpha)i}$ from (6.5.5) can be written in the adapted basis

$$\begin{aligned} d\delta y^{(\alpha)i} = {} & \{d\underset{(\alpha)}{M}{}^i_j - \underset{(1)}{N}{}^m_j d\underset{(\alpha-1)}{M}{}^i_m - \ldots - \underset{(\alpha-1)}{N}{}^i_j d\underset{(1)}{M}{}^i_j\} \wedge dx^j + \\ & \{d\underset{(\alpha-1)}{M}{}^i_j - \underset{(1)}{N}{}^m_j d\underset{(\alpha-2)}{M}{}^i_m - \ldots - \underset{(\alpha-2)}{N}{}^i_j d\underset{(1)}{M}{}^i_j\} \wedge dy^{(1)j} + \\ & + \ldots + \{d\underset{(2)}{M}{}^i_j - \underset{(1)}{N}{}^m_j d\underset{(1)}{M}{}^i_m\} \wedge dy^{(\alpha-2)j} + d\underset{(1)}{M}{}^i_j \wedge dy^{(\alpha)j}. \end{aligned} \tag{6.5.10}$$

Identifying to $d\delta y^{(\alpha)i}$ for (6.5.6) we have

$$\begin{aligned} \underset{(\alpha 0)}{P}{}^i_j &= d\underset{(\alpha)}{M}{}^i_j - \underset{(1)}{N}{}^m_j d\underset{(\alpha-1)}{M}{}^i_m - \ldots - \underset{(\alpha-1)}{N}{}^m_j d\underset{(1)}{M}{}^i_m, \\ \underset{(\alpha 1)}{P}{}^i_j &= d\underset{(\alpha-1)}{M}{}^i_j - \underset{(1)}{N}{}^m_j d\underset{(\alpha-2)}{M}{}^i_m - \ldots - \underset{(\alpha-2)}{N}{}^m_j d\underset{(1)}{M}{}^i_m, \\ & \cdots\cdots\cdots\cdots\cdots\cdots\cdots\cdots \\ \underset{(\alpha,\alpha-1)}{P} &= d\underset{(1)}{M}{}^i_j. \end{aligned} \tag{6.5.11}$$

Taking into account that, with respect to the adapted basis the 1-form $d\underset{(\alpha)}{M}{}^i_j$ are given by

$$d\underset{(\alpha)}{M}{}^i_j = \frac{\delta \underset{(\alpha)}{M}{}^i_j}{\delta x^h}\delta x^h + \sum_{\gamma=1}^{k-1} \frac{\delta \underset{(\alpha)}{M}{}^i_j}{\delta y^{(\gamma)h}}\delta y^{(\gamma)h} + \frac{\delta \underset{(\alpha)}{M}{}^i_j}{\delta p_h}\delta p_h. \tag{6.5.12}$$

Substituting (6.5.11) and identifying to (6.5.8) we completely determine the coefficients of the Lie brackets.

For instance, from $P_{\alpha,\alpha-1}$, (6.5.8) $_{\alpha}$) and $\underset{(\alpha,\alpha-1)}{P}$ (6.5.11) we obtain

$$\underset{(\alpha-1,\alpha)}{B^i_{mj}} = -\frac{\delta \underset{(\alpha)}{M^i_j}}{\delta x^m};\ \underset{(\gamma,\alpha-1)}{\overset{(\alpha)}{C^i_{mj}}} = -\frac{\delta \underset{(\alpha)}{M^i_j}}{\delta y^{(\gamma)h}};\ \underset{(\alpha-1,\alpha)}{C^{im}_j} = -\frac{\delta \underset{(\alpha)}{M^i_j}}{\delta p_m}.$$

6.6 The Almost Product Structure $\mathbb{P}$. The Almost Contact Structure $\mathbb{F}$

The $\mathcal{F}(T^{*k}M)$-linear mapping $\mathbb{P}: \mathcal{X}(T^{*p}M) \to \mathcal{X}(T^{*p}M)$ defined by

$$\mathbb{P}(X^H) = X^H, \mathbb{P}(X^{V_\alpha}) = -X^{V_\alpha}, \mathbb{P}(X^{W_k}) = -X^{W_k}, (\alpha = 1, ..., k-1) \tag{6.6.1}$$

determines an almost product structure on the manifold $T^{*k}M$. It is given by means of a nonlinear connection N.

We have

$$\begin{aligned} &\mathbb{P}\circ\mathbb{P} = I,\\ &\mathbb{P} = I - 2(v_1 + ... + v_{k-1} + w_k),\\ &\text{rank}\mathbb{P} = (k+1)n. \end{aligned} \tag{6.6.2}$$

Theorem 6.6.1 *A nonlinear connection N on $T^{*k}M$ is characterized by the existence of an almost product structure $\mathbb{P}$ on $T^{*k}M$ whose eingenspaces corresponding to the eingenvalue -1 coincides with the linear space of the vertical distribution V on $T^{*k}M$.*

The proof is same as in the case $k = 2$ (see the book [115]).

Theorem 6.6.2 *The almost product structure $\mathbb{P}$, defined by (6.6.1) is integrable if and only if the horizontal distribution N is integrable.*

The proof is exactly as in Prop. 9.8.1 of the book [115].

Another important structure on $T^{*k}M$ is determined by the $\mathcal{F}(T^{*k}M)-$ linear mapping

$$\begin{aligned} \mathbb{F}\left(\frac{\delta}{\delta x^i}\right) = -\frac{\partial}{\partial y^{(k-1)i}}, \mathbb{F}\left(\frac{\delta}{\delta y^{(\alpha)i}}\right) = 0, (\alpha = 1, .., k-2),\\ \mathbb{F}\left(\frac{\partial}{\partial y^{(k-1)}}\right) = \frac{\delta}{\delta x^i}, \mathbb{F}\left(\frac{\delta}{\delta p_i}\right) = 0, \end{aligned} \tag{6.6.3}$$

where $\left(\frac{\delta}{\delta x^i}, ..., \frac{\delta}{\delta p_i}\right)$ is the adapted basis of a nonlinear connection N and of the vertical distribution V. Now, is not difficult to prove

Theorem 6.6.3 *The mapping* $\mathbb{F}$ *has the following properties:*

1° $\mathbb{F}$ *is globally defined on* $\widetilde{T^{*k}M}$.

2° $\mathbb{F}$ *is a tensor field of type* $(1,1)$:

$$F = -\frac{\partial}{\partial y^{(k-1)i}} \otimes dx^i + \frac{\delta}{\delta x^i} \otimes \delta y^{(k-1)i}, \tag{6.6.4}$$

3° $Ker\mathbb{F} = N_1 \oplus ... \oplus N_{k-2} \oplus W_k,\ Im\mathbb{F} = N_0 \oplus V_{k-1},$

4° $\text{rank}\mathbb{F} = 2n,$

5° $\mathbb{F}^3 + \mathbb{F} = 0.$

Looking at 5°, we can say that $\mathbb{F}$ is an *almost* $(k-1)n$*-contact structure* determined by the nonlinear connection N.

The Nijenhuis tensor of structure $\mathbb{F}$ is expressed by:

$$\mathcal{N}_{\mathbb{F}}(X,Y) = \mathbb{F}^2[X,Y] + [\mathbb{F}X, \mathbb{F}Y] - \mathbb{F}[\mathbb{F}X, Y] - \mathbb{F}[X, \mathbb{F}Y]$$

and the condition of normality of the structure $\mathbb{F}$ is as follows:

(6.6.5)

$$\mathcal{N}_{\mathbb{F}}(X,Y) + \sum_{i=1}^{n}[\sum_{\alpha=1}^{k-1} d(\delta y^{(\alpha)i}(X,Y) + d(\delta p_i)(X,Y)] = 0, \quad \forall X,Y \in \mathcal{X}(T^{*k}M).$$

Using the formulas (6.4.2) and (6.4.5) we can obtain the explicit form of the last equation.

6.7 The Riemannian Structure $\mathbb{G}$ on $T^{*k}M$

Let $\mathbb{G}$ be a Riemannian structure on the manifold $T^{*k}M$. $\mathbb{G}$ determines uniquely a nonlinear connection N on $T^{*k}M$. N is the ortoghonal distribution, with respect to $\mathbb{G}$, to the vertical distribution V.

In the case when the base manifold M is paracompact, Theorem 4.1.2 affirms that the manifolds $T^{*k}M$ is paracompact, too. So on $T^{*k}M$ there exists a Riemannian structure $\mathbb{G}$. Consequently, we have:

Theorem 6.7.1 *If the base manifold* M *is paracompact then on the manifold* $T^{*k}M$ *there exist nonlinear connections* N.

Let $\mathbb{G}$ be a Riemannian structure on $T^{*k}M$ and N the nonlinear connection, whose distribution is orthogonal to the vertical distribution V. The problem is to determine the local coefficients $\underset{(1)}{N^i_j}, ..., \underset{(k-1)}{N^i_j}, N_{ij}$ of N by means of the local coefficients of $\mathbb{G}$:

$$\overset{(00)}{g_{ij}} = \mathbb{G}(\frac{\partial}{\partial x^i}, \frac{\partial}{\partial x^j}), \overset{(0\alpha)}{g_{ij}} = \mathbb{G}(\frac{\partial}{\partial x^i}, \frac{\partial}{\partial y^{(\alpha)j}}),$$

$$\overset{(0k)}{g_i^j} = \mathbb{G}(\frac{\partial}{\partial x^i}, \frac{\partial}{\partial p_j}), ,..., \overset{(k,k)}{g^{ij}} = \mathbb{G}(\frac{\partial}{\partial p_i}, \frac{\partial}{\partial p_j}). \tag{6.7.1}$$

It follows that $\overset{(k-1,k-1)}{g_{ij}}$ and $\overset{(k,k)}{g^{ij}}$ are d-tensor fields symmetric and positively defined. They are coefficients of the restrictions of $\mathbb{G}$ to the distributions V_{k-1} and W_k.

The coefficients of the nonlinear connection N enter in the adapted basis $\{\frac{\delta}{\delta x^i}\}$ to the distribution $N = N_0$, (6.2.4), in the adapted basis $\{\frac{\delta}{\delta y^{(1)i}}\}, ..., \{\frac{\delta}{\delta y^{(k-1)i}}\}$ to the distribution $N_1, ..., V_{k-1}$ and in the adapted basis $\{\frac{\delta}{\delta p_i}\}$ to the vertical distributions W_k. They are uniquely determined by the conditions that each of the distributions $\{N_1, ..., N_{k-2}\}$ is orthogonal to V_{k-1} with respect to $\mathbb{G}$ and N_0 is orthogonal to V.

Indeed, the conditions of orthogonality

$$\mathbb{G}(\frac{\delta}{\delta y^{(k-2)i}}, \frac{\delta}{\delta y^{(k-1)j}}) = 0$$

give us

$$\overset{(k-2,k-1)}{g_{ij}} - \underset{(1)}{N_i^m} \overset{(k-1,k-1)}{g_{mj}} = 0.$$

But $rank(\overset{(k-1,k-1)}{g_{ij}}) = n$. So the previous equation uniquely gives the coefficients $\underset{(1)}{N_i^m}$.

The following equations

$$\mathbb{G}(\frac{\delta}{\delta y^{(k-3)i}}, \frac{\delta}{\delta y^{(k-1)j}}) = 0$$

lead to the equations

$$\overset{(k-3,k-1)}{g_{ij}} - \underset{(1)}{N_i^m} \overset{(k-2,k-1)}{g_{mj}} - \underset{(2)}{N_i^m} \overset{(k-1,k-1)}{g_{mj}} = 0.$$

These equations uniquely determine the coefficients $\underset{(2)}{N_i^m}$, etc.

Now we prove the following result.

Theorem 6.7.2 *Any Riemannian structure G on the manifold $T^{*k}M$ determines on this manifold a Riemannian almost contact structure $(\overset{\circ}{\mathbb{G}}, \mathbb{F})$.*

Proof. The structure $\mathbb{G}$ determine a nonlinear connection N with the local coefficients $(\underset{(1)}{N_i^m}, ..., \underset{(k-1)}{N_i^m}, N_{ij})$. With these coefficients we construct the adapted basis
$(\frac{\delta}{\delta x^i}, \frac{\delta}{\delta y^{(1)i}}, ..., \frac{\delta}{\delta y^{(k-1)j}}, \frac{\delta}{\delta p_j})$ to N and V_1. The restrictions of G to the distribution V_{k-1} and W_k give us the symmetric, positively defined d-tensor fields

$$g_{ij} = \mathbb{G}(\frac{\delta}{\delta y^{(k-1)i}}, \frac{\delta}{\delta y^{(k-1)j}}),\ h^{ij} = \mathbb{G}(\frac{\partial}{\partial p_i}, \frac{\partial}{\partial p_j})$$

Therefore $\mathbb{G}$ determines on the manifold $T^{*k}M$ the Riemannian structure

$$(6.7.3)\qquad \begin{aligned} \overset{\circ}{\mathbb{G}} &= g_{ij}dx^i \otimes dx^j + g_{ij}\delta y^{(1)i} \otimes \delta y^{(1)j} + ...+ \\ &+ g_{ij}\delta y^{(k-1)i} \otimes \delta y^{(k-1)i} + h^{ij}\delta p_i \otimes \delta p_j. \end{aligned}$$

Consider the almost $(k-1)n$-contact structure $\mathbb{F}$ determined by N. It is given by (6.6.3). Consequently, $\mathbb{F}$ is determined only by the Riemannian structure $\mathbb{G}$.

Therefore the pair $(\overset{\circ}{\mathbb{G}}, \mathbb{F})$ is a Riemannian almost contact structure determined only by $\mathbb{G}$.

Of course, the equation

$$\overset{\circ}{\mathbb{G}}(\mathbb{F}X, Y) = -\overset{\circ}{\mathbb{G}}(X, \mathbb{F}Y),$$

is verified on the adapted basis to N and V. q.e.d

Remark 6.7.1 *We shall use the Riemannian structure $\overset{\circ}{\mathbb{G}}$, in the case of Hamilton space of order k, for $h^{ij} = g^{ij}$*

6.8 The Riemannian Almost Contact Structure $(\overset{\vee}{\mathbb{G}}, \overset{\vee}{\mathbb{F}})$

Consider a d-tensor field g_{ij} on $\widetilde{T^{*k}M}$, symmetric and positively defined. It follows $\det \|g_{ij}\| > 0$ on $\widetilde{T^{*k}M}$. Let N be an apriori given nonlinear connection on the manifold $T^{*k}M$ and the adapted basis with respect to N and V : $(\frac{\delta}{\delta x^j}, \frac{\delta}{\delta y^{(1)j}}, ..., \frac{\delta}{\delta y^{(k-1)j}}, \frac{\delta}{\delta p_j})$.

As usual, the dual basis is $(dx^i \delta y^{(1)i}, ..., \delta y^{(k-1)i}, \delta p_i)$. The N-lift of the d-tensor fields g_{ij} at every point $u \in T^{*k}M$ is defined by

$$(6.8.1)\qquad \begin{aligned} \overset{\vee}{\mathbb{G}} = g_{ij}dx^i \otimes dx^j + g_{ij}\delta y^{(1)i} \otimes \delta y^{(1)j} + ... + g_{ij}\delta y^{(k-1)i} \otimes \delta y^{(k-1)i} + \\ + g^{ij}\delta p_i \otimes \delta p_j, \end{aligned}$$

where g^{ij} is the contravariant tensor of g_{ij}. We have:

Proposition 6.8.1 *$\overset{\vee}{\mathbb{G}}$ is a Riemannian structure on $T^{*k}M$.*

Indeed, $\overset{\vee}{\mathbb{G}}$ is a tensor field on the manifold $\widetilde{T^{*k}M}$ covariant symmetric and positively defined.

The geometrical object fields N and g_{ij} allows to define the $\mathcal{F}(\widetilde{T^{*k}M})$-linear mapping

$$\overset{\vee}{\mathbb{F}}: \mathcal{X}(\widetilde{T^{*k}M}) \to \mathcal{X}(\widetilde{T^{*k}M})$$

by

$$\overset{\vee}{\mathbb{F}}\left(\frac{\delta}{\delta x^i}\right) = -g_{ij}\frac{\delta}{\delta p_j}, \overset{\vee}{\mathbb{F}}\left(\frac{\delta}{\delta y^{(\alpha)i}}\right) = 0, \alpha = 1, ..., k, \overset{\vee}{\mathbb{F}}\left(\frac{\delta}{\delta p_i}\right) = g^{ij}\frac{\delta}{\delta x^j} \tag{6.8.2}$$

Taking into account the rule of transformations of the d-tensor g_{ij} and of the vector fields $\frac{\delta}{\delta x^j}, ..., \frac{\delta}{\delta p_j}$ it follows that $\overset{\vee}{\mathbb{F}}$ has a geometrical meaning.

We have also:

Theorem 6.8.1 *1° $\overset{\vee}{\mathbb{F}}$ is a tensor field on $\widetilde{T^{*k}M}$ of type (6.1.1)*

2° In adapted basis $\overset{\vee}{\mathbb{F}}$ is expressed by

$$\overset{\vee}{\mathbb{F}} = -g_{ij}\frac{\delta}{\delta p_j} \otimes dx^i + g^{ij}\frac{\delta}{\delta x^i} \otimes \delta p_j, \tag{6.8.3}$$

3° $Ker\, \overset{\vee}{\mathbb{F}} = N_1 \oplus ... \oplus N_{(k-1)}$, $Im\, \overset{\vee}{\mathbb{F}} = N_0 \oplus W_k$,

4° $rank \overset{\vee}{\mathbb{F}} = 2n$,

5° $\overset{\vee}{\mathbb{F}}^3 + \overset{\vee}{\mathbb{F}} = 0$.

By means of (6.8.2) the proof is immediate.

Consequently, $\overset{\vee}{\mathbb{F}}$ is an almost $(k-1)n$-contact structure on $T^{*k}M$ determined by the nonlinear connection N and d-tensor g_{ij}. Of course, it is useful in the case of Hamilton spaces of order k.

The conditions of normality of the structure $\overset{\vee}{\mathbb{F}}$ can be written exactly as in (6.6.5).

Between the Riemannian structure $\overset{\vee}{\mathbb{G}}$ and the structure $\overset{\vee}{\mathbb{F}}$ there is a strongly relation.

Theorem 6.8.2 *1° The pair $(\overset{\vee}{\mathbb{G}}, \overset{\vee}{\mathbb{F}})$ is a Riemannian almost contact structure determined only by the pair (N, g_{ij})*

2° The associated 2-form is

$$\theta = \delta p_i \wedge dx^i.$$

3° If the coefficients N_{ij} of N are symmetric, then θ reduces to the presymplectic structure

$$\theta = dp_i \wedge dx^i.$$

Proof:

1° The following formula $\check{\mathbb{G}}(\check{\mathbb{F}}X, Y) = -\check{\mathbb{G}}(X, \check{\mathbb{F}}Y)$ can be verified on the adapted basis, using (6.8.2).

2° $\theta(X, Y) = \check{\mathbb{G}}(\check{\mathbb{F}}X, Y)$ is satisfied, too.

3° From $N_{ij} = N_{ji}$ and $\delta p_i = dp_i - N_{ji}dx^j$ we deduce $\delta p_i \wedge dx^i = dp_i \wedge dx^i$, q.e.d.

Chapter 7

Linear Connections on the Manifold $T^{*k}M$

In the chapter 10 of the book 'The Geometry of Hamilton and Lagrange Spaces', [115], we studied the notions of d-tensor algebra and N -linear connections on $T^{*2}M$. The corresponding theory will be extended in the present section to the case $k > 2$.

7.1 The Algebra of Distinguished Tensor Fields

Let N be a nonlinear connection on the manifold $T^{*k}M$. It determines, at every point $u \in T^{*k}M$, the direct decomposition of the linear space $T_u(T^{*k}M)$:

$$T_u(T^{*k}M) = N_{0,u} \oplus N_{1,u} \oplus ... \oplus N_{k-2,u} \oplus V_{k-1,u} \oplus W_{k,u}. \tag{7.1.1}$$

A vector field $X \in \mathcal{X}(T^{*k}M)$ and an one form $\omega \in \mathcal{X}^*(T^{*k}M)$ can be uniquely written in the form

$$\begin{cases} X = X^H + X^{V_1} + ... + X^{V_{k-1}} + X^{W_k}, \\ \omega = \omega^H + \omega^{V_1} + ... + \omega^{V_{k-1}} + \omega^{W_k}. \end{cases} \tag{7.1.2}$$

Clearly if $h, v_1, v_2, ..., v_{k-1}, w_k$ are the projectors determined by the decomposition (7.1.1), we have

$$\begin{gathered} X^H = hX,\ X^{V_\alpha} = v_\alpha X,\ (\alpha = 1, ..., k-1),\ X^{W_k} = w_k X, \\ \omega^H = \omega \circ h,\ \omega^{V_\alpha} = \omega \circ v_\alpha,\ (\alpha = 1, ..., k-1),\ \omega^{W_k} = \omega \circ w_k. \end{gathered} \tag{7.1.3}$$

Definition 7.1.1 *A distinguished tensor field (briefly d-tensor field) on the manifold $T^{*k}M$ of type (r,s) is a d-tensor field T of type (r,s) on $T^{*k}M$ with the property:*

$$(7.1.4)\quad \begin{aligned} &T(\overset{1}{\omega},...,\overset{r}{\omega},\underset{1}{X},...,\underset{s}{X}) = T(\overset{1}{\omega}^{H},...,\overset{1}{\omega}^{W_k},\underset{1}{X}^{H},...,\underset{s}{X}^{W_k}) \\ &\forall \overset{1}{\omega},...,\overset{r}{\omega} \in \mathcal{X}^*(T^{*k}M), \forall \underset{1}{X},...,\underset{s}{X} \in \mathcal{X}(T^{*k}M) \end{aligned}$$

For instance, every components $X^H, X^{V_1}, X^{V_{k-1}}, X^{W_k}$ of a vector field X is a d -vector field and every component of an 1-form $\omega^H, \omega^{V_1}, \omega^{V_{k-1}}, \omega^{W_k}$ is a d-1-form field.

In the adapted basis $(\frac{\delta}{\delta x^i}, \frac{\delta}{\delta y^{(1)i}}, ..., \frac{\delta}{\delta y^{(k-1)i}}, \frac{\delta}{\delta p_i})$ to the decomposition (7.1.1) and in its dual basis $(dx^i, \delta y^{(1)i}, ..., \delta y^{(k-1)i}, \delta p_i)$, given by (4.1.10) and (4.2.2), a d-tensor field T of type (r,s) can be written in the form

$$(7.1.5)\quad T(u) = T^{i_1...i_r}_{j_1...j_s}(u)\frac{\delta}{\delta x^{i_1}} \otimes ... \otimes \frac{\delta}{\delta p_{j_s}} \otimes dx^{j_1} \otimes ... \otimes \delta p_{i_r}, \forall u \in T^{*k}M$$

It follows that the set $\{1, \frac{\delta}{\delta x^i}, \frac{\delta}{\delta y^{(1)i}}, ..., \frac{\delta}{\delta y^{(k-1)i}}, \frac{\delta}{\delta p_i}\}$ generates the algebra of the d-tensor fields over the ring of functions $\mathcal{F}(T^{*k}M)$.

With respect to a local transformation of the coordinates on $T^{*k}M$, the local coefficients $T^{i_1...i_r}_{j_1...j_s}$ of T are transformed by the classical rule

$$(7.1.6)\quad \widetilde{T}^{i_1...i_r}_{j_1...j_s} = \frac{\partial \widetilde{x}^{i_1}}{\partial x^{h_1}} \cdots \frac{\partial \widetilde{x}^{i_r}}{\partial x^{h_r}} \frac{\partial x^{k_1}}{\partial \widetilde{x}^{j_1}} \cdots \frac{\partial x^{k_s}}{\partial \widetilde{x}^{j_s}} T^{h_1...h_r}_{k_1...k_s}.$$

For instance, if $f \in \mathcal{F}(T^{*k}M)$, then each set of functions $\frac{\delta f}{\delta x^i}, \frac{\delta f}{\delta y^{(1)i}}, ..., \frac{\delta f}{\delta y^{(k-1)i}}$, $(i = 1, ..., n)$ is a d-covector field, and $\frac{\delta f}{\delta p_i}$ is a d-vector field.

7.2 N-Linear Connections

The N-linear connections, for $k = 2$, were studied in chapter 10, of the book [115]. Their definition for $k \geq 2$ is as follows:

Definition 7.2.1 *A linear connection D on the manifold $T^{*k}M$ is called an N-linear connection if:*

(1) D preserves by parallelism the distributions $N_0, N_1, ..., N_{k-2}, V_{k-1}, W_k$.

(2) The $k-1$ tangent structure J is absolutely parallel with respect to D.

(3) The presymplectic structure θ is absolutely parallel with respect D.

Directly from the definition we can establish, without difficulties the following characterization of an N-linear connection

Theorem 7.2.1 *A linear connection D is an N-linear connection on the manifold $T^{*k}M$ if and only if:*

(1) D preserves by parallelism the distributions $N_0, N_1, ..., N_{k-2}, V_{k-1}$ and W_k.

(2)

$$D_X(JY^H) = J(D_XY^H),\ D_X(JY^{V_\alpha}) = J(D_XY^{V_\alpha}), \quad (\alpha = 1, ..., k-1) \tag{7.2.1}$$
$$D_X(JY^{W_k}) = J(D_XY^{W_k}),\ \forall X, Y \in X(T^{*k}M).$$

(3)

$$D\theta = 0. \tag{7.2.2}$$

We remark that the equalities $D_X(JY^{V_{k-1}}) = J(D_XY^{V_{k-1}})$ and $D_X(JY^{W_k}) = J(D_XY^{W_k})$ are trivial, since $J(Y^{V_{k-1}}) = 0$, and $J(Y^{W_k}) = 0$.

We obtain also

Theorem 7.2.2 *For any N- linear connection D we have*

$$D_Xh = D_Xv_\alpha = D_Xw_k = 0, (\alpha = 1, ..., k-1), \tag{7.2.3}$$

$$D_X\mathbb{P} = 0, D_X\mathbb{F} = 0. \tag{7.2.4}$$

Indeed, from $(D_Xh)(Y) = D_X(hY) - hD_XY$ if $Y = Y^H$ we obtain

$$(D_Xh)(Y^H) = D_XY^H - hD_XY^H = 0$$

and for $Y = Y^{V_\alpha}$,

$$(D_Xh)(Y^{V_\alpha}) = D_X(hY^{V_\alpha}) - hD_XY^{V_\alpha} = 0.$$

Also $(D_Xh)(Y^{W_k}) = 0$. That means $(D_Xh)(Y) = 0, \forall Y \in \chi((T^{*k}M)$. Similarly, we prove the other equalities (7.2.3).

Now, taking into account the formula (6.6.2), we deduce $D_X\mathbb{P} = 0$. The last equality (7.2.4) can be proved by the formula $(D_X\mathbb{F})(Y) = D_X\mathbb{F}(Y) - \mathbb{F}(D_XY)$, using the local expression of D_XY (see the section 5, from this chapter).

Let us consider the vector field X written in the form (7.1.2). From the linearity of the operator D_XY with respect to X we deduce

$$D_XY = D_{X^H}Y + D_{X^{V_1}}Y + ... + D_{X^{V_{k-1}}}Y + D_{X^{W_k}}Y. \tag{7.2.5}$$

Here appears $(k+1)$ new operators of derivation in the d–tensor algebra, defined by

$$D^H_X = D_{X^H}, D^{V_1}_X = D_{X^{V_1}}, ..., D^{V_{k-1}}_X = D_{X^{V_{k-1}}}, D^{W_k}_X = D_{X^{W_k}}. \tag{7.2.6}$$

These operators are not covariant derivations in the d-tensor algebra, since

$$D^H_Xf = X^Hf \neq Xf, D^{V_\alpha}_Xf = X^{V_\alpha}f \neq Xf, (\alpha = 1, ..., k),$$
$$D^{W_k}_Xf = X^{W_k}f \neq Xf.$$

However, they have similar properties with the covariant derivatives.

¿From (7.2.5) and (7.2.6) we deduce

(7.2.7) $$D_XY = D^H_XY + D^{V_1}_XY + ... + D^{V_{k-1}}_XV + D^{W_k}_XY.$$

By means of Theorem 7.2.2, we have:

Theorem 7.2.3 *The operators $D^H_X, D^{V_1}_X, ..., D^{V_{k-1}}_X$ and $D^{W_k}_X$ have the following properties:*

1) Each operator $D^H_X, D^{V_1}_X, ..., D^{V_{k-1}}_X$ and $D^{W_k}_X$ maps a vector field that belongs to one of the distributions $N_0, N_1, ..., N_{k-2}, V_{k-1}$ and W_k into a vector field that belongs to the same distribution,

2) $D^H_Xf = X^Hf,\ D^{V_\alpha}_Xf = X^{V_\alpha}f,\ (\alpha = 1, ..., k),\ D^{W_k}_Xf = X^{W_k}f,$

3) $D^H_X(fY) = X^H(fY) + fD^H_XY,\ D^{V_\alpha}_X(fY) = X^{V_\alpha}(fY) + fD^{V_\alpha}_XY,$

$(\alpha = 1, ..., k-1),\ D^{W_k}_X(fY) = X^{W_k}(fY) + fD^{W_k}_XY,$

4) $D^H_X(Y + Z) = D^H_XY + D^H_XZ,\ D^{V_\alpha}_X(Y + Z) = D^{V_\alpha}_XY + D^{V_\alpha}_XZ,$

$(\alpha = 1, ..., k-1),\ D^{W_k}_X(Y + Z) = D^{W_k}_XY + D^{W_k}_XZ,$

5) $D^H_{X+Y} = D^H_X + D^H_Y,\ D^{V_\alpha}_{X+Y} = D^{V_\alpha}_X + D^{V_\alpha}_Y,\ (\alpha = 1, ...k-1),$

$D^{W_k}_{X+Y} = D^{W_k}_X + D^{W_k}_Y,$

6) $D^H_{fX} = fD^H_X; D^{V_\alpha}_{fX} = fD^{V_\alpha}_Y, (\alpha = 1, ..., k-1), D^{W_k}_{fX} = fD^{W_k}_Y,$

7) $D^H_X(JY) = JD^H_XY,\ D^{V_\alpha}_X(JY) = JD^{V_\alpha}_XY,\ (\alpha = 1, ..., k-1)$

$D^{W_k}_X(JY) = JD^{W_k}_XY,$

8) $D^H_X\theta = 0, D^{V_\alpha}_X\theta = 0, (\alpha = 1, ..., k-1), D^{W_k}_X\theta = 0.$

9) For any open set $U \subset T^{*k}M$ the following properties hold:

$(D^H_XY)_{|U} = D^H_{X_{|U}}Y_{|U}, (D^{V_\alpha}_XY)_{|U} = D^{V_\alpha}_{X_{|U}}Y_{|U}, (\alpha = 1, ..., k-1),$

$(D^{W_k}_XY)_{|U} = D^{W_k}_{X_{|U}}Y_{|U}.$

The proof of this theorem can be done by the classical methods, [115].

The operators $D^H_X, D^{V_\alpha}_X, D^{W_k}_X$ will be called the operators of h- , v_α- and w_k-covariant derivation.

The action of these operators over the 1-form fields ω are given by

$$(D^H_X\omega)(Y) = X^H\omega(Y) - \omega(D^H_X Y),$$

(7.2.8)
$$(D^{V_\alpha}_X\omega)(Y) = X^{V_\alpha}\omega(Y) - \omega(D^{V_\alpha}_X Y), (\alpha = 1, ...k-1),$$

$$(D^{W_k}_X\omega)(Y) = X^{W_k}\omega(Y) - \omega(D^{W_k}_X Y).$$

Of course, the action of the operators $D^H_X, D^{V_\alpha}_X, D^{W_k}_X$ can be extended to any tensor fields, particularly to any d-tensor field on $T^{*k}M$.

For instance, if the d-tensor T verifies (7.1.4) we have

(7.2.9)
$$(D^H_X T)(\overset{1}{\omega}{}^H, ..., \overset{1}{\omega}{}^{W_k}, .\underset{1}{X}{}^H, ..., \underset{s}{X}{}^{W_k}) = X^H T(\overset{1}{\omega}{}^H, ..., \overset{1}{\omega}{}^{W_k}, \underset{1}{X}{}^H, ..., \underset{s}{X}{}^{W_k})$$

$$-T(D^H_X \overset{1}{\omega}{}^H, ..., \overset{1}{\omega}{}^{W_k}, \underset{1}{X}{}^H, ..., \underset{s}{X}{}^{W_k}) - ... -$$

$$-T(\overset{1}{\omega}{}^H, ..., \overset{r}{\omega}{}^{W_k}, .\underset{1}{X}{}^H, ..., D^H_X \underset{s}{X}{}^{W_k}).$$

Now, let us consider a parametrized smooth curve $\gamma : t \in I \to \gamma(t) \in \widetilde{T^{*k}M}$ having the image in a domain of local chart.

Its tangent vector field $\dot{\gamma} = \dfrac{d\gamma}{dt}$ is uniquely written in the form

(7.2.10)
$$\dot{\gamma} = \overset{.}{\gamma}{}^H + \overset{.}{\gamma}{}^{V_1} + ... + \overset{.}{\gamma}{}^{V_{k-1}} + \overset{.}{\gamma}{}^{W_k}.$$

If the curve γ is analytically given by (6.3.2), then $\dot{\gamma}$, in the adapted basis is given by (6.3.20"), (6.3.21). The horizontal curves are defined by Theorem 6.3.2 and the autoparallel curves of the nonlinear connection N are given by the equations (6.3.22) in the conditions

$$y^{(1)i} = \frac{dx^i}{dt}, ..., y^{(k-1)i} = \frac{1}{(k-1)!}\frac{d^{k-1}x^i}{dt^{k-1}}.$$

A vector field $Y(\gamma(t))$, along curve γ has the covariant derivative

(7.2.11)
$$D_{\dot{\gamma}}Y = D^H_{\dot{\gamma}}Y + D^{V_1}_{\dot{\gamma}}Y + ... + D^{V_{k-1}}_{\dot{\gamma}}Y + D^{W_k}_{\dot{\gamma}}Y.$$

The vector field $Y(\gamma(t))$ is *parallel* along curve γ if

(7.2.12)
$$D_{\dot{\gamma}}Y = 0.$$

The curve γ is autoparallel with respect to the N- linear connection D if $D_{\dot{\gamma}}\dot{\gamma} = 0$. This equation is equivalent to

(7.2.13)
$$D^H_{\dot{\gamma}}\dot{\gamma} + D^{V_1}_{\dot{\gamma}}\dot{\gamma} + ... + D^{V_{k-1}}_{\dot{\gamma}}\dot{\gamma} + D^{W_k}_{\dot{\gamma}}\dot{\gamma} = 0$$

In a next section we shall express this equation in an adapted basis.

7.3 The Torsion and Curvature of an N-Linear Connection

The torsion tensor field $\mathbb{T}$ of an N-linear connection D is expressed as usually by

$$\mathbb{T}(X,Y) = D_X Y - D_Y X - [X,Y], \forall X,Y \in \mathcal{X}(T^{*k}M). \tag{7.3.1}$$

$\mathbb{T}$ can by characterized by the vector fields

$$\begin{aligned}\mathbb{T}(X^H,Y^H) &= h\mathbb{T}(X^H,Y^H) + v_1\mathbb{T}(X^H,Y^H) + ... + v_{k-1}\mathbb{T}(X^H,Y^H) + \\ &+ w_k\mathbb{T}(X^H,Y^H),\end{aligned}$$

$$\begin{aligned}\mathbb{T}(X^H,Y^{V_\alpha}) &= h\mathbb{T}(X^H,Y^{V_\alpha}) + v_1\mathbb{T}(X^H,Y^{V_\alpha}) + ... + v_{k-1}\mathbb{T}(X^H,Y^{V_\alpha}) + \\ &+ w_k\mathbb{T}(X^H,Y^{V_\alpha}),\end{aligned}$$

$$\begin{aligned}\mathbb{T}(X^H,Y^{W_k}) &= h\mathbb{T}(X^H,Y^{W_k}) + v_1\mathbb{T}(X^H,Y^{W_k}) + ...+ \\ &+ v_{k-1}\mathbb{T}(X^H,Y^{W_k}) + w_k\mathbb{T}(X^H,Y^{W_k}),\end{aligned} \tag{7.3.2}$$

$$\begin{aligned}\mathbb{T}(X^{V_\alpha},Y^{V_\beta}) &= h\mathbb{T}(X^{V_\alpha},Y^{V_\beta}) + v_1\mathbb{T}(X^{V_\alpha},Y^{V_\beta}) + ... + v_{k-1}\mathbb{T}(X^{V_\alpha},Y^{V_\beta}) + \\ &+ w_k\mathbb{T}(X^{V_\alpha},Y^{V_\beta}),\end{aligned}$$

$$\alpha \leq \beta; \alpha,\beta = 1,2,...,k-1,$$

$$\begin{aligned}\mathbb{T}(X^{V_\alpha},Y^{W_k}) &= h\mathbb{T}(X^{V_\alpha},Y^{W_k}) + v_1\mathbb{T}(X^{V_\alpha},Y^{W_k}) + ... + v_{k-1}\mathbb{T}(X^{V_\alpha},Y^{W_k}) + \\ &+ w_k\mathbb{T}(X^{V_\alpha},Y^{W_k}),\end{aligned}$$

$$\begin{aligned}\mathbb{T}(X^{W_k},Y^{W_k}) = h\mathbb{T}(X^{W_k},Y^{W_k}) + v_1\mathbb{T}(X^{W_k},Y^{W_k}) + ... + \\ +v_{k-1}\mathbb{T}(X^{W_k},Y^{W_k}) + w_k\mathbb{T}(X^{W_k},Y^{W_k}).\end{aligned}$$

Since D preserves by parallelism the distributions $N_0, N_1, ..., N_{k-2}, V_{k-1}$ and W_k we deduce

Proposition 7.3.1 *The following properties of the torsion $\mathbb{T}$ of the N-linear connection D hold:*

$$hT(X^{V_{k-1}}, Y^{V_{k-1}}) = 0, hT(X^{W_k}, X^{W_k}) = 0. \tag{7.3.3}$$

Indeed, the distributions V_{k-1} and W_k being integrable, the equations (7.3.3) are verified.

Now, using the formula (7.3.1), we can write the expression of $\mathbb{T}(X,Y)$ in the form

$$\mathbb{T}(X,Y) = h\mathbb{T}(X,Y) + v_1\mathbb{T}(X,Y) + ... + v_{k-1}\mathbb{T}(X,Y) + w_k\mathbb{T}(X,Y), \tag{7.3.4}$$

taking into account the components of X and Y from the decomposition (7.1.2).

The curvature of the N-linear connection D is given by

$$\mathbb{R}(X,Y)Z = (D_XD_Y - D_YD_X)Z - D_{[X,Y]}Z, \ \forall X, Y, Z \in \mathcal{X}(T^{*k}M) \tag{7.3.5}$$

Taking into account the decomposition of vector fields X, Y, Z in the form (7.1.2) we can write the curvature $\mathbb{R}$ in a similar manner as the torsion $\mathbb{T}$.

The definition 7.2.1 allows to prove, without difficulties

Proposition 7.3.2 *The following properties hold:*

$$J(\mathbb{R}(X,Y)Z) = \mathbb{R}(X,Y)JZ; \ D_X\theta = 0. \tag{7.3.5'}$$

Consequently, we have

Theorem 7.3.1 *The curvature $\mathbb{R}$ has the properties:*

1° The essential components of R are

$$\mathbb{R}(X,Y)Z^H, \mathbb{R}(X,Y)Z^{V_\alpha}, (\alpha = 1, ...k-1), \mathbb{R}(X,Y)Z^{W_k}. \tag{7.3.6}$$

2° The vector field $\mathbb{R}(X,Y)Z^H$ belongs to the horizontal distribution $N = N_0$.

3° The vector field $\mathbb{R}(X,Y)Z^{V_\alpha}, (\alpha = 1, ...k-1)$ belongs to the distribution N_α.

4° The vector field $\mathbb{R}(X,Y)Z^{W_k}$ belongs to the distribution W_k.

5° The following equations hold:

$$\begin{aligned}
&v_\alpha\{\mathbb{R}(X,Y)Z^H\} = 0, (\alpha = 1, ...k-1), w_k\{\mathbb{R}(X,Y)Z^H\} = 0,\\
&v_\alpha\{\mathbb{R}(X,Y)Z^{V_\beta}\} = 0, (\alpha \neq \beta; \alpha, \beta = 1, ...k-1),\\
&v_\alpha\{\mathbb{R}(X,Y)Z^{W_k}\} = 0,\\
&h\{\mathbb{R}(X,Y)Z^{V_\alpha}\} = 0, (\alpha = 1, ...k-1), h\{\mathbb{R}(X,Y)Z^{W_k}\} = 0.
\end{aligned} \tag{7.3.7}$$

Of course we can express the d-tensor of curvature by means of the operators $D^H_X, D^{V_\alpha}_X, D^{W_k}_X$. They will be written in the adapted basis in a next section.

Proposition 7.3.3 *The Ricci identities of the N-linear connection D are:*

$$[D_X, D_Y]Z^H = \mathbb{R}(X,Y)Z^H + D_{[X,Y]}Z^H,$$

$$[D_X, D_Y]Z^{V_\alpha} = \mathbb{R}(X,Y)Z^{V_\alpha} + D_{[X,Y]}Z^{V_\alpha}, (\alpha = 1, ..., k-1), \tag{7.3.8}$$

$$[D_X, D_Y]Z^{W_k} = \mathbb{R}(X,Y)Z^{W_k} + D_{[X,Y]}Z^{W_k}.$$

Let us consider the Liouville vector fields $\overset{1}{\Gamma}, ..., \overset{k-1}{\Gamma}$ and C^* from Theorem 4.2.1 and let us apply the previous Proposition.

Theorem 7.3.2 *For any N-linear connection D the following identities hold:*

$$(7.3.9)\quad \begin{aligned} &[D_X, D_Y]\overset{\alpha}{\Gamma} = \mathbb{R}(X,Y)\overset{\alpha}{\Gamma} + D_{[X,Y]}\overset{\alpha}{\Gamma}, (\alpha = 1, ..., k-1),\\ &[D_X, D_Y]C^* = \mathbb{R}(X,Y)C^* + D_{[X,Y]}C^*. \end{aligned}$$

Using the considerations from this chapter we can establish the Bianchi identities of an N-linear connection D, by means of the operators D^H_X, $D^{V_\alpha}_X$, $D^{W_k}_X$ taking into account the classical identities

$$(7.3.10)\quad \begin{aligned} &\sum_{(X,Y,Z)} \{(D_X\mathbb{T})(Y,Z) - \mathbb{R}(X,Y)Z + \mathbb{T}(\mathbb{T}(X,Y),Z)\} = 0,\\ &\sum_{(X,Y,Z)} \{(D_X\mathbb{R})(U,Y,Z) - \mathbb{R}(\mathbb{T}(X,Y),Z)U\} = 0, \end{aligned}$$

where $\sum_{(X,Y,Z)}$ means the cyclic sum.

7.4 The Coefficients of a N-Linear Connection

A N-linear connection D is characterized by its coefficients in the adapted basis $(\frac{\delta}{\delta x^i}, \frac{\delta}{\delta y^{(\alpha)i}}, \frac{\delta}{\delta p_i})$. As we shall see these coefficients obey particular rules of transformations with respect to the change of local coordinates on the manifold $T^{*k}M$.

Taking into account Theorem 7.2.1 we can prove:

Theorem 7.4.1 *We have:*

1° An N-linear connection D can be uniquely represented in the adapted basis in the following form:

(7.4.1)

$$D_{\frac{\delta}{\delta x^j}}\frac{\delta}{\delta x^i} = H^s_{ij}\frac{\delta}{\delta x^s}; D_{\frac{\delta}{\delta x^j}}\frac{\delta}{\delta y^{(\alpha)i}} = H^s_{ij}\frac{\delta}{\delta y^{(\alpha)s}}, (\alpha = 1, ...k-1);$$

$$D_{\frac{\delta}{\delta x^j}}\frac{\delta}{\delta p_i} = -H^i_{sj}\frac{\delta}{\delta p_s};$$

$$D_{\frac{\delta}{\delta y^{(\alpha)j}}}\frac{\delta}{\delta x^i} = \underset{(\alpha)}{C^s_{ij}}\frac{\delta}{\delta x^s}; D_{\frac{\delta}{\delta y^{(\alpha)j}}}\frac{\delta}{\delta y^{(\beta)i}} = \underset{(\alpha)}{C^s_{ij}}\frac{\delta}{\delta y^{(\beta)s}}, (\alpha, \beta = 1, ...k-1);$$

$$D_{\frac{\delta}{\delta y^{(\alpha)j}}}\frac{\delta}{\delta p_i} = \underset{(\alpha)}{C^i_{sj}}\frac{\delta}{\delta p_s};$$

$$D_{\frac{\delta}{\delta p_j}}\frac{\delta}{\delta x^i} = C^{js}_i\frac{\delta}{\delta x^s}; D_{\frac{\delta}{\delta p_j}}\frac{\delta}{\delta y^{(\alpha)i}} = C^{js}_i\frac{\delta}{\delta y^{(\alpha)s}}, (\alpha = 1, ...k-1);$$

$$D_{\frac{\delta}{\delta p_i}}\frac{\delta}{\delta p_j} = -C^{ij}_s\frac{\delta}{\delta p_s}.$$

2° With respect to the transformation (4.1.2), the coefficients H^i_{jk} obey the rule of transformation

$$\widetilde{H}^i_{rs}\frac{\partial \widetilde{x}^r}{\partial x^j}\frac{\partial \widetilde{x}^s}{\partial x^h} = \frac{\partial \widetilde{x}^i}{\partial x^r}H^r_{jh} - \frac{\partial^2 \widetilde{x}^i}{\partial x^j \partial x^h} \tag{7.4.2}$$

3° The system of functions $\underset{(\alpha)}{C^i_{jh}}, C^{jh}_i, (\alpha = 1, ..., k-1)$ are d-tensor fields of type (1,2) and (2,1), respectively.

Proof: According to the definition 2.1, we can write

$$D_{\frac{\delta}{\delta x^j}}\frac{\delta}{\delta x^i} = \underset{(0)}{H^s_{ij}}\frac{\delta}{\delta x^s}; D_{\frac{\delta}{\delta x^j}}\frac{\delta}{\delta y^{(1)i}} = \underset{(1)}{H^s_{ij}}\frac{\delta}{\delta y^{(1)s}}, ...,$$

$$D_{\frac{\delta}{\delta x^j}}\frac{\delta}{\delta y^{(k-1)i}} = \underset{(k-1)}{H^s_{ij}}\frac{\delta}{\delta y^{(k-1)s}}; D_{\frac{\delta}{\delta x^j}}\frac{\delta}{\delta p_i} = \underset{(k)}{H^i_{js}}\frac{\delta}{\delta p_s};$$

Applying the mapping J and looking to (4.3.1) and to Theorem 7.2.3, we obtain

$$D_{\frac{\delta}{\delta x^j}}(J\frac{\delta}{\delta x^i}) = J(\underset{(0)}{H^s_{ij}}\frac{\delta}{\delta x^s}), D_{\frac{\delta}{\delta x^j}}(J\frac{\delta}{\delta y^{(1)i}}) = J(\underset{(1)}{H^s_{ij}}\frac{\delta}{\delta y^{(1)s}}), ...,$$

$$D_{\frac{\delta}{\delta x^j}}(J\frac{\delta}{\delta y^{(k-2)i}}) = J(\underset{(k-1)}{H^s_{ij}}\frac{\delta}{\delta y^{(k-1)s}}).$$

Consequently,

$$\underset{(1)}{H^s_{ij}} = \underset{(0)}{H^s_{ij}}, \underset{(2)}{H^s_{ij}} = \underset{(1)}{H^s_{ij}}, ..., \underset{(k-1)}{H^s_{ij}} = \underset{(k-2)}{H^s_{ij}}$$

and, from $D_{\frac{\delta}{\delta x^j}}\theta = 0$, we have $\underset{(0)}{H^s_{ij}} = \underset{(k)}{H^s_{ij}}$. It follows that the formulae of the first line of (7.4.1) are valid. Similarly we prove the other equalities of (7.4.1). By a direct calculus we prove that 2° and 3° from this theorem hold. q.e.d

The systems of functions

$$D\Gamma(N) = \{H^i_{jh}, \underset{(\alpha)}{C^i_{jh}}, C^{jh}_i\}, (\alpha = 1, ..., k-1) \tag{7.4.3}$$

is the system of coefficients of the N-linear connection D in the adapted basis.

The converse of the previous theorem holds,too.

Theorem 7.4.2 *If the system of functions (7.4.3) are apriori given over every domain of a local chart on the manifold $T^{*k}M$, having the rule of transformation mentioned in the previous theorem, then there exists an unique N-linear connection D whose local coefficients are just the system of given functions.*

Corollary 7.4.1 *The following formulae hold:*

(7.4.4)

$$D_{\frac{\delta}{\delta x^j}} dx^i = -H^i_{js}dx^s; D_{\frac{\delta}{\delta x^j}} \delta y^{(\alpha)i} = -H^i_{js}\delta y^{(\alpha)s}, (\alpha = 1, ...k-1);$$

$$D_{\frac{\delta}{\delta x^j}} \delta p_i = H^s_{ij}\delta p_s,$$

$$D_{\frac{\delta}{\delta y^{(\alpha)j}}} dx^i = -\underset{(\alpha)}{C^i_{js}}dx^s; D_{\frac{\delta}{\delta y^{(\alpha)j}}} \delta y^{(\beta)i} = -H^i_{js}\delta y^{(\beta)s}, (\alpha, \beta = 1, ...k-1);$$

$$D_{\frac{\delta}{\delta y^{(\alpha)j}}} \delta p_i = \underset{(\alpha)}{C^s_{ij}}\delta p_s;$$

$$D_{\frac{\delta}{\delta p_j}} dx^i = -C^{ij}_s dx^s; D_{\frac{\delta}{\delta p_j}} \delta y^{(\alpha)i} = C^{ij}_s \delta y^{(\alpha)s}, (\alpha = 1, ...k-1);$$

$$D_{\frac{\delta}{\delta p_j}} \delta p_i = C^{js}_i \delta p_s.$$

Indeed, the formula (7.4.1) and the conditions of duality between the vector fields from the adapted basis and its dual 1-form basis lead to the formula (7.4.4).

7.5 The h-, v_α- and w_k-Covariant Derivatives in Local Adapted Basis

In the adapted basis a tensor field T can be written in the form (7.1.5):

(7.5.1) $$T = T^{i_1...i_r}_{j_1...j_s} \frac{\delta}{\delta x^{i_1}} \otimes ... \otimes \frac{\delta}{\delta p_{j_s}} \otimes dx^{j_1} \otimes ... \otimes \delta p_{i_r}.$$

Applying the operator of covariant derivation D_X for $X = X^H = X^i \frac{\delta}{\delta x^i}$ and taking into account the formulae (7.4.1), (7.4.4) and the properties of the operator $D^H_X T = X^i D^H_{\frac{\delta}{\delta x^j}} T$, expressed in Theorem 7.2.3, we deduce

(7.5.2) $$D^H_X T = X^m T^{i_1...i_r}_{j_1...j_s|m} \frac{\delta}{\delta x^{i_1}} \otimes ... \otimes \frac{\delta}{\delta p_{j_s}} \otimes dx^{j_1} \otimes ... \otimes \delta p_{i_r},$$

where

(7.5.2') $$T^{i_1...i_r}_{j_1...j_s|m} = \frac{\delta T^{i_1...i_r}_{j_1...j_s}}{\delta x^m} + T^{hi_2...i_r}_{j_1...j_s} H^{i_1}_{hm} + ... + T^{i_1...h}_{j_1...j_s} H^{i_r}_{hm} -$$
$$-T^{i_1...i_r}_{h...j_s} H^h_{j_1m} - ... - T^{i_1...i_r}_{j_1...h} H^h_{j_sm}.$$

The operator "$_|$" is called the h-covariant derivative with respect to $D\Gamma(N)$.

Now, we put $X = X^{V_\alpha} = X^i \frac{\delta}{\delta y^{(\alpha)i}}, (\alpha = 1, ..., k)$. From (7.5.1) we deduce

(7.5.3) $$D^{V_\alpha}_X T = X^m T^{i_1...i_r}_{j_1...j_s} \overset{(\alpha)}{|}_m \frac{\delta}{\delta x^{i_1}} \otimes ... \otimes \frac{\delta}{\delta p_{j_s}} \otimes dx^{j_1} \otimes ... \otimes \delta p_{i_r}$$
$$\alpha = 1, ..., k-1.$$

where

(7.5.3') $$T^{i_1...i_r}_{j_1...j_s} \overset{(\alpha)}{|}_m = \frac{\delta T^{i_1...i_r}_{j_1...j_s}}{\delta y^{(\alpha)m}} + T^{hi_2...i_r}_{j_1...j_s} \underset{(\alpha)}{C^{i_1}_{hm}} + ... + T^{i_1...h}_{j_1...j_s} \underset{(\alpha)}{C^{i_r}_{hm}} -$$
$$-T^{i_1...i_r}_{h...j_s} \underset{(\alpha)}{C^h_{j_1m}} - ... - T^{i_1...i_r}_{j_1...h} \underset{(\alpha)}{C^h_{j_sm}}, (\alpha = 1, ..., k-1).$$

The operator $\overset{(\alpha)}{|}$ is called the v_α -covariant derivative with respect to $D\Gamma(N)$.

Finally, taking $X = X^{W_k} = X_i \frac{\delta}{\delta p_i}$, then $D^{W_k}_X T$ has the form:

(7.5.4) $$D^{W_k}_X T = X_m T^{i_1...i_r}_{j_1...j_s}|^m \frac{\delta}{\delta x^{i_1}} \otimes ... \otimes \frac{\delta}{\delta p_{j_s}} \otimes dx^{j_1} \otimes ... \otimes \delta p_{i_r},$$

where

(7.5.4') $$T^{i_1...i_r}_{j_1...j_s}|^m = \frac{\delta T^{i_1...i_r}_{j_1...j_s}}{\delta p_m} + T^{hi_2...i_r}_{j_1...j_s} C_h^{i_1m} + ... + T^{i_1...h}_{j_1...j_s} C_h^{i_rm} -$$
$$-T^{i_1...i_r}_{h...j_s} C^{hm}_{j_1} - ... - T^{i_1...i_r}_{j_1...h} C^{hm}_{j_s}.$$

It is not hard to prove

Proposition 7.5.1 *The following properties hold:*

$$T^{i_1...i_r}_{j_1...j_s|m},\ T^{i_1...i_r}_{j_1...j_s} \overset{(\alpha)}{|}_m,\ T^{i_1...i_r}_{j_1...j_s}|^m, (\alpha = 1, ..., k-1)$$

are d-tensor fields. The first two are of type $(r, s+1)$ *and the last one is of type* $(r+1, s)$.

As an application, the d-tensor field g_{ij} has the h-, v_α- and w_k- covariant derivatives with respect to the N-linear connection with the coefficients $D\Gamma(N)$, given by:

$$(7.5.5)\qquad \begin{cases} g_{ij|k} = \dfrac{\delta g_{ij}}{\delta x^m} - H^h_{im} g_{hj} - H^h_{jm} g_{ih}, \\ g_{ij} \overset{(\alpha)}{|}_m = \dfrac{\delta g_{ij}}{\delta y^{(\alpha)m}} - \underset{(\alpha)}{C^h_{im}} g_{hj} - \underset{(\alpha)}{C^h_{jm}} g_{ih}, \quad (\alpha = 1, .., k-1), \\ g_{ij}|^m = \dfrac{\delta g_{ij}}{\delta p_m} - C^{hm}_i g_{hj} - C^{hm}_j g_{ih}. \end{cases}$$

In a next chapter we shall determine the coefficients $D\Gamma\,(N)$ from the conditions that g_{ij} is covariant constant with respect D.

Proposition 7.5.2 *The operators* $_|$, $\overset{(\alpha)}{|}$ *and* $|$ *have the properties:*

$$1^\circ\ f_{|m} = \frac{\delta f}{\delta x^m}, f \overset{(\alpha)}{|}_m = \frac{\delta f}{\delta y^{(\alpha)m}}, (\alpha = 1, ..., k-1), f|^m = \frac{\delta f}{\delta p_m}$$

2° These operators are distributive with respect to the addition of the d-tensor of the same type.

3° They commute with the operation of contraction

4° They verify the Leibniz rule to the tensor product.

As an application we study the $(\overset{(\alpha)}{z})$ -deflection tensor of $D\Gamma\,(N)$. They are defined by:

$$(7.5.6)\qquad \overset{(\alpha)}{D^i_j} = \overset{(\alpha)}{z^i}_{|j},\ \overset{(\alpha\beta)}{D^i_j} = \overset{(\alpha)}{z^i} \overset{(\beta)}{|}_j, (\alpha, \beta = 1, ..., k-1),\ \overset{(\alpha)}{D^{ij}} = \overset{(\alpha)}{z^i}|^j,$$

where $\overset{(\alpha)}{z^i}$, $(\alpha = 1, ..., k-1)$ are the Liouville d-vector fields.

Evidently, they have the following expressions

$$(7.5.6')\qquad \begin{aligned} &\overset{(\alpha)}{D^i_j} = \frac{\delta \overset{(\alpha)}{z^i}}{\delta x^j} + \overset{(\alpha)}{z^m} H^i_{mj},\quad \overset{(\alpha\beta)}{D^i_j} = \frac{\delta \overset{(\alpha)}{z^i}}{\delta y^{(\beta)j}} + \overset{(\alpha)}{z^m} \underset{(\beta)}{C^i_{mj}}, \\ &\overset{(\alpha)}{D^{ij}} = \frac{\delta \overset{(\alpha)}{z^i}}{\delta p_j} + \overset{(\alpha)}{z^m} C^{ij}_m. \end{aligned}$$

Similarly, we introduce the (p)-deflection tensors by

$$\Delta_{ij} = p_{i|j},\quad \overset{(\alpha)}{\delta}_{ij} = p_i \overset{(\alpha)}{|}_j,\quad \delta^j_i = p_i|^j. \tag{7.5.7}$$

We deduce:

$$\Delta_{ij} = -p_h H^h_{ij},\quad \overset{(\alpha)}{\delta}_{ij} = -p_h \underset{(\alpha)}{C^h_{ij}},\quad \delta^j_i = \delta^j_i - p_h C^{hj}_i .$$

The deflection tensors will be used in some important identities determined by the Ricci identities, applied to the Liouville d-tensor fields $\overset{(\alpha)}{z^i}$, and to the d-covector p_i.

Some remarks:

1° In the adapted basis we can prove the equation

$$D_X \mathbb{F} = 0,\ \forall X \in X(T^{*k}M)$$

2° A Berwald connection is an N-linear connection D with the coefficients

$$B\Gamma(N) = (B^i_{jk}, 0, ..., 0, 0) \tag{7.5.8}$$

where B^i_{jh} has the same rule of transformation as H^i_{jh} from (7.4.2) and is determined by the nonlinear connection N.

We have:

Theorem 7.5.1 *Any nonlinear connection N, with the coefficients $(\underset{(1)}{N^i_j}, ..., \underset{(k-1)}{N^i_j}, N_{ij})$ determines the following Berwald connections:*

$$B\Gamma(N) = (\dot{\partial}^i N_{ij}, 0, ..., 0, 0) \tag{7.5.9}$$

$$\underset{(1)}{B\Gamma}(N) = \left(\frac{\delta \underset{(1)}{N^i_j}}{\delta y^{(1)h}}, 0, ..., 0, 0\right) \tag{7.5.9'}$$

Proof: The first one is given by the last formula (7.1.6) applying the derivation $\dot{\tilde{\partial}}^i = \dfrac{\partial \widetilde{x}^i}{\partial x^s}\dot{\partial}^s$ and (7.5.9') is obtained from the first formula (7.1.6), applying the derivation
$\dfrac{\delta}{\delta \widetilde{y}^{(1)h}} = \dfrac{\partial x^m}{\partial \widetilde{x}^h}\dfrac{\delta}{\delta y^{(1)m}}$. q.e.d.

If the base manifold M is paracompact, then the manifold $T^{*k}M$ is paracompact, too. Consequently, on $T^{*k}M$ there exist nonlinear connections. Therefore, we have:

Theorem 7.5.2 *If the base manifold M is paracompact, then on the manifold $T^{*k}M$ there exist N-linear connections D.*

Indeed, on $T^{*k}M$ there exist nonlinear connections N. Applying the Theorem 7.5.1 the conclusion follows. Q.E.D.

7.6 Ricci Identities. Local Expressions of d-Tensor of Curvature and Torsion. Bianchi Identities.

Let D be an N-linear connection with the local coefficients

$$D\Gamma(N) = (H^i_{jh}, \underset{(1)}{C}{}^i_{jh}, ..., \underset{(k-1)}{C}{}^i_{jh}, C^{jh}_i) \tag{7.6.1}$$

The Ricci identities for a d-vector field X^i can be deduced from the formulae (7.3.8) written in the adapted basis. But we can obtain them by a straighforward calculus.

Theorem 7.6.1 *For any N-linear connection D and any d-vector field X^i the following Ricci formulae hold:*

$$\begin{aligned} X^i_{|j|h} - X^i_{|h|j} &= X^m R_m{}^i{}_{jh} - X^i_{|m} T^m_{jh} - \{X^i \overset{(1)}{|}_m \underset{(01)}{R}{}^m_{jh} + ... + \\ &+ X^i \overset{(k-1)}{|}_m \underset{(0,k-1)}{R}{}^m_{jh} + X^i|^m \underset{(0)}{R}{}_{mjh}\}, \end{aligned} \tag{7.6.2$_1$}$$

(7.6.2$_2$)

$$\begin{aligned} X^i_{|j}\overset{(\alpha)}{|}_h - X^i\overset{(\alpha)}{|}_{h|j} &= X^m \underset{(\alpha)}{P}{}_m{}^i{}_{jh} - X^i_{|m} \underset{(\alpha)}{C}{}^m_{jh} - \{X^i \overset{(1)}{|}_m \underset{(01)}{R}{}^m_{jh} + X^i \overset{(\alpha)}{|}_m H^m_{jh}\} - \\ &-\{X^i \overset{(1)}{|}_m \underset{(\alpha,1)}{B}{}^m_{jh} + ... + X^i \overset{(k-1)}{|}_m \underset{(\alpha,k-1)}{B}{}^m_{jh} + X^i|^m \underset{(\alpha)}{B}{}_{mjh}\}, \end{aligned}$$

$$\begin{aligned} X^i_{|j}|^h - X^i|^h{}_{|j} &= X^m P_m{}^i{}_j{}^h - X^i_{|m} C^{mh}_j - X^i|^m H_{mj}{}^h - \\ &- \left\{X^i \overset{(1)}{|}_m \underset{(1)}{B}{}^{mh}_j + \cdots + X^i \overset{(k-1)}{|}_m \underset{(k-1)}{B}{}^{mh}_j + X^i|_m \underset{(0)}{B}{}^h_{mj}\right\}, \end{aligned} \tag{7.6.2$_3$}$$

(7.6.2$_4$)

$$\begin{aligned} X^i \overset{(\alpha)}{|}_j \overset{(\beta)}{|}_h - X^i \overset{(\beta)}{|}_h \overset{(\alpha)}{|}_j &= X^m \underset{(\alpha\beta)}{S}{}_m{}^i{}_{jh} - X^i \overset{(\alpha)}{|}_m \underset{(\beta)}{C}{}^m_{jh} + X^i \overset{(\beta)}{|}_m \underset{(\alpha)}{C}{}^m_{hj} - \\ &- \left\{X^i \overset{(1)}{|}_m \underset{(\alpha\beta)}{\overset{(1)}{C}}{}^m_{jh} + \cdots + X^i \overset{(k-1)}{|}_m \underset{(\alpha\beta)}{\overset{(k-1)}{C}}{}^m_{jh} + X^i|_m \underset{(\alpha\beta)}{B}{}_{mjh}\right\}, \end{aligned}$$

(7.6.2₅)

$$X^i \overset{(\alpha)}{|}_j |^h - X^i|^h \overset{(\alpha)}{|}_j = X^m \underset{(\alpha)}{S}{}_m{}^i{}^h{}_j - X^i|^m \underset{(\alpha)}{C}{}^h_{mj} + X^i \overset{(\alpha)}{|}_m C^{mh}_j -$$
$$- \left\{ X^i \overset{(1)}{|}_m \underset{(\alpha 1)}{C}{}^{mh}_j + \cdots + X^i \overset{(k-1)}{|}_m \underset{(\alpha,k-1)}{C}{}^{mh}_j + X^i|^m \underset{(\alpha)}{C}{}^h_{mj} \right\},$$

(7.6.2₆)
$$X^i|^j|^h - X^i|^h|^j = X^m S_m{}^{ijh} + X^i|^m S^{jh}_m,$$

where all terms in $\underset{(0\alpha)}{R}{}^i{}_{jh}$, $\underset{(0)}{R}{}_{mjh}$, $\underset{(\alpha\beta)}{B}{}^i{}_{jh}$, $\underset{(\alpha)}{B}{}_{ijh}$, $\underset{(0)}{B}{}_{m}{}^h_j$ *are known from the Lie brackets (6.5.1).*

The coefficients $D\Gamma(N)$ are given in (7.6.1). Supplementary we put:

(7.6.3)
$$T^i{}_{jh} = H^i_{jh} - H^i_{hj}, \quad S_i{}^{jh} = C_i^{jh} - C_i^{hj}.$$

Here $R^i{}_{mjh}$, ..., are called *d-tensor of curvature of* D and they have the expressions:

(7.6.4)

$$\begin{cases} R_m{}^i{}_{jh} = \dfrac{\delta H^i_{mj}}{\delta x^h} - \dfrac{\delta H^i_{mh}}{\delta x^j} + H^s_{mj}H^i_{sh} - H^s_{mh}H^i_{sj} + \underset{(1)}{C}{}^i{}_{ms}\underset{(01)}{R}{}^s{}_{jh} + \\ \quad + \cdots + \underset{(k-1)}{C}{}^i{}_{ms}\underset{(0,k-1)}{R}{}^s{}_{jh} + C^{is}_m \underset{(0)}{R}{}_{sjh}, \\ \underset{(\alpha)}{P}{}_m{}^i{}_{jh} = \dfrac{\delta H^i_{mj}}{\delta y^{(\alpha)h}} - \dfrac{\delta \underset{(\alpha)}{C}{}^i{}_{mh}}{\delta x^j} + H^s_{mj}\underset{(\alpha)}{C}{}^i{}_{sh} - \underset{(\alpha)}{C}{}^s{}_{mh}H^i_{sj} + \underset{(1)}{C}{}^i{}_{ms}\underset{(\alpha 1)}{B}{}^s{}_{jh} + \\ \quad + \cdots + \underset{(k-1)}{C}{}^i{}_{ms}\underset{(\alpha,k-1)}{B}{}^s{}_{jh} + C^{is}_m \underset{(\alpha)}{B}{}_{sjh}, \\ P_m{}^i{}_j{}^h = \dfrac{\delta H^i_{mj}}{\delta p_h} - \dfrac{\delta C^{ih}_m}{\delta x^j} + H^s_{mj}C^{ih}_s - C^{sh}_m H^i_{sj} + \underset{(1)}{C}{}^i{}_{ms}\underset{(1)}{B}{}^{sh}_j + \\ \quad + \cdots + \underset{(k-1)}{C}{}^i{}_{ms}\underset{(k-1)}{B}{}^{sh}_j + C^{is}_m \underset{(0)}{B}{}_s{}^h{}_j. \end{cases}$$

and
(7.6.5)

$$\begin{cases} \underset{(\alpha\beta)}{S}{}_m{}^i{}_{jh} = \dfrac{\delta \underset{(\alpha)}{C}{}^i_{mj}}{\delta y^{(\beta)h}} - \dfrac{\delta \underset{(\beta)}{C}{}^i_{mh}}{\delta y^{(\alpha)j}} + \underset{(\alpha)}{C}{}^s_{mj}\underset{(\beta)}{C}{}^i_{sh} - \underset{(\beta)}{C}{}^s_{mh}\underset{(\alpha)}{C}{}^i_{sj} + \underset{(1)}{C}{}^i_{ms}\underset{(\alpha\beta)}{\overset{(1)}{C}}{}^s_{jh} + \\ + \cdots + \underset{(k-1)}{C}{}^i_{ms}\underset{(\alpha\beta)}{\overset{(k-1)}{C}}{}^s_{jh} + C^{is}_m \underset{(\alpha\beta)}{B}{}_{sjh}, \ (\alpha \leq \beta, \alpha, \beta = 1, ..., k-1) \\ \underset{(\alpha)}{S}{}_m{}^i{}_j{}^h = \dfrac{\delta \underset{(\alpha)}{C}{}^i_{mj}}{\delta p_h} - \dfrac{\delta C^{ih}_m}{\delta y^{(\alpha)j}} + \underset{(\alpha)}{C}{}^s_{mj}C^{ih}_s - C^s_{mh}\underset{(\alpha)}{C}{}^i_{sj} + \underset{(1)}{C}{}^i_{ms}\underset{(\alpha 1)}{C}{}_j^{sh} + \\ + \cdots + \underset{(k-1)}{C}{}^i_{ms}\underset{(\alpha,k-1)}{C}{}^i_{sj} + C^{is}_m\underset{(\alpha)}{C}{}^h_{sj}, \\ S_m{}^{ijh} = \dfrac{\delta C^{ij}_m}{\delta p_h} - \dfrac{\delta C^{ih}_m}{\delta p_j} + C^{sj}_m C^{ih}_s - C^{sh}_m C^{ij}_s. \end{cases}$$

As usually, we extend the Ricci identities (7.6.2) for any d-tensor field $T^{i_1...i_r}_{j_1...j_s}$.

For instance, if $g^{ij}(x, y^{(1)}, ..., y^{(k-1)}, p)$ is a d-tensor field, the Ricci identities of g^{ij}, with respect to the N-linear connection $D\Gamma(N)$, are:

(7.6.6)

$$g^{ij}{}_{|h|m} - g^{ij}{}_{|m|h} = g^{sj}R_s{}^i{}_{hm} - g^{is}R_s{}^j{}_{hm} - g^{ij}{}_{|s}T^s{}_{hm} - \{g^{ij}\overset{(1)}{|}_s \underset{(01)}{R}{}^s{}_{hm} +$$

$$+ \cdots + g^{ij}\overset{(k-1)}{|}_s \underset{(0,k-1)}{R}{}^s{}_{hm} + g^{ij}|^s \underset{(0)}{R}{}_{shm}\},$$

....

$$g^{ij}|^h|^m - g^{ij}|^m|^h = g^{sj}S_s{}^{ihm} + g^{is}S_s{}^{jhm} + g^{ij}|^s S_s{}^{hm}.$$

In particular if $D\Gamma(N)$ satisfies the supplementary conditions:

(7.6.7) $$g^{ij}{}_{|h} = 0, \quad g^{ij}\overset{(\alpha)}{|}_h = 0, \quad g^{ij}|^h = 0, \quad (\alpha = 1, ..., k-1),$$

then the Ricci identities (7.6.6) give us:

(7.6.8)

$$\begin{aligned} &g^{sj}R_s{}^i{}_{hm} + g^{is}R_s{}^j{}_{hm} = 0, \\ &g^{sj}\underset{(\alpha)}{P}{}_s{}^i{}_{hm} + g^{is}\underset{(\alpha)}{P}{}_s{}^j{}_{hm} = 0, \\ &\text{.......................................}, \\ &g^{sj}S_s{}^{ihm} + g^{is}S_s{}^{jhm} = 0. \end{aligned}$$

Some important identities are obtained applying the Ricci identities to the d-covector p_i and to the Liouville vector fields $z^{(\alpha)i}$, $(\alpha = 1, ..., k-1)$.

Theorem 7.6.2 *Any N-linear connection $D\Gamma(N)$ satisfies the following identities:*

(7.6.9)

$$\begin{aligned}
\Delta_{ij|h}-\Delta_{ih|j} &= -p_sR_i{}^s{}_{jh}-\Delta_{is}T^s{}_{jh}-\\
&- \left\{\overset{(1)}{\not\delta}_{is}\underset{(01)}{R}{}^s{}_{jh}+\cdots+\overset{(k-1)}{\not\delta}_{is}\underset{(0,k-1)}{R}{}^s{}_{jh}+\not\delta^s_i\underset{(0)}{R}{}_{sjh}\right\},\\
\Delta_{ij}\overset{(\alpha)}{|}_h-\overset{(\alpha)}{\not\delta}_{ih|j} &= -p_s\underset{(\alpha)}{P}{}_i{}^s{}_{jh}-\Delta_{is}\underset{(\alpha)}{C}{}^s_{jh}-\overset{(\alpha)}{\not\delta}_{is}H^s_{jh}-\\
&- \left\{\overset{(1)}{\not\delta}_{is}\underset{(\alpha1)}{B}{}^s{}_{jh}+\cdots+\overset{(k-1)}{\not\delta}_{is}\underset{(\alpha,k-1)}{B}{}^s{}_{jh}+\not\delta^s_i\underset{(\alpha)}{B}{}_{sjh}\right\},\\
\Delta_{ij}|^h-\not\delta_i{}^h{}_{|j} &= -p_sP_i{}^s{}_j{}^h-\Delta_{is}C^{sh}_j-\not\delta^s_iH^h_{sj}-\\
&- \left\{\overset{(1)}{\not\delta}_{is}\underset{(1)}{B}{}^{sh}_j+\cdots+\overset{(k-1)}{\not\delta}_{is}\underset{(k-1)}{B}{}^{sh}_j+\not\delta^s_i\underset{(0)}{B}{}_{sj}{}^h\right\},\\
\overset{(\alpha)}{\not\delta}_{ij}\overset{(\beta)}{|}_h-\overset{(\beta)}{\not\delta}_{ih}\overset{(\alpha)}{|}_j &= -p_s\underset{(\alpha\beta)}{S}{}_i{}^s{}_{jh}-\overset{(\alpha)}{\not\delta}_{is}\underset{(\beta)}{C}{}^s{}_{jh}-\overset{(\beta)}{\not\delta}_{is}\underset{(\alpha)}{C}{}^s{}_{hj}-\\
&- \left\{\overset{(1)}{\not\delta}_{is}\underset{(\alpha\beta)}{\overset{(1)}{C}}{}^s{}_{jh}+\cdots+\overset{(k-1)}{\not\delta}_{is}\underset{(\alpha\beta)}{\overset{(k-1)}{C}}{}^s{}_{jh}+\not\delta^s_i\underset{(\alpha\beta)}{B}{}_{sjh}\right\},\\
\overset{(\alpha)}{\not\delta}_{ij}|^h-\not\delta^h_i\overset{(\beta)}{|}_j &= -p_s\underset{(\alpha)}{S}{}_i{}^s{}_j{}^h-\not\delta^s_i\underset{(\alpha)}{C}{}_s{}^h{}_j-\overset{(\alpha)}{\not\delta}_{is}C^{sh}_j-\\
&- \left\{\overset{(1)}{\not\delta}_{is}\underset{(\alpha1)}{C}{}^{sh}_j+\cdots+\overset{(k-1)}{\not\delta}_{is}\underset{(\alpha,k-1)}{C}{}^{sh}_j+\not\delta^s_i\underset{(\alpha)}{C}{}_{sj}{}^h\right\},\\
\not\delta_i{}^j|^h-\not\delta_i{}^h|^j &= -p_sS_i{}^{sjh}+\not\delta^s_iS^{jh}_s.
\end{aligned}$$

The similar identities are obtained applying the Ricci identities to the Liouville vector fields $z^{(\alpha)i}$.

Theorem 7.6.3 *Any N-linear connection $D\Gamma(N)$ satisfies the following identities, obtained from (7.6.2) for $X^i=z^{(\alpha)i}$:*

(7.6.10)

$$\begin{aligned}
\overset{(\alpha)}{D}{}^i{}_{j|h}-\overset{(\alpha)}{D}{}^i{}_{h|j} &= z^{(\alpha)s}R_s{}^i{}_{jh}-\overset{(\alpha)}{D}{}^i_sT^s{}_{jh}-\\
&- \{\overset{(\alpha1)}{D}{}^i{}_s\underset{(0)}{R}{}^s{}_{jh}+\cdots+\overset{(\alpha,k-1)}{D}{}^i{}_s\underset{(0,k-1)}{R}{}^s{}_{jh}+\overset{(\alpha)}{D}{}^{is}\underset{(0)}{R}{}_{sjh}\},\\
\cdots\cdots\cdots\cdots &\;\cdots\;\cdots\cdots\cdots\cdots\cdots\cdots\cdots\cdots\cdots\cdots\\
\overset{(\alpha)}{D}{}^{ij}|^h-\overset{(\alpha)}{D}{}^{ih}|^j &= z^{(\alpha)s}S_s{}^{ijh}-\overset{(\alpha)}{\not\delta}{}^{is}S_s{}^{jh},\ (\alpha=1,\dots,k-1).
\end{aligned}$$

Evidently, the whole theory from the present section is simplified if $D\Gamma(N)=B\Gamma(N)=(B^i_{jh},0,\dots,0,0)$ is a Berwald connection, taking $X^i{}_{\|j}=\dfrac{\delta X^i}{\delta x^j}+X^hB^i_{hj}$, $X^i\overset{(\alpha)}{\|}_j=\dfrac{\delta X^i}{\delta y^{(\alpha)j}}$, $X^i\|^j=\dfrac{\delta X^i}{\delta p_j}$, '$_\|$', '$\overset{(\alpha)}{\|}$' and '$\|$' being the $h-$, $v_\alpha-$ and w_k-covariant derivatives with respect to $B\Gamma(N)$.

The Bianchi identities for the N-linear connection $D\Gamma(N)$ can be obtained using the methods described above. Namely, applying the Ricci identities to d-tensor field $X^i{}_{|j}$ we obtain:

$$\left(X^i{}_{|j}\right)_{|h|m}-\left(X^i{}_{|j}\right)_{|m|h}=X^r{}_{|j}R_r^i{}_{hm}-X^i{}_{|r}R_j^i{}_{hm}-X^i{}_{|r}T^r_{hm}-$$
$$\left(X^i{}_{|j}\overset{(1)}{|}_r\underset{(01)}{R}{}^r{}_{hm}+\cdots+X^i{}_{|j}\overset{(k-1)}{|}_r\underset{(0,k-1)}{R}{}^r{}_{hm}+X^i{}_{|j}|^r\underset{(01)}{R}{}_{rhm}\right).$$

If we cyclically permute the indices j, h, m and add the identitiessuch obtained we determine a first set of Bianchi identities:

(7.6.11)

$$\underset{(jhm)}{S}\left\{R_j{}^i{}_{hm}-T^i{}_{jh|m}-T^i{}_{jr}T^r{}_{hm}+\sum_{\alpha=1}^{k-1}\underset{(\alpha)}{C}{}^i_{jr}\underset{(0\alpha)}{R}{}^r{}_{hm}+C_j^{ir}\underset{(0)}{R}{}_{rhm}\right\}=0,$$

$$\underset{(jhm)}{S}\left\{R_s^i{}_{jh|m}+R_s{}^i{}_{jr}T^r{}_{hm}-\sum_{\alpha=1}^{k-1}\underset{(\alpha)}{P}{}_s{}^i{}_{jr}\underset{(0\alpha)}{R}{}^r{}_{hm}-P_s{}^i{}_j{}^r\underset{(0)}{R}{}_{rhm}\right\}=0,$$

$$\underset{(jhm)}{S}\left\{\underset{(0\alpha)}{R}{}^i{}_{jh|m}-\underset{(0\alpha)}{R}{}^i{}_{jr}T^r{}_{hm}+\underset{(\alpha)}{P}{}^{is}_{jr}\underset{(0\alpha)}{R}{}^r{}_{hm}+\underset{(\alpha)}{B}{}^{ir}_j\underset{(0)}{R}{}_{rhm}\right\}=0,$$

$$\underset{(jhm)}{S}\left\{\underset{(0)}{R}{}_{sjh|m}-\underset{(0)}{R}{}_{sjr}T^r{}_{hm}+\sum_{\alpha=1}^{k-1}\underset{(\alpha)}{B}{}_{sjr}\underset{(0\alpha)}{R}{}^r{}_{hm}+\underset{(0)}{P}{}^r{}_{sj}\underset{(0)}{R}{}_{rhm}\right\}=0.$$

(7.6.12) $$\underset{(\alpha)}{P}{}^i{}_{jr}=H^i_{jr}+\underset{(\alpha)}{B}{}^i{}_{jr},\ (\alpha=1,...,k-1).$$

In a similar way we have the second set of Bianchi identities:

(7.6.13)
$$\underset{(jhm)}{S}\left\{S_s{}^{ijh}|^m-S_s{}^{ijr}S_r{}^{hm}\right\}=0,$$
$$\underset{(ihm)}{S}\left\{S_s{}^{ihm}-S_s{}^{ih}|^m-S_s{}^{ir}S_r{}^{hm}\right\}=0,$$

where $\underset{(jhm)}{S}$ is the cyclic sum.

Similarly, we can get the other Bianchi identities.

7.7 Parallelism of the Vector Fields on the Manifold $T^{*k}M$

Consider a N-linear connection D on the manifold $T^{*k}M$ with the coefficients $D\Gamma(N)=(H^i{}_{jh},\ \underset{(\alpha)}{C}{}^i{}_{jh},\ C_i{}^{jh})$, $(\alpha=1,...,k-1)$ in the adapted basis $\left(\frac{\delta}{\delta x^i},\frac{\delta}{\delta y^{(\alpha)i}},\frac{\delta}{\delta p_i}\right)$.

A smooth curve $\gamma: I\to T^{*k}M$ having the image in a domain of a local chart is given by

(7.7.1) $$x^i=x^i(t),\ y^{(\alpha)i}=y^{(\alpha)i}(t),\ p_i=p_i(t),(\alpha=1,...,k-1),\ t\in I.$$

The tangent vector field $\dot{\gamma}=\dfrac{d\gamma}{dt}$ can be written by means of (7.2.10), in the form:

$$\dot{\gamma}=\frac{dx^i}{dt}\frac{\delta}{\delta x^i}+\frac{\delta y^{(1)i}}{dt}\frac{\delta}{\delta y^{(1)i}}+\cdots+\frac{\delta y^{(k-1)i}}{dt}\frac{\delta}{\delta y^{(k-1)i}}+\frac{\delta p_i}{dt}\frac{\partial}{\partial p_i} \tag{7.7.2}$$

where, from (6.3.2),

(7.7.3)

$$\begin{cases}\dfrac{\delta y^{(1)i}}{dt}=\dfrac{dy^{(1)i}}{dt}+\underset{(1)}{M}{}^i_j\dfrac{dx^j}{dt},\\ \dotfill\\ \dfrac{\delta y^{(k-1)i}}{dt}=\dfrac{dy^{(k-1)i}}{dt}+\underset{(1)}{M}{}^i_j\dfrac{dy^{(k-2)j}}{dt}+\cdots+\underset{(k-2)}{M}{}^i_j\dfrac{dy^{(1)j}}{dt}+\underset{(k-1)}{M}{}^i_j\dfrac{dx^j}{dt},\\ \dfrac{\delta p_i}{dt}=\dfrac{dp_i}{dt}-N_{ji}\dfrac{dx^j}{dt}.\end{cases}$$

Let us denote

$$D_{\dot{\gamma}}X=\frac{DX}{dt},\quad DX=\frac{DX}{dt}dt,\ \forall X\in\mathcal{X}(T^{*k}M). \tag{7.7.4}$$

DX is called *the covariant differential of the vector field* X and $\dfrac{DX}{dt}$ is *the covariant differential of* X *along curve* γ.

If X is written in the form

$$\begin{aligned}X &= X^H+X^{V_1}+\cdots+X^{V_{k-1}}+X^{W_k}=\\ &= \overset{(0)i}{X}\frac{\delta}{\delta x^i}+\overset{(1)i}{X}\frac{\delta}{\delta y^{(1)i}}+\cdots+\overset{(k-1)i}{X}\frac{\delta}{\delta y^{(k-1)i}}+X_i\frac{\delta}{\delta p_i}\end{aligned} \tag{7.7.4'}$$

and $\dot{\gamma}$ from (7.7.2) is $\dot{\gamma}=\dot{\gamma}^H+\dot{\gamma}^{V_1}+\cdots+\dot{\gamma}^{V_{k-1}}+\dot{\gamma}^{W_k}$ we have

$$D_{\dot{\gamma}}=D^H_{\dot{\gamma}}+D^{V_1}_{\dot{\gamma}}+\cdots+D^{V_{k-1}}_{\dot{\gamma}}+D^{W_k}_{\dot{\gamma}}.$$

Then DX has the final form:

$$\begin{aligned}DX &= \left(d\overset{(0)i}{X}+\overset{(0)s}{X}\omega^i_s\right)\frac{\delta}{\delta x^i}+\left(d\overset{(1)i}{X}+\overset{(1)s}{X}\omega^i_s\right)\frac{\delta}{\delta y^{(1)i}}+\cdots+\\ &+\left(d\overset{(k-1)i}{X}+\overset{(k-1)s}{X}\omega^i_s\right)\frac{\delta}{\delta y^{(k-1)i}}+(dX_i-X_s\omega^s_i)\frac{\delta}{\delta p_i},\end{aligned} \tag{7.7.5}$$

where

$$\omega^i_j=H^i_{js}dx^s+\underset{(1)}{C}{}^i_{js}\delta y^{(1)s}+\cdots+\underset{(k-1)}{C}{}^i_{js}\delta y^{(k-1)s}+C^{is}_j\delta p_s. \tag{7.7.6}$$

The differential forms ω^i_j are called *the* 1 *-forms connection of the connection* D.

Putting

$$(7.7.7)\qquad \frac{\omega^i_j}{dt} = H^i_{js}\frac{dx^s}{dt} + \underset{(1)}{C}{}^i_{js}\frac{\delta y^{(1)s}}{dt} + \cdots + \underset{(k-1)}{C}{}^i_{js}\frac{\delta y^{(k-1)s}}{dt} + C^{is}_j\frac{\delta p_s}{dt}.$$

the covariant differential $\frac{DX}{dt}$ along curve γ is

(7.7.8)

$$\begin{aligned}\frac{DX}{dt} &= \left(\frac{d\overset{(0)i}{X}}{dt} + \overset{(0)s}{X}\frac{\omega^i_s}{dt}\right)\frac{\delta}{\delta x^i} + \left(\frac{d\overset{(1)i}{X}}{dt} + \overset{(1)s}{X}\frac{\omega^i_s}{dt}\right)\frac{\delta}{\delta y^{(1)i}} + \cdots + \\ &+ \left(\frac{d\overset{(k-1)i}{X}}{dt} + \overset{(k-1)s}{X}\frac{\omega^i_s}{dt}\right)\frac{\delta}{\delta y^{(k-1)i}} + \left(\frac{dX_i}{dt} - X_s\frac{\omega^s_i}{dt}\right)\frac{\delta}{\delta p_i}.\end{aligned}$$

The theory of the parallelism of vector fields along curve γ, presented in the chapter 10, section 7 of the book [115], in the case $k = 2$, can be extended without difficulties for $k > 2$.

Consequently, we define the notion of parallelism of a vector field $X(\gamma(t))$ along curve γ, by the differential equation $\frac{DX}{dt} = 0$. We obtain:

Theorem 7.7.1 *The vector field X, given by (7.7.4') is parallel along the parametrized curve γ, with respect to the N-linear connection D, if and only if its components $\overset{(0)i}{X}$, $\overset{(\alpha)i}{X}$, X_i ($\alpha = 1, ..., k-1$) are solutions of the differential equations*

$$(7.7.9)\qquad \begin{aligned}&\frac{d\overset{(0)i}{X}}{dt} + \overset{(0)s}{X}\frac{\omega^i_s}{dt} = 0,\\ &\frac{d\overset{(\alpha)i}{X}}{dt} + \overset{(\alpha)s}{X}\frac{\omega^i_s}{dt} = 0,\ (\alpha = 1, ..., k-1),\\ &\frac{dX_i}{dt} - X_s\frac{\omega^s_i}{dt} = 0.\end{aligned}$$

By means of the formula (7.7.8), the proof of the previous theorem is immediate.

The vector field $X \in \mathcal{X}(T^{*k}M)$ is called *absolute parallel* with respect to the N-linear connection D if the equation $DX = 0$ holds for any curve γ. This equations $DX = 0$, $\forall\gamma$ is equivalent to the integrability of the following system of Pffaf equations

(7.7.10)

$$d\overset{(0)i}{X} + \overset{(0)s}{X}\omega^i_s = 0,\ d\overset{(\alpha)i}{X} + \overset{(\alpha)s}{X}\omega^i_s = 0,\ (\alpha = 1, ..., k-1),\ dX_i - X_s\omega^s_i = 0.$$

But the previous system is equivalent to
(7.7.10')

$$\overset{(0)i}{X}{}_{|j}=\overset{(\alpha)i}{X}{}_{|j}=0,\ (\alpha=1,...,k-1),\ X_{i|j}=0,$$

$$\overset{(0)i(\beta)}{X}\ \overset{}{|}_{j}=\overset{(\alpha)i(\beta)}{X}\ |_{j}=0,\ (\alpha=1,...,k-1),\ X_i\overset{(\beta)}{|}_{j}=0,\ (\beta=1,...,k-1),$$

$$\overset{(0)i}{X}|^{j}=\overset{(\alpha)i}{X}|^{j}=0,\ (\alpha=1,...,k-1),\ X_i|^{j}=0,$$

which must be integrable.

Using the Ricci identities (7.6.2) the system (7.7.10') is integrable if and only if the coordinates $(\overset{(0)i}{X},\overset{(\alpha)i}{X},X_i)$ of the vector field X satisfy the following equations:
(7.7.11)

$$\overset{(\alpha)s}{X}R_s{}^i{}_{jh}=0,\ \overset{(\alpha)s}{X}\underset{(\beta)^s}{P}{}^i{}_{jh}=0,\ \overset{(\alpha)s}{X}P_s{}^i{}_j{}^h=0,\ \overset{(\alpha)s}{X}\underset{(\beta\gamma)^s}{S}{}^i{}_{jh}=0,$$

$$\overset{(\alpha)s}{X}\underset{(\beta)^s}{S}{}^i{}_j{}^h=0,\ \overset{(\alpha)s}{X}S_s{}^{ijh}=0,\ (\alpha=0,1,...,k-1;\beta,\gamma=1,...,k-1)$$

and

(7.7.11')

$$X^sR_s{}^i{}_{jh}=0,\ X_s\underset{(\beta)^i}{P}{}^s{}_{jh}=0,\ X_sP_i{}^s{}_j{}^h=0,\ X_s\underset{(\beta\gamma)^i}{S}{}^s{}_{jh}=0,$$

$$X_sS_i{}^s{}_j{}^h=0,\ X_sS_i{}^{sjh}=0,(\beta,\gamma=1,...,k-1).$$

The manifold $T^{*k}M$ is called *with absolute parallelism of vectors, with respect to D* if any vector field on $T^{*k}M$ is absolute parallel.

In this case the systems of equations (7.7.11), (7.7.11') are verified for any vector field X with the coefficients $(\overset{(0)i}{X},\overset{(\alpha)i}{X},X_i)$ in the adapted cobasis.

We obtain:

Theorem 7.7.2 *The manifold $T^{*k}M$ is with absolute parallelism of vectors, with respect to the N-linear connection D if and only if all curvature d-tensors of D vanish, i.e.*
$R_m{}^i{}_{jh}=0,\ \underset{(\alpha)}{P}{}_m{}^i{}_{jh}=0,\ P_m{}^i{}_j{}^h=0,\ \underset{(\alpha\beta)}{S}{}_m{}^i{}_{jh}=0,\ \underset{(\alpha)}{S}{}_m{}^i{}_j{}^h=0,S_m{}^{ijh}=0,\ \alpha,\ \beta=1,\ ...,\ k-1.$

The previous theory can be applied to investigate the autoparallel curves with respect to a N-linear connection D.

The parametrized curve $\gamma:t\in I\to\gamma(t)\in T^{*k}M$, is an autoparellel curve with respect to D if $D_{\dot\gamma}\dot\gamma=0$.

By means of (7.7.2), (7.7.8) we obtain

$$D_{\dot{\gamma}}\dot{\gamma} = \frac{D\dot{\gamma}}{dt} =$$

$$(7.7.12)\quad = \left(\frac{d^2x^i}{dt^2} + \frac{dx^s}{dt}\frac{\omega^i_s}{dt}\right)\frac{\delta}{\delta x^i} + \left(\frac{d}{dt}\frac{\delta y^{(1)i}}{dt} + \frac{\delta y^{(1)s}}{dt}\frac{\omega^i_s}{dt}\right)\frac{\delta}{\delta y^{(1)i}} + \cdots +$$

$$+ \left(\frac{d}{dt}\frac{\delta y^{(k-1)i}}{dt} + \frac{\delta y^{(k-1)s}}{dt}\frac{\omega^i_s}{dt}\right)\frac{\delta}{\delta y^{(k-1)i}} + \left(\frac{d}{dt}\frac{\delta p_i}{dt} - \frac{\delta p_s}{dt}\frac{\omega^s_i}{dt}\right)\frac{\delta}{\delta p_i}.$$

Theorem 7.7.3 *A smooth parametrized curve γ, (7.7.1) is an autoparallel curve with respect to the N-linear connection D if and only if the functions $x^i(t)$, $y^{(\alpha)i}(t)$, $p_i(t)$, $(\alpha = 1, ..., k-1)$, $t \in I$ verify the following system of differential equations:*

$$(7.7.13)\quad \begin{cases} \dfrac{d^2x^i}{dt^2} + \dfrac{dx^s}{dt}\dfrac{\omega^i_s}{dt} = 0, \\ \dfrac{d}{dt}\dfrac{\delta y^{(\alpha)i}}{dt} + \dfrac{\delta y^{(\alpha)s}}{dt}\dfrac{\omega^i_s}{dt} = 0, \ (\alpha = 1, ..., k-1), \\ \dfrac{d}{dt}\dfrac{\delta p_i}{dt} - \dfrac{\delta p_s}{dt}\dfrac{\omega^s_i}{dt} = 0. \end{cases}$$

Of course, the theorem of existence and uniqueness for the autoparallel curves can be formulated taking into account the system of differential equations (7.7.13).

We recall that γ is an horizontal curve if $\dot{\gamma} = \dot{\gamma}^H$. The horizontal curves are characterized by

$$(7.7.14)\quad x^i = x^i(t),\ \frac{\delta y^{(\alpha)i}}{dt} = 0,\ (\alpha = 1, ..., k-1),\ \frac{\delta p_i}{dt} = 0.$$

Definition 7.7.1 *A horizontal path of an N-linear connection D is an horizontal autoparallel curve γ, with respect to D.*

So, a horizontal path γ is characterized by $D_{\dot{\gamma}^H}\dot{\gamma}^H = 0$. Taking into account (7.7.13) we get:

Theorem 7.7.4 *The horizontal paths of an N-linear connection D are characterized by the system of differential equations*

$$(7.7.15)\quad \frac{d^2x^i}{dt^2} + H^i_{jh}\frac{dx^j}{dt}\frac{dx^h}{dt} = 0,\ \frac{\delta y^{(\alpha)i}}{dt} = 0,\ (\alpha = 1, ..., k-1),\ \frac{\delta p_i}{dt} = 0.$$

Indeed, (7.7.14) implies $\dfrac{\omega^i_j}{dt} = H^i_{jh}\dfrac{dx^h}{dt}$. But (7.7.13) gives us the mentioned equations (7.7.15).

A parametrized curve $\gamma : I \to T^{*k}M$ is called *v_α-vertical at the point $x_0 \in M$* if $\dot{\gamma} = \dot{\gamma}^{V_\alpha}$. It is analytically given by

$$x^i = x^i_0,\ y^{(\alpha)i} = y^{(\alpha)i}(t),\ y^{(\beta)i}(t) = 0,\ \beta \neq \alpha,\ p_i = 0,\ t \in I.$$

A v_α-vertical path γ, with respect to D is defined by $D_{\dot{\gamma}^{V_\alpha}} \dot{\gamma}^{V_\alpha} = 0$.

In this case, the equations (7.7.13) are as follows

$$\text{(7.7.16)} \quad \begin{aligned} &\frac{dx^i}{dt} = 0,\ \frac{dy^{(\beta)i}}{dt} = 0,\ (\beta \neq \alpha),\ \frac{dp_i}{dt} = 0 \text{ and} \\ &\frac{d}{dt}\frac{\delta y^{(\alpha)i}}{dt} + \underset{(\alpha)}{C}{}^{i}_{sj}\frac{dy^{(\alpha)s}}{dt}\frac{dy^{(\alpha)j}}{dt} = 0 \end{aligned}$$

Similarly, a w_k-vertical curve γ at the point $x_0 \in M$ is defined by the condition $\dot{\gamma} = \dot{\gamma}^{W_k}$. Analytically it is expressed by

$$x^i = x^i_0,\ y^{(\alpha)i} = y^{(\alpha)i}(t) = 0,\ (\alpha = 1, ..., k-1),\ p_i = p_i(t),\ t \in I.$$

A w_k-path γ, with respect to D has the property $D_{\dot{\gamma}^{W_k}} \dot{\gamma}^{W_k} = 0$.

The w_k-paths, with respect to the N-linear connection D are characterized by

$$\text{(7.7.17)} \quad \begin{aligned} &\frac{dx^i}{dt} = \frac{dy^{(1)i}}{dt} = \cdots = \frac{dy^{(k-1)i}}{dt} = 0, \\ &\frac{d^2p_i}{dt^2} - C_i^{jm}(x_0, 0, ..., 0, p)\frac{dp_j}{dt}\frac{dp_m}{dt} = 0. \end{aligned}$$

In the case when D is a Berwald N-linear connection the previous theory is a simple one.

7.8 Structure Equations of a N-Linear Connection

Let us consider a N-linear connection D with the coefficients $D\Gamma(N)$ in the adapted basis $\left(\frac{\delta}{\delta x^i}, \frac{\delta}{\delta y^{(1)i}}, ..., \frac{\delta}{\delta y^{(k-1)i}}, \frac{\delta}{\delta p_i}\right)$.

It is not difficult to prove:

Lemma 7.8.1 *1^0 Each of the following object fields*

$$d(dx^i) - dx^m \wedge \omega^i_m;\ d(\delta y^{(\alpha)i}) - \delta y^{(\alpha)m} \wedge \omega^i_m,\ (\alpha = 1, ..., k-1);\ d(\delta p_i) + \delta p_m \wedge \omega^m_i$$

is a d-vector field, except the last one which is a d-covector field.

2^0 The geometrical object field

$$d\omega^i_j - \omega^m_j \wedge \omega^i_m$$

is a d-tensor field of type $(1, 1)$.

Using this lemma we obtain a fundamental result in the geometry of the manifold $T^{*k}M$ and implicitly in the geometry of higher order Hamilton spaces.

Theorem 7.8.1 *For any N-linear connection D with the coefficients $D\Gamma(N) = \left(H^i_{jh}, \underset{(1)}{C}{}^i_{jh}, ..., \underset{(k-1)}{C}{}^i_{jh}, C^{jh}_i\right)$ the following structure equations hold:*

$$(7.8.1)\qquad \begin{aligned} d(\delta x^i) - dx^m \wedge \omega^i_m &= -\overset{(0)}{\Omega}{}^i, \\ d(\delta y^{(\alpha)i}) - dy^{(\alpha)m} \wedge \omega^i_m &= -\overset{(\alpha)}{\Omega}{}^i, (\alpha = 1, ..., k-1), \\ d(\delta p_i) + \delta p_m \wedge \omega^m_i &= -\Omega_i \end{aligned}$$

and

$$(7.8.2)\qquad d\omega^i_j - \omega^m_j \wedge \omega^i_m = -\Omega^i_j,$$

where $\overset{(0)}{\Omega}{}^i$, $\overset{(1)}{\Omega}{}^i$, $\overset{(k-1)}{\Omega}{}^i$ and Ω_i are the 2-forms of torsion
(7.8.3)

$$\begin{aligned} \overset{(0)}{\Omega}{}^i &= dx^j \wedge \left\{\frac{1}{2}T^i_{jm}dx^m + \sum_{\alpha=1}^{k-1} \underset{(\alpha)}{C}{}^i_{jm}\delta y^{(\alpha)m} + C^{im}_j \delta p_m\right\}, \\ \overset{(\alpha)}{\Omega}{}^i &= dx^j \wedge \underset{(\alpha 0)}{P}{}^i{}_j + \sum_{\gamma=1}^{k-1} \delta y^{(\gamma)j} \wedge \underset{(\alpha\gamma)}{P}{}^i{}_j + \\ &+ \delta y^{(\alpha)j} \wedge \left\{H^i_{jm}dx^m + \sum_{\gamma=1}^{k-1} \underset{(\gamma)}{C}{}^i_{jm}\delta y^{(\gamma)m} + C^{im}_j \delta p_m\right\}, \ (\alpha = 1, ..., k-1) \\ \Omega_i &= dx^j \wedge \left\{\frac{1}{2}\underset{(0)}{R}{}_{ijm}dx^m + \sum_{\gamma=1}^{k-1} \underset{(\gamma)}{B}{}_{ijm}\delta y^{(\gamma)m} + \underset{(0)}{B}{}_{ij}{}^m \delta p_m\right\} - \\ &- \delta p_j \wedge \left\{H^j_{im}dx^m + \sum_{\gamma=1}^{k-1} \underset{(\gamma)}{C}{}^j_{im}\delta y^{(\gamma)m} + C^{jm}_i \delta p_m\right\}, \end{aligned}$$

$\underset{(\alpha 0)}{P}{}^i{}_j$, ..., $\underset{(\alpha,k-1)}{P}{}^i{}_j$ being given by (5.4.8) and where Ω^i_j are the 2-forms of curvature:

$$(7.8.4)\qquad \begin{aligned} \Omega^i_j &= \frac{1}{2}R_j{}^i{}_{hm}dx^h \wedge dx^m + \sum_{\gamma=1}^{k-1} \underset{(\gamma)}{P}{}_j{}^i{}_{hm}dx^h \wedge \delta y^{(\gamma)m} + \\ &+ P_j{}^i{}_h{}^m dx^h \wedge \delta p_m + \sum_{\alpha\leq\beta}^{k-1}\sum_{\beta=1}^{k-1} \underset{(\alpha\beta)}{S}{}_j{}^i{}_{hm}\delta y^{(\alpha)h} \wedge \delta y^{(\beta)m} + \\ &+ \sum_{\gamma=1}^{k-1} \underset{(\gamma)}{S}{}_j{}^i{}_h{}^m \delta y^{(\gamma)h} \wedge \delta p_m + \frac{1}{2}S_i{}^{jhm}\delta p_h \wedge \delta p_m. \end{aligned}$$

Indeed, by means of the exterior differential $d\delta y^{(\alpha)i}$ from (7.4.7), (7.4.8) and ω^i_j from (7.7.6) we get the formulas (7.8.3) and (7.8.4).

These formulas have a very simple form in the case of Berwald connection, where $\underset{(\alpha)}{C}{}^{i}{}_{jh} = 0$, $C_j^{ih} = 0$.

The structure equations will be used in a theory of submanifold of the Hamilton spaces, studied in chapter 9.

Chapter 8

Hamilton Spaces of Order $k \geq 1$

The Hamilton spaces of order 1 and 2 were investigated in the chapter 5 and 12 of the book [115]. In the present chapter we study the natural extension of this notion to order $k \geq 1$.

A Hamilton space of order k is a pair $H^{(k)n} = (M, H(x, y^{(1)}, ..., y^{(k-1)}, p))$ in which M is a real n -dimensional manifold and $H : T^{*k}M \to \mathbf{R}$ is a regular Hamiltonian function on the manifold $T^{*k}M = T^{(k-1)}M \times_M T^*M$.

The geometry of the spaces $H^{(k)n}$ can be developed step by step following the same ideas as in the cases $k = 1$ or $k = 2$ and using the geometrical theory of the manifold $T^{*k}M$ described in the last three chapters. Of course, $T^{*k}M$ being the dual of T^kM, the geometry of the Hamilton spaces of order k, $H^{(k)n} = (M, H)$, appears as dual of the geometry of Lagrange spaces of order k, $L^{(k)n} = (M, L)$, via a Legendre mapping.

Therefore, in this chapter we study the notion of Hamilton space $H^{(k)n} = (M, H)$, the canonical presymplectic structure and canonical Poisson structure, Legendre mappings, the nonlinear connection and canonical metrical connection. We end with the Riemannian almost contact model of this space.

8.1 The Spaces $H^{(k)n}$

Let us consider the dual bundle $(T^{*k}M, \pi^{*k}, M)$. The local coordinates of a point $u = (x, y^{(1)}, ..., y^{(k-1)}, p)$, $u \in T^{*k}M$, will

be denoted as usually by $(x^i, y^{(1)i}, ..., y^{(k-1)i}, p_i)$; (x^i) being the coordinates of the particle x, $y^{(1)i}$, ..., $y^{(k-1)i}$ are seen as the coordinates of the accelerations of order 1, ..., $k-1$, respectively and p_i are the momenta. The coordinate transformations on $T^{*k}M$ are given by (4.1.2), (4.1.3).

On the manifold $T^{*k}M$ there are the vertical distributions $V_{k-1} \subset V_{k-2} \subset \cdots \subset V_1 \subset V$ and a vertical distribution W_k such that $V_u = V_{1,u} \oplus W_{k,u}$, $\forall u \in T^{*k}M$.

Also, on the manifold $T^{*k}M$ there exist the Liouville vector fields $\overset{(1)}{\Gamma}, ...,$ $\overset{(k-1)}{\Gamma}$ and the Hamilton vector field C^*, linearly independent, expressed by

$$(8.1.1)\quad \begin{cases} \overset{(1)}{\Gamma} = y^{(1)i}\dfrac{\partial}{\partial y^{(k-1)i}}, \\ \overset{(2)}{\Gamma} = y^{(1)i}\dfrac{\partial}{\partial y^{(k-2)i}} + 2y^{(2)i}\dfrac{\partial}{\partial y^{(k-1)i}}, \\ \cdots\cdots\cdots\cdots\cdots\cdots\cdots\cdots\cdots\cdots \\ \overset{(k-1)}{\Gamma} = y^{(1)i}\dfrac{\partial}{\partial y^{(1)i}} + 2y^{(2)i}\dfrac{\partial}{\partial y^{(2)i}} + \cdots + (k-1)y^{(k-1)i}\dfrac{\partial}{\partial y^{(k-1)i}}. \end{cases}$$

and

$$(8.1.1') \qquad C^* = p_i\frac{\partial}{\partial p_i}.$$

Theorem 4.2.1 stipulates that these vector fields are globally defined on the total space of the dual bundle.

The function

$$(8.1.2) \qquad \varphi = p_i y^{(1)i}$$

is a scalar function on $T^{*k}M$.

A Hamiltonian is a scalar function $H : (x, y^{(1)}, ..., y^{(k-1)}, p) \in T^{*k}M \to H(x, y^{(1)}, ..., y^{(k-1)}, p) \in \mathbf{R}$. 'Scalar' means that H does not depend on the changing of coordinates on $T^{*k}M$.

As we know, the Hamiltonian H is differentiable if it is differentiable on the manifold $\widetilde{T^{*k}M} = T^{*k}M \setminus \{0\}$ (where $\mathbf{0}$ is the null section of the projection π^{*k}) and H is continuous on the null section. Evidently,

$$\widetilde{T^{*k}M} = \left\{(x, y^{(1)}, ..., y^{(k-1)}, p) \in T^{*k}M | y^{(1)}, ..., y^{(k-1)}, p \text{ are not all null}\right\}.$$

The null section $\mathbf{0} : M \to T^{*k}M$, having the property $\pi^{*k} \circ \mathbf{0} = 1_M$ can be identified with the manifold M.

Definition 8.1.1 *A regular Hamiltonian $H : T^{*k}M \to \mathbf{R}$ is a differentiable Hamiltonian whose Hessian with respect to the momenta p_i, with the entries:*

$$(8.1.3) \qquad g^{ij}(x, y^{(1)}, ..., y^{(k-1)}, p) = \frac{1}{2}\frac{\partial^2 H}{\partial p_i \partial p_j}$$

is nondegenerate on the manifold $\widetilde{T^{*k}M}$.

Of course, g^{ij} from (8.1.3) is a d-tensor field, contravariant of order 2, symmetric.

The condition of regularity is expressed by

$$rank\|g^{ij}\| = n, \quad on\ \widetilde{T^{*k}M}. \tag{8.1.3'}$$

If the base manifold M is paracompact, then the manifold $T^{*k}M$ is paracompact, too and on $T^{*k}M$ there exist regular Hamiltonians.

The d-tensor field g^{ij} being nonsingular on $\widetilde{T^{*k}M}$ there exists a d-tensor field g_{ij} covariant of order 2, symmetric, uniquely determined, at every point $u \in \widetilde{T^{*k}M}$, by

$$g_{ij}g^{jk} = \delta_i^k. \tag{8.1.4}$$

Definition 8.1.2 *An Hamilton space of order k is a pair* $H^{(k)n} = (M, H(x, y^{(1)}, ..., y^{(k-1)}, p))$, *where* M *is a real* n *-dimensional manifold and* H *is a differentiable regular Hamiltonian having the property that the d-tensor field* g^{ij} *has a constant signature on* $\widetilde{T^{*k}M}$.

As usually, H is called *the fundamental function* and g^{ij} *the fundamental tensor* of the space $H^{(k)n}$.

In the case when the fundamental tensor g^{ij} is positively defined, then the condition of regularity (8.1.3') is verified.

Theorem 8.1.1 *Assuming that the base manifold M is paracompact, then there exists on* $\widetilde{T^{*k}M}$ *a regular Hamiltonian H such that the pair* (M, H) *is a Hamilton space of order k.*

Proof: Let $F^{(k-1)n} = (M, F(x, y^{(1)}, ..., y^{(k-1)}))$ be a Finsler space of order $k-1$ on the manifold $T^{k-1}M$, where $T^{k-1}M = \pi^{*k}_{k-1}(T^{*k}M)$, having $\gamma_{ij}(x, y^{(1)}, ..., y^{(k-1)})$ as fundamental tensor. The manifold M being paracompact, the space $F^{(k-1)n}$ exists.

Then, the function

$$H(x, y^{(1)}, ..., y^{(k-1)}, p) = \alpha\gamma^{ij}(x, y^{(1)}, ..., y^{(k-1)})p_ip_j, \ (\alpha \in \mathbf{R}, \alpha > 0),$$

is well defined in every point $(x, y^{(1)}, ..., y^{(k-1)}, p) \in \widetilde{T^{*k}M}$ and it is a fundamental function for a Hamilton space of order k. Its fundamental tensor field is $\alpha\gamma^{ij}$. Q.E.D.

One of the important d-tensor field derived from the fundamental function H of the space $H^{(k)n}$ is:

$$C^{ijh} = -\frac{1}{2}\overset{\cdot h}{\partial}g^{ij} = -\frac{1}{4}\overset{\cdot i}{\partial}\overset{\cdot j}{\partial}\overset{\cdot h}{\partial}H, \quad \left(\overset{\cdot i}{\partial} = \frac{\partial}{\partial p_i}\right). \tag{8.1.5}$$

Proposition 8.1.1 *We have:*

1⁰ C^{ijh} *is a totally symmetric d-tensor field;*

2⁰ C^{ijh} *vanishes if and only if the fundamental tensor* g^{ij} *does not depend on the momenta* p_i.

Other geometrical object fields which are entirely determined by the fundamental function H are the coefficients $C_i^{\ jh}$ of the w_k -covariant derivation, given by

$$C_i^{\ jh} = -\frac{1}{2} g_{is} \left(\overset{\cdot j}{\partial} g^{sh} + \overset{\cdot h}{\partial} g^{js} - \overset{\cdot s}{\partial} g^{jh} \right). \tag{8.1.6}$$

Proposition 8.1.2 *1⁰ $C_i^{\ jh}$ are the components of a d-tensor fields of type* $(2,1)$.

2⁰ They depend on the fundamental function H only.

3⁰ They are symmetric in the indices j,h.

4⁰ The formula

$$C_i^{\ jh} = g_{is} C^{sjh}$$

holds.

5⁰ The w_k-covariant derivative of the fundamental tensor g^{ij}, vanishes:

$$g^{ij}|^h = 0. \tag{8.1.7}$$

The proof is not difficult.

8.2 The k-Tangent Structure J and the Adjoint k-Tangent Structure J^*

For the Hamilton space of order k, $H^{(k)n} = (M,H)$, the structures J and J^* defined on the manifold $T^{*k}M$ in the section 3, Ch.4 have some special properties.

The k-tangent structure is the mapping:

$$\begin{gathered} J\left(\frac{\partial}{\partial x^i}\right) = \frac{\partial}{\partial y^{(1)i}}, \dots, J\left(\frac{\partial}{\partial y^{(k-2)i}}\right) = \frac{\partial}{\partial y^{(k-1)i}}, \\ J\left(\frac{\partial}{\partial y^{(k-1)i}}\right) = 0, J\left(\overset{\cdot i}{\partial}\right) = 0. \end{gathered} \tag{8.2.1}$$

Locally, it is expressed by (4.3.2):

$$J = \frac{\partial}{\partial y^{(1)i}} \otimes dx^i + \frac{\partial}{\partial y^{(2)i}} \otimes dy^{(1)i} + \dots + \frac{\partial}{\partial y^{(k-1)i}} \otimes dy^{(k-2)i}. \tag{8.2.2}$$

The main properties of J are explicitly given in Theorem 4.3.1.

Let X be a vector field on $T^{*k}M$, locally expressed by

$$X = \overset{(0)i}{X} \frac{\partial}{\partial x^i} + \overset{(1)i}{X} \frac{\partial}{\partial y^{(1)i}} + \dots + \overset{(k-1)i}{X} \frac{\partial}{\partial y^{(k-1)i}} + X_i \overset{\cdot i}{\partial}. \tag{8.2.3}$$

Here, $\overset{\cdot i}{\partial} = \dfrac{\partial}{\partial p_i}$.

Consider the following vector fields $\overset{0}{X}, \overset{1}{X}, ..., \overset{k-1}{X}$

(8.2.3')
$$\overset{1}{X}= JX,\ \overset{2}{X}= J^2X,\ ...,\ \overset{k-1}{X}= J^{k-1}X.$$

Taking into account (8.2.3), these vector fields have the form:

(8.2.4)
$$\begin{aligned}\overset{1}{X} &= \overset{(0)i}{X}\frac{\partial}{\partial y^{(1)i}} + \cdots + \overset{(k-2)i}{X}\frac{\partial}{\partial y^{(k-1)i}},\\ \overset{2}{X} &= \overset{(0)i}{X}\frac{\partial}{\partial y^{(2)i}} + \cdots + \overset{(k-3)i}{X}\frac{\partial}{\partial y^{(k-1)i}},\\ &\cdots\cdots\cdots\cdots\cdots\cdots\cdots\cdots\\ \overset{k-1}{X} &= \overset{(0)i}{X}\frac{\partial}{\partial y^{(k-1)i}}.\end{aligned}$$

Now, the adjoint J^* of J is defined by

(8.2.5) $J^*(dx^i) = 0, J^*(dy^{(1)}) = dx^i, ..., J^*(dy^{(k-1)i}) = dy^{(k-2)i}, J^*(dp_i) = 0.$

J^* is the following d-tensor field of type $(1,1)$:

(8.2.6)
$$J^* = dx^i \otimes \frac{\partial}{\partial y^{(1)i}} + dy^{(1)i} \otimes \frac{\partial}{\partial y^{(2)i}} + \cdots + dy^{(k-2)i} \otimes \frac{\partial}{\partial y^{(k-1)i}}.$$

J^* is an integrable structure and *rank* $J^* = (k-1)n$.
If ω is an 1-form field on the manifold $T^{*k}M$ and

(8.2.7)
$$\omega = \overset{(0)}{\omega}_i\, dx^i + \cdots + \overset{(k-1)}{\omega}_i\, dy^{(k-1)i} + \omega^i dp_i,$$

then by means of J^* we obtain a number of $k-1$ 1-forms on $T^{*k}M$:

(8.2.8)
$$\overset{1}{\omega}= J^*\omega,\ ...,\ \overset{k-1}{\omega}= J^{*(k-1)}\omega.$$

The vertical differential operators d_0, ..., d_{k-2} are introduced in §4.3 by

(8.2.9)
$$d_0 = J^{*(k-1)}d, ..., d_{k-2} = J^*d,$$

where d is the operator of differentiation on the manifold $T^{*k}M$. We know from §4.3 that these operators are the antiderivations of degree 1 in the exterior algebra $\Lambda(T^{*k}M)$.

The following formula hold:

(8.2.10)
$$d \circ d = 0,\ d_\alpha \circ d_\alpha = 0,\ (\alpha = 0, 1, ..., k-2).$$

We get:

Proposition 8.2.1 *For any Hamilton space of order k, $H^{(k)n} = (M, H)$ the 1-forms (8.2.8) of the form dH are given by*

$$
\begin{aligned}
& d_0 H = \overset{(0)}{p}_i \, dx^i, \\
(8.2.11) \qquad & d_1 H = \overset{(1)}{p}_i \, dx^i + \overset{(0)}{p}_i \, dy^{(1)i}, \\
& \dots\dots\dots\dots\dots\dots\dots\dots\dots\dots \\
& d_{k-2} H = \overset{(k-2)}{p}_i \, dx^i + \overset{(k-3)}{p}_i \, dy^{(1)i} + \dots + \overset{(0)}{p}_i \, dy^{(k-2)i}.
\end{aligned}
$$

where

$$
(8.2.11') \qquad \overset{(0)}{p}_i = \frac{\partial H}{\partial y^{(k-1)i}}, \ \overset{(1)}{p}_i = \frac{\partial H}{\partial y^{(k-2)i}}, \ \dots, \ \overset{(k-2)}{p}_i = \frac{\partial H}{\partial y^{(1)i}}.
$$

Proposition 8.2.2 *The following 2-forms depend only on the fundamental function H of the Hamilton space $H^{(k)n}$:*

$$
\begin{aligned}
& dd_0 H = d \, \overset{(0)}{p}_i \wedge dx^i, \\
(8.2.12) \qquad & dd_1 H = d \, \overset{(1)}{p}_i \wedge dx^i + d \, \overset{(0)}{p}_i \wedge dy^{(1)i}, \\
& \dots\dots\dots\dots\dots\dots\dots\dots\dots\dots \\
& dd_{k-2} H = d \, \overset{(k-2)}{p}_i \wedge dx^i + \dots + d \, \overset{(0)}{p}_i \wedge dy^{(k-2)i}.
\end{aligned}
$$

We have, also:

$$
(8.2.13) \qquad d_\alpha \circ d_\alpha H = 0, \ \forall \alpha = 0, 1, \dots, k-2.
$$

Clearly, $dd_\alpha H = 0, \ \alpha = 0, \dots, k-2$ are closed 2-forms.

As we know, the operator

$$
(8.2.14) \qquad d_{k-1} = \frac{\partial}{\partial x^i} dx^i + \frac{\partial}{\partial y^{(1)i}} dy^{(1)i} + \dots + \frac{\partial}{\partial y^{(k-1)i}} dy^{(k-1)i}
$$

is not a vertical differentiation.

Proposition 8.2.3 *We have:*

1^0 $d_{k-1}H$ is not an 1-form;

2^0 Under a change of local coordinate on $T^{*k}M$ we have

$$
d_{k-1} H = d_{k-1} \widetilde{H} + \widetilde{\dot{\partial}^j} \widetilde{H} \frac{\partial \widetilde{p}_j}{\partial x^i} dx^i;
$$

3^0 $J^* \circ d_{k-1} = d_{k-2}$.

8.3 The Canonical Poisson Structure of the Hamilton Space $H^{(k)n}$

Consider a Hamilton space $H^{(k)n} = (M, H)$. As we know from the section 4 ch.4 on the manifold $T^{*k}M$ there exists a canonical Poisson structure $\{,\}_{k-1}$. Besides this natural Poisson structure on the submanifold Σ_0 of $T^{*k}M$ there exist a remarkable Poisson structure of the space $H^{(k)n} = (M, H)$.

Let $(T^{*k}M, \overline{\pi}^*, T^*M)$ be the bundle introduced in §1, ch.4. The projection $\overline{\pi}^*$ is given by $\overline{\pi}^*(x, y^{(1)}, ..., y^{(k-1)}, p) = (x, p)$. The canonical section $\sigma_0 : (x, p) \in T^*M \to (x, 0, ..., 0, p) \in T^{*k}M$ has the image $\Sigma_0 = Im\ \sigma_0$, a submanifold of the manifold $T^{*k}M$. The canonical presymplectic structure $\theta = dp_i \wedge dx^i$ has its restriction θ_0 to Σ_0, given by $\theta_0 = dp_i \wedge dx^i$ in every point $(x, p) \in \Sigma_0$. The equations of the submanifold Σ_0 being $y^{(\alpha)i} = 0$, $(\alpha = 1, ..., k-1)$, then (x^i, p_i) are the coordinate of the points $(x, p) \in \Sigma_0$.

Theorem 8.3.1 *The pair* (Σ_0, θ_0) *is a symplectic manifold.*

Proof: Indeed, $\theta_0 = dp_i \wedge dx^i$ is a closed 2-form and $rank\ ||\theta_0|| = 2n = \dim \Sigma_0$. Q.E.D.

In a point $u = (x, p) \in \Sigma_0$ the tangent space $T_u\Sigma_0$ has the natural basis $\left(\dfrac{\partial}{\partial x^i}, \dfrac{\partial}{\partial p_i}\right)_u$, $(i = 1, ..., n)$ and natural cobasis $\left(dx^i, dp_i\right)_u$.

Let us consider $\mathcal{F}(\Sigma_0)$-module $\mathcal{X}(\Sigma_0)$ of vector fields and $\mathcal{F}(\Sigma_0)$-module $\mathcal{X}^*(\Sigma_0)$ of covector fields on the submanifold Σ_0.

The following $\mathcal{F}(\Sigma_0)$-linear mapping
$S_{\theta_0} : \mathcal{X}(\Sigma_0) \to \mathcal{X}^*(\Sigma_0)$ defined by

$$S_{\theta_0}(X) = i_X\theta_0, \forall X \in \mathcal{X}(\Sigma_0) \tag{8.3.4}$$

gives us

$$S_{\theta_0}\left(\frac{\partial}{\partial x^i}\right) = -dp_i,\ S_{\theta_0}\left(\frac{\partial}{\partial p_i}\right) = dx^i. \tag{8.3.4'}$$

These equalities have as a consequence:

Proposition 8.3.1 *The mapping* S_{θ_0} *is an isomorphism.*

The Hamilton space $H^{(k)n} = (M, H)$, allows to consider the restriction H_0 of the fundamental function H to the submanifold Σ_0, $H_0(x, p) = H(x, 0, ..., 0, p)$. Therefore the pair $(M, H_0(x, p))$ is a classical Hamilton space (cf. [115]) with fundamental tensor field

$$g^{ij}(x, 0, ..., 0, p) = \frac{1}{2}\dot{\partial}^i\dot{\partial}^j H_0.$$

By means of the last proposition it follows:

Proposition 8.3.2 *1⁰ There exists a unique vector field $X_{H_0} \in \mathcal{X}(\Sigma_0)$ such that*

$$S_{\theta_0}(X_{H_0}) = i_{X_{H_0}}\theta_0 = -dH_0.$$

2⁰ X_{H_0} is given by

$$X_{H_0} = \frac{\partial H_0}{\partial p_i}\frac{\partial}{\partial x^i} - \frac{\partial H_0}{\partial x^i}\frac{\partial}{\partial p_i}. \tag{8.3.5}$$

Theorem 8.3.2 *The integral curves of the vector field X_{H_0} are given by the Σ_0-canonical equations*

$$\frac{dx^i}{dt} = \frac{\partial H_0}{\partial p_i},\ \frac{dp_i}{dt} = -\frac{\partial H_0}{\partial x^i},\ y^{(\alpha)i} = 0,\ (\alpha = 1, ..., k-1). \tag{8.3.6}$$

For two functions $f,\ g \in \mathcal{F}(\Sigma_0)$, let X_f and X_g be the corresponding Hamilton vector fields given by

$$i_{X_f}\theta_0 = -df,\ i_{X_g}\theta_0 = -dg.$$

Theorem 8.3.3 *The following formula holds*

$$\{f,g\}_0 = \theta_0(X_f, X_g),\ \forall f,g \in \mathcal{F}(\Sigma_0) \tag{8.3.7}$$

Proof: We have
$\theta(X_f, X_g) = (i_{X_f}\theta_0)(X_g) = S_{\theta_0}(X_f)(X_g) = -df(X_g) = -X_g f =$

$$= \left(\frac{\partial f}{\partial x^i}\frac{\partial g}{\partial p_i} - \frac{\partial f}{\partial p_i}\frac{\partial g}{\partial x^i}\right) = \{f,g\}_0\ .\ \text{Q.E.D.}$$

Remark 1⁰ The previous theory can be extended to the other Poisson structures $\{,\}_\alpha$, $(\alpha = 1, ..., k-1)$ (cf. [4]).

2⁰ The triple $(T^{*k}M, H(x, y^{(1)}, ..., y^{(k-1)}, p), \theta)$ is an Hamiltonian system in which θ is a presymplectic structure. Therefore we can apply Gotay's method (cf. M. de Leon and Gotay, [115]) taking into account and the considerations from the previous section, §8.2.

The equations (8.3.6) are particular. For a Hamilton space $H^{(k)n} = (M, H)$, the integral of action, (see Ch.5):

$$I(c) = \int_0^1 [p_i\frac{dx^i}{dt} - \frac{1}{2}H(x, \frac{dx}{dt}, ..., \frac{1}{(k-1)!}\frac{d^{k-1}}{dt^{k-1}}, p)]dt$$

leads, via the variational problem, to the fundamental equations of the space $H^{(k)n}$, i.e. the Hamilton-Jacobi equations

$$\begin{aligned} &\frac{dx^i}{dt} = \frac{1}{2}\frac{\partial H}{\partial p_i}, \\ &\frac{dp_i}{dt} = -\frac{1}{2}[\frac{\partial H}{\partial x^i} - \frac{d}{dt}\frac{\partial H}{\partial y^{(1)i}} + \cdots + (-1)^{k-1}\frac{d^{k-1}}{dt^{k-1}}\frac{\partial H}{\partial y^{(k-1)i}}]. \end{aligned} \tag{8.3.8}$$

The energies of order k-1, $\mathcal{E}^{k-1}(H)$ of the considered space has the expression

$$\mathcal{E}^{k-1}(H) = I^{k-1}(H) - \frac{1}{2!}\frac{d}{dt}I^{k-2}(H) + \cdots + (-1)^{k-2}\frac{1}{(k-1)!}\frac{d^{k-2}}{dt^{k-2}}I^1(H) - H$$

and we have:

Theorem 8.3.4 *For a Hamilton space $H^{(k)n} = (M, H)$ the energy of order $k-1$, $\mathcal{E}^{k-1}(H)$ is constant along every solution curve of the Hamilton-Jacobi equations.*

8.4 Legendre Mapping Determined by a Lagrange Space $L^{(k)n} = (M, L)$

Let $L^{(k)n} = (M, L(x, y^{(1)}, ..., y^{(k)}))$ be a Lagrange space of order k. It determines a local diffeomorphism $\varphi : \widetilde{T^kM} \to \widetilde{T^{*k}M}$ which preserves the fibres. The mapping φ transforms the canonical k-semispray S, (1.2.5), of $L^{(k)n}$ in the dual k-semispray S_ξ, where $\xi = \varphi^{-1}$ and determines a nonlinear connection N^* on $T^{*k}M$. But φ does not transform the regular Lagrangian L in a regular Hamiltonian, in the case $k > 1$. We need some supplementary geometrical object fields for getting a regular Hamiltonian H from the fundamental function L of the space $L^{(k)n}$. In this section we investigate the above mentioned problems.

A point $(x, y^{(1)}, ..., y^{(k)})$ of the manifold T^kM will be denoted by $(y^{(0)}, y^{(1)}, ..., y^{(k)})$ and its coordinates by $(y^{(0)i}, ..., y^{(k)i})$.

The fundamental tensor of the space $L^{(k)n} = (M, L(y^{(0)}, ..., y^{(k)}))$ will be given by

$$a_{ij} = \frac{1}{2}\frac{\partial^2 L}{\partial y^{(k)i}\partial y^{(k)j}}. \tag{8.4.1}$$

Proposition 8.4.1 *The mapping $\varphi : u = (y^{(0)}, y^{(1)}, ..., y^{(k)}) \in \widetilde{T^kM} \to u^* = (x, y^{(1)}, ..., y^{(k-1)}, p) \in \widetilde{T^{*k}M}$ given by*

$$\begin{cases} x^i = y^{(0)i}, \ y^{(1)i} = y^{(1)i}, \ ..., \ y^{(k-1)i} = y^{(k-1)i}, \\ p_i = \dfrac{1}{2}\dfrac{\partial L}{\partial y^{(k)i}}. \end{cases} \tag{8.4.2}$$

is a local diffeomorphism, which preserves the fibres.

Proof: The mapping φ is differentiable on the manifold $\widetilde{T^kM}$ and its Jacobian has the determinant equal to $\det ||a_{ij}|| \neq 0$. Of course, $\pi^k(y^{(0)}, ..., y^{(k)}) = \pi^{*k} \circ \varphi(y^{(0)}, ..., y^{(k)}) = y^{(0)}$.

The diffeomorphism φ is called *the Legendre mapping* (or *Legendre transformation*).

We denote

$$p_i = \frac{1}{2}\frac{\partial L}{\partial y^{(k)i}} = \varphi_i(y^{(0)}, y^{(1)}, ..., y^{(k)}). \tag{8.4.2'}$$

Clearly, φ_i is a d-covector field on $\widetilde{T^kM}$.

The local inverse diffeomorphism $\xi = \varphi^{-1} : \widetilde{T^{*k}M} \to \widetilde{T^kM}$ is expressed by

$$\begin{cases} y^{(0)i} = x^i,\ y^{(1)i} = y^{(1)i},\ ...,\ y^{(k-1)i} = y^{(k-1)i}, \\ y^{(k)i} = \xi^i(x, y^{(1)}, ..., y^{(k-1)}, p). \end{cases} \tag{8.4.3}$$

With respect to a change of local coordinates on the manifold $T^{*k}M$, ξ^i is transformed exactly as the variables $y^{(k)i}$ from the formulas (1.1.2).

The mappings φ and ξ satisfy the conditions

$$\xi \circ \varphi = 1_{\hat{U}}, \varphi \circ \xi = 1_{\overset{\vee}{U}}, \hat{U} = (\pi^{*k})^{-1}(U), \overset{\vee}{U} = (\pi^k)^{-1}(U), U \subset M.$$

Therefore we have the following identities

(8.4.4)

$$a_{ij}(y^{(0)}, ..., y^{(k)}) = \frac{\partial \varphi_i}{\partial y^{(k)j}};\ a^{ij}(x, y^{(1)}, ..., y^{(k-1)}, \xi(x, y^{(1)}, ..., y^{(k-1)}, p)) = \frac{\partial \xi^i}{\partial p_j}$$

and

(8.4.5)

$$\begin{cases} \dfrac{\partial \varphi_i}{\partial x^j} = -a_{is}\dfrac{\partial \xi^s}{\partial x^j};\ \dfrac{\partial \varphi_i}{\partial y^{(\alpha)j}} = -a_{is}\dfrac{\partial \xi^s}{\partial y^{(\alpha)j}},\ (\alpha = 1, ..., k-1);\ \dfrac{\partial \varphi_i}{\partial y^{(k)j}} = a_{ij}; \\ \dfrac{\partial \xi^i}{\partial x^j} = -a^{is}\dfrac{\partial \varphi_s}{\partial x^j};\ \dfrac{\partial \xi^i}{\partial y^{(\alpha)j}} = -a^{is}\dfrac{\partial \varphi_s}{\partial y^{(\alpha)j}},\ (\alpha = 1, ..., k-1);\ \dfrac{\partial \xi^i}{\partial p_j} = a^{ij}. \end{cases}$$

The differential $d\varphi$ of the mapping φ, $\varphi_* : T_u(T^kM) \to T_{\varphi(u)}(T^{*k}M)$ in the natural basis is expressed by

$$\begin{aligned} &\varphi_*\left(\frac{\partial}{\partial y^{(0)i}}|_u\right) = \frac{\partial}{\partial x^i}|_{u^*} + \frac{\partial \varphi_m}{\partial x^i}\frac{\partial}{\partial p_m}|_{u^*},\quad u^* = \varphi(u) \\ &\varphi_*\left(\frac{\partial}{\partial y^{(\alpha)i}}|_u\right) = \frac{\partial}{\partial y^{(\alpha)i}}|_{u^*} + \frac{\partial \varphi_m}{\partial y^{(\alpha)i}}\frac{\partial}{\partial p_m}|_{u^*},\quad (\alpha = 1, ..., k-1), \\ &\varphi_*\left(\frac{\partial}{\partial y^{(k)i}}|_u\right) = a_{im}\frac{\partial}{\partial p_m}|_{u^*}. \end{aligned} \tag{8.4.6}$$

Theorem 8.4.1 *The Legendre mapping $\varphi : T^kM \to T^{*k}M$ transforms the canonical k-semispray S of the Lagrange space $L^{(k)n}$*

(8.4.7)

$$S = y^{(1)i}\frac{\partial}{\partial y^{(0)i}} + 2y^{(2)i}\frac{\partial}{\partial y^{(1)i}} + \cdots + ky^{(k)i}\frac{\partial}{\partial y^{(k-1)i}} - (k+1)G^i(y^{(0)}, ..., y^{(k)})\frac{\partial}{\partial y^{(k)i}}$$

in the dual k-semispray S^*_ξ on $T^{*k}M$:

$$\begin{aligned} S^*_\xi &= y^{(1)i}\frac{\partial}{\partial x^i} + 2y^{(2)i}\frac{\partial}{\partial y^{(1)i}} + \cdots + (k-1)y^{(k-1)i}\frac{\partial}{\partial y^{(k-2)i}} + \\ &+ k\xi^i(x, y^{(1)}, ..., y^{(k-1)}, p)\frac{\partial}{\partial y^{(k-1)}} + \eta_i(x, y^{(1)}, ..., y^{(k-1)}, p)\frac{\partial}{\partial p_i}, \end{aligned} \tag{8.4.8}$$

with the coefficients ξ^i from (8.4.3) and

$$\begin{aligned} \eta_i = -a_{is}&\left(\frac{\partial \xi^s}{\partial x^r}y^{(1)r} + \cdots + (k-1)\frac{\partial \xi^s}{\partial y^{(k-1)r}}\xi^r + \right. \\ &\left. + (k+1)G^s(x, y^{(1)}, ..., y^{(k-1)}, \xi(x, y^{(1)}, ..., y^{(k-1)}, p))\right) \end{aligned} \tag{8.4.9}$$

Proof: If $X \in \mathcal{X}(\widetilde{T^kM})$ have the local expression $X = X^{(0)i}\dfrac{\partial}{\partial y^{(0)i}} + X^{(1)i}\dfrac{\partial}{\partial y^{(1)i}} + \cdots + X^{(k)i}\dfrac{\partial}{\partial y^{(k)i}}$ in every point $u \in \widetilde{T^kM}$, then

$X^* = (\varphi_* X)(u^*) = X^{(0)i}(u^*)\varphi_*\dfrac{\partial}{\partial x^i} + \cdots + X^{(k)i}(u^*)\varphi_*\dfrac{\partial}{\partial y^{(k)i}} =$

$= X^{(0)i}(u^*)\dfrac{\partial}{\partial x^i}|_{u^*} + \cdots + X^{(k-1)i}(u^*)\dfrac{\partial}{\partial y^{(k-1)i}}|_{u^*} +$

$+ \left(X^{(0)m}(u^*)\dfrac{\partial \varphi_i}{\partial x^m} + \cdots + X^{(k-1)m}(u^*)\dfrac{\partial \varphi_i}{\partial y^{(k-1)m}}\right)\dfrac{\partial}{\partial p_i}|_{u^*} +$

$+ X^{(k)m}(u^*)a_{mi}(u^*)\dfrac{\partial}{\partial p_i}|_{u^*}$ in the points $u^* = \varphi(u)$.

Consequently, $S^*_\xi = \varphi_* S$ holds.

Now, applying Theorem 5.3.1 we obtain without difficulties:

Theorem 8.4.2 *The Legendre mapping $\varphi : T^kM \to T^{*k}M$ determined by a Lagrange space of order k, $L^{(k)n}$, transforms the canonical k -semispray S in the dual k-semispray S^*_ξ with the coefficients ξ^i, η_i. S^*_ξ determines locally a nonlinear connection N^* on $T^{*k}M$ with dual coefficients:*

$$\underset{(1)}{M^*}{}^i_j = -\frac{\partial \xi^i}{\partial y^{(1)j}}, ..., \underset{(k-1)}{M^*}{}^i_j = -\frac{\partial \xi^i}{\partial y^{(k-1)j}} \tag{8.4.10}$$

and

$$N^*_{ij} = \frac{\delta \eta_i}{\delta y^{(1)j}}, \tag{8.4.10'}$$

where the operators $\dfrac{\delta}{\delta y^{(1)i}}$ are obtained by means of the coefficients, (8.4.10).

We may ask wether by means of the Legendre mapping φ we can transfer the fundamental function L of the Lagrange space $L^{(k)n}$ to a fundamental function H of a Hamilton space of order k, $H^{(k)}$, like in the classical case of $k = 1$.

Remarking that the energy of order k, $\mathcal{E}_c^k(L)$ of the space $L^{(k)n}$ is not a regular Lagrangian on T^kM, the Legendre transformation φ cannot apply $\mathcal{E}_c^k(L)$ in a regular Hamiltonian on $T^{*k}M$. Therefore it is necessary to look for another way to solve this problem.

Let us consider a fixed nonlinear connection $\overset{\circ}{N}$ on the submanifold $T^{k-1}M$ in T^kM. The dual coefficients of $\overset{\circ}{N}$ are denoted by $\underset{(1)}{\overset{\circ}{M}}{}^i_j(x, y^{(1)}, ..., y^{(k-1)})$, ..., $\underset{(k-1)}{\overset{\circ}{M}}{}^i_j(x, y^{(1)}, ..., y^{(k-1)})$. It is not difficult (cf. (1.4.7)) to prove that

$$kz^{(k)i} = ky^{(k)i} + (k-1)\underset{(1)}{\overset{\circ}{M}}{}^i_m y^{(k-1)m} + \cdots + \underset{(k-1)}{\overset{\circ}{M}}{}^i_m y^{(1)m} \tag{8.4.11}$$

is a d-vector field at every point $u = (x, y^{(1)}, ..., y^{(k)})$.

The Legendre mapping

$\varphi : u \in T^kM \to u^* \in T^{*k}M, u^* = (x, y^{(1)}, ..., y^{(k-1)}, p)$ transform $z^{(k)i}$

in a d-vector field $\overset{\vee}{z}{}^{(k)i}$ at u^*, given by

$$k\overset{\vee}{z}{}^{(k)i} = k\xi^i + (k-1)\underset{(1)}{\overset{\circ}{M}}{}^i_m y^{(k-1)m} + \cdots + \underset{(k-1)}{\overset{\circ}{M}}{}^i_m y^{(1)m} \tag{8.4.12}$$

Let us consider the following Hamiltonian

(8.4.13)

$$H(x, y^{(1)}, ..., y^{(k-1)}, p) = 2p_i \overset{\vee}{z}{}^{(k)i} - L(x, y^{(1)}, ..., y^{(k-1)}, \xi(x, y^{(1)}, ..., y^{(k-1)}, p)).$$

Clearly, H is a differentiable Hamiltonian function on $T^{*k}M$.

Theorem 8.4.3 *The Hamiltonian function H, (8.4.13), is the fundamental function of a Hamilton space of order k, $H^{(k)n}$ and its fundamental tensor field is*

(8.4.14)

$$g^{ij}(x, y^{(1)}, ..., y^{(k-1)}, p) = a^{ij}(x, y^{(1)}, ..., y^{(k-1)}, \xi(x, y^{(1)}, ..., y^{(k-1)}, p)),$$

a_{ij} being the fundamental tensor field of the Lagrange space of order k, $L^{(k)n} = (M, L)$.

Proof: From (8.4.13) we have

$$\frac{1}{2}\dot{\partial}^j H = \overset{\vee}{z}{}^{(k)j} + p_m \dot{\partial}^j \overset{\vee}{z}{}^{(k)m} - \frac{1}{2}\frac{\partial L}{\partial y^{(k)m}} \dot{\partial}^j \xi^{(k)m} =$$

$$= \overset{\vee}{z}{}^{(k)j} + p_m \dot{\partial}^j \overset{\vee}{z}{}^{(k)m} - p_m \dot{\partial}^j \overset{\vee}{z}{}^{(k)m} = \overset{\vee}{z}{}^{(k)j}.$$

Consequently,

$$g^{ij}(x, y^{(1)}, ..., y^{(k-1)}, p) = \frac{1}{2}\dot{\partial}^i\dot{\partial}^j H = \frac{1}{2}\dot{\partial}^i \overset{\vee}{z}{}^{(k)j} =$$

$$= \frac{1}{2} \overset{\cdot i}{\partial} \xi^{(k)j} = a^{ij}(x, y^{(1)}, ..., y^{(k-1)}, \xi).$$

It follows that H is a regular Hamiltonian and its fundamental tensor g^{ij} has a constant signature on $T^{*k}M$. Q.E.D.

The Hamilton space of order k, $H^{(k)n} = (M, H)$ is called the *dual* of the Lagrange space of order k, $L^{(k)n} = (M, L)$.

Remarks. All geometrical object fields of the space $H^{(k)n}$, which are derived from the fundamental tensor field g^{ij} do not depend on the apriori fixed nonlinear connection $\overset{\circ}{N}$. Consequently, the geometry of the Hamilton space $H^{(k)n} = (M, H)$ can be constructed by means of the dual k-semispray S^*_ξ and the fundamental tensor field g^{ij} from (8.4.14).

8.5 Legendre Mapping Determined by a Hamilton Space of Order k

Let us consider a converse for the previous problem: being given a Hamilton space of order k, $H^{(k)n} = (M, H(x, y^{(1)}, ..., y^{(k-1)}, p))$ let us determine its *dual* like a Lagrange space of order k, $L^{(k)n} = (M, L(x, y^{(1)}, ..., y^{(k)}))$.

In this case, we start from the space $H^{(k)n}$ and try to determine the local diffeomorphism in the form of φ^{-1} from (8.4.3). But $y^{(k)i}$ is not a vector field on T^kM and we can not define it only by the d-vector field $\overset{\cdot i}{\partial} H$. As in the previous section we assume that a nonlinear connection $\overset{\circ}{N}$ on the submanifold $T^{k-1}M$ in T^kM is apriori fixed. The dual coefficients of $\overset{\circ}{N}$ being $\underset{(1)}{\overset{\circ}{M}}{}^i_j, ..., \underset{(k-1)}{\overset{\circ}{M}}{}^i_j$ they are functions of the points $(x, y^{(1)}, ..., y^{(k-1)})$ from $T^{k-1}M$.

Then the d-vector field $z^{(k)i}$, given by (8.4.11) on T^kM can be considered.

The mapping

$\overset{*}{\xi}$: $u^* = (x, y^{(1)}, ..., y^{(k-1)}, p) \in T^{*k}M \to u = (y^{(0)}, y^{(1)}, ..., y^{(k-1)}, y^{(k)}) \in T^kM$, defined as follows:

$$y^{(0)i} = x^i, y^{(1)i} = y^{(1)i}, ..., y^{(k-1)i} = y^{(k-1)i},$$

(8.5.1)

$$y^{(k)i} = \overset{*i}{\xi}(x, y^{(1)}, ..., y^{(k-1)}, p),$$

where $\overset{*i}{\xi}$ is expressed from the formula:

$$k \overset{*i}{\xi}(x, y^{(1)}, ..., y^{(k-1)}, p) + (k-1) \underset{(1)}{\overset{\circ}{M}}{}^i_m (x, y^{(1)}, ..., y^{(k-1)}) y^{(k-1)m} +$$

(8.5.2)

$$+ \cdots + \underset{(k-1)}{\overset{\circ}{M}}{}^i_m (x, y^{(1)}, ..., y^{(k-1)}) y^{(1)m} = \frac{k}{2} \overset{\cdot i}{\partial} H.$$

Theorem 8.5.1 *The mapping $\overset{*}{\xi}$ is a local diffeomorphism, which preserves the fibres of $T^{*k}M$ and T^kM.*

Proof: The determinant of the Jacobian of $\overset{*}{\xi}$ is $\det \|g^{ij}(x, y^{(1)}, ..., y^{(k-1)}, p)\| \neq 0$, since $\dot{\partial}^j \overset{*}{\xi}{}^i = g^{ij}$ and $\pi^{*k} = \pi^k \circ \overset{*}{\xi}$. q.e.d.

The formulae (8.5.1) and (8.5.2) imply:

$$\dot{\partial}^i \overset{*}{\xi}{}^j = g^{ij} \tag{8.5.3}$$

$$z^{(k)i}(x, y^{(1)}, ..., y^{(k-1)}, \overset{*}{\xi}) = \frac{1}{2} \dot{\partial}^i H(x, y^{(1)}, ..., y^{(k-1)}, p), \tag{8.5.4}$$

g^{ij} being the fundamental tensor field of space $H^{(k)n}$.

The inverse mapping $\overset{*}{\varphi}$: $u = (y^{(0)}, y^{(1)}, ..., y^{(k-1)}, y^{(k)}) \in T^kM \to u^* = (x, y^{(1)}, ..., y^{(k-1)}, p) \in T^{*k}M$ of the Legendre transformation $\overset{*}{\xi}$ can be written as follows:

$$\begin{aligned} &x^i = y^{(0)i}, y^{(1)i} = y^{(1)i}, ..., y^{(k-1)i} = y^{(k-1)i}, \\ &p_i = \overset{*}{\varphi}{}^i (y^{(0)}, y^{(1)}, ..., y^{(k-1)}, y^{(k)}). \end{aligned} \tag{8.5.5}$$

¿From here we deduce

$$\frac{\partial \overset{*}{\varphi}_i}{\partial y^{(k)j}} = g_{ij}(x, y^{(1)}, ..., y^{(k-1)}, \overset{*}{\varphi}(u)), \tag{8.5.6}$$

where g_{ij} is the covariant tensor of the fundamental tensor g^{ij} of $H^{(k)n}$.

The Legendre transformation $\overset{*}{\xi}$ maps the regular Hamiltonian H into the differentiable Lagrangian $L = \overset{*}{\xi}(H)$:

$$\begin{aligned} &L(x, y^{(1)}, ..., y^{(k-1)}, y^{(k)}) = 2p_i z^{(k)i} - H(x, y^{(1)}, ..., y^{(k-1)}, p), \\ &p_i = \overset{*}{\varphi}_i (y^{(0)}, y^{(1)}, ..., y^{(k-1)}, y^{(k)}). \end{aligned} \tag{8.5.7}$$

Theorem 8.5.2 *The Lagrangian L from (8.5.7) is a regular one. Its fundamental tensor field is $g_{ij}(x, y^{(1)}, ..., y^{(k-1)}, \overset{*}{\varphi}{}^i (x, y^{(1)}, ..., y^{(k-1)}, y^{(k)}))$.*

Indeed, $\dfrac{\partial z^{(k)i}}{\partial y^{(k)j}} = \delta^i_j$ implies $\dfrac{1}{2}\dfrac{\partial L}{\partial y^{(k)i}} = \dfrac{\partial p_m}{\partial y^{(k)i}} z^{(k)m} + p_i - \dfrac{1}{2} \dot{\partial}^m H \dfrac{\partial p_m}{\partial y^{(k)i}}$.
Taking into account (8.5.4) it follows

$$p_i = \frac{1}{2}\frac{\partial L}{\partial y^{(k)i}}, \quad p_i = \overset{*}{\varphi}_i (x, y^{(1)}, ..., y^{(k-1)}, y^{(k)}). \tag{8.5.8}$$

By means of (8.5.6) it follows

$$a_{ij}(u) = \frac{1}{2}\frac{\partial^2 L}{\partial y^{(k)i}\partial y^{(k)j}}(u) = g_{ij}(x, y^{(1)}, ..., y^{(k-1)}, \overset{*}{\varphi}(u)), \tag{8.5.9}$$

where $u = (x, y^{(1)}, ..., y^{(k)}) \in T^k M$. Q.E.D.

The Lagrange space of order k, $L^{(k)n} = (M, L)$ is called the *dual* of the Hamilton space of order k.

Remarks 1^0 Of course, the space $L^{(k)n} = (M, L)$ is locally determined.

2^0 The relations (8.5.8) give us the geometrical meaning of the mapping $\overset{*}{\varphi}$, the inverse of the bundle morphism $\overset{*}{\xi}$.

It follows

Theorem 8.5.3 *We have the following local properties:*

Consider the Legendre transformation determined by the Lagrange space of order k, $L^{(k)n} = (M, L)$. Then, the dual Hamilton space of order k, $H^{(k)n} = (M, H)$, $L = \overset{*}{\xi}(H)$ is defined by the local diffeomorphism $\overset{*}{\varphi}$, the inverse of the local diffeomorphism $\overset{*}{\xi}$ and $H = \overset{*}{\varphi}(L)$ is locally given by

$$\begin{aligned} H(x, y^{(1)}, ..., y^{(k-1)}, p) &= 2p_i z^{(k)}(x, y^{(1)}, ..., y^{(k-1)}, y^{(k)}) - \\ &\quad - L(x, y^{(1)}, ..., y^{(k-1)}, y^{(k)}), \\ y^{(k)i} &= \overset{*}{\xi}{}^{i}(x, y^{(1)}, ..., y^{(k-1)}, p). \end{aligned} \tag{8.5.10}$$

Indeed, as the Legendre transformation $\varphi : T^k M \to T^{*k} M$ is defined by $\varphi : (x, y^{(1)}, ..., y^{(k-1)}, y^{(k)}) \in T^k M \to (x, y^{(1)}, ..., y^{(k-1)}, p) \in T^{*k} M$, with $p_i = \frac{1}{2}\frac{\partial L}{\partial y^{(k)i}} = \overset{*}{\varphi}(x, y^{(1)}, ..., y^{(k)})$. It follows, locally, $\varphi = \overset{*}{\varphi}$, and $\varphi^{-1} = \overset{*}{\xi}$. But H from (8.5.10) is $H = \overset{*}{\varphi}(L)$. So, locally, $H = \overset{*}{\varphi}(\overset{*}{\xi} H)$ and $L = \overset{*}{\xi}(\overset{*}{\varphi} L)$ .Q.E.D.

8.6 The Canonical Nonlinear Connection of the Space $H^{(k)n}$

There exists a nonlinear connection N^* on the manifold $T^{*k}M$, which is determined only by a Hamilton space of order k, $H^{(k)n} = (M, H)$.

Such a property holds in the case of Lagrange spaces of order k, $L^{(k)n} = (M, L)$, (see ch. 2). Namely, $L^{(k)n}$ determines a canonical k -semispray S and S allows to construct a nonlinear connection N, which depends only on the fundamental function L (Theorem 2.5.2.). Since $L^{(k)n}$ defines the Legendre mapping φ from (8.4.2), (8.4.2'), then $\varphi^{-1} = \xi$ determines a bundle morphism and S is transformed in a k -dual semispray $S^*_\xi = \varphi_*(S)$. Theorem 8.4.2. shows that S^*_ξ determine a nonlinear connection N^* on $T^{*k}M$, which depends on the fundamental function L of the space $L^{(k)n}$.

Considering the Hamilton space $H^{(k)n} = (M, H)$, the dual of $L^{(k)n}$, its fundamental function $H = \varphi(L)$ is built locally by means of L and by an apriori given nonlinear connection $\overset{\circ}{N}$ on the submanifold $T^{k-1}M$. The connection N^* can be considered the $\overset{\circ}{N}$-canonical nonlinear connection of the space $H^{(k)n}$.

Assuming now that a Hamilton space of order k, $H^{(k)n} = (M, H)$ is considered, we construct a bundle morphism $\overset{*}{\xi}$, (8.5.1), (8.5.2), by means of H and $\overset{\circ}{N}$, and we determine locally the Lagrange space $L^{(k)n} = (M, L)$, where $L = \overset{*}{\xi}(H)$ is from (8.5.7). It is the *dual* space of $H^{(k)n}$. The Legendre transformation $\overset{*}{\varphi} = \overset{*}{\xi}^{-1}$, with $\overset{*}{\varphi}_i = \dfrac{1}{2}\dfrac{\partial L}{\partial y^{(k)i}}$, transforms the k-semispray S of L from (8.4.7) in the dual k-semispray S^*_ξ from (8.4.8):

$$(8.6.1)\quad S^*_\xi = y^{(1)i}\frac{\partial}{\partial x^i} + \cdots + (k-1)y^{(k-1)i}\frac{\partial}{\partial y^{(k-2)i}} + k\,\overset{*}{\xi}{}^i\,\frac{\partial}{\partial y^{(k-1)i}} + \eta_i\frac{\partial}{\partial p_i},$$

where $\overset{*}{\xi}{}^i$ is given by (8.5.2) and its coefficients are

$$(8.6.2)\quad \begin{aligned}\eta_i = -g_{is}\Big[&\frac{\partial \overset{*}{\xi}{}^s}{\partial x^r}y^{(1)r} + \cdots + (k-1)\frac{\partial \overset{*}{\xi}{}^s}{\partial y^{(k-1)r}}\overset{*}{\xi}{}^r + \\ &(k+1)G^s(x, ..., y^{(k-1)}, \overset{*}{\xi}(u^*))\Big]\end{aligned}$$

So, we have obtained:

Theorem 8.6.1 *The dual coefficients of the $\overset{\circ}{N}$-canonical nonlinear connection N^* of the Hamilton space of order k, $H^{(k)n}$, are expressed as follows:*

$$(8.6.3)\quad \underset{(1)}{\overset{*}{M}}{}^i_j = -\frac{\partial \overset{*}{\xi}{}^i}{\partial y^{(1)j}}, ..., \underset{(k-1)}{\overset{*}{M}}{}^i_j = -\frac{\partial \overset{*}{\xi}{}^i}{\partial y^{(k-1)j}}$$

and

$$(8.6.3')\quad \overset{*}{N}_{ij} = \frac{\delta \eta_i}{\delta y^{(1)j}},$$

where the operator $\dfrac{\delta}{\delta y^{(1)j}}$ is constructed by means of the coefficients $\underset{(1)}{N}{}^i_j, ..., \underset{(k)}{N}{}^i_j$ given by the dual coefficients (8.6.3).

The previous theory was presented in the book [115], ch. 12, in the case $k = 2$.

It is useful to prove the existence of a nonlinear connection canonically determined by the fundamental function of the Hamilton space $H^{(k)n}$.

8.7 Canonical Metrical N-Linear Connection of the Space $H^{(k)n}$

For a Hamilton space of order k, $H^{(k)n} = (M, H)$, we consider the canonical nonlinear connection N determined in the previous section.

We consider also the adapted basis $\left(\frac{\delta}{\delta x^i}, \frac{\delta}{\delta y^{(1)i}}, \dots, \frac{\delta}{\delta y^{(k-1)i}}, \dot{\partial}^i\right)$ and its dual basis $(\delta x^i, \delta y^{(1)i}, ..., \delta y^{(k-1)i}, \delta p_i)$ determined by N and by the distribution W_k (cf. ch. 6).

We can define a N-linear connection D by its coefficients $D\Gamma(N) = (H^i_{jh}, \underset{(\alpha)}{C}{}^i_{jh}, C^{jh}_i)$, $(\alpha = 1, 2, ..., k-1)$. Therefore the h-, v_α-,w_k-operators of covariant derivations with respect to D, denoted by '$_|$', '$\overset{(\alpha)}{|}$', '$|$', can be applied to the d-tensor fields on the manifold $T^{*k}M$.

Therefore if $g^{ij} = \frac{1}{2}\dot{\partial}^i\dot{\partial}^j H$ is the fundamental tensor of the Hamilton space $H^{(k)n}$, we have

$$
\begin{aligned}
&g^{ij}{}_{|h} = \frac{\delta g^{ij}}{\delta x^h} + g^{mj}H^i_{mh} + g^{im}H^j_{mh},\\
(8.7.1)\qquad &g^{ij}\overset{(\alpha)}{|}_h = \frac{\delta g^{ij}}{\delta y^{(\alpha)h}} + g^{mj}\underset{(\alpha)}{C}{}^i_{mh} + g^{im}\underset{(\alpha)}{C}{}^j_{mh},\ (\alpha = 1, ..., k-1),\\
&g^{ij}|^h = \frac{\partial g^{ij}}{\partial p_h} + g^{mj}C^{ih}_m + g^{im}C^{jh}_m.
\end{aligned}
$$

A N-linear connection $D\Gamma(N)$ is called *compatible* with the fundamental tensor g^{ij} of the Hamilton space of order k, $H^{(k)n} = (M, H)$, (or it is *metrical*) if g^{ij} is covariant constant (or absolute parallel) with respect to $D\Gamma(N)$, i.e.

$$
(8.7.2)\qquad g^{ij}{}_{|h} = 0,\ g^{ij}\overset{(\alpha)}{|}_h = 0,\ g^{ij}|^h = 0.
$$

The previous equations have a geometrical meaning. Indeed, if the tensor field g^{ij} is positively defined and ω_i is a d-covector field, then

$$
||\omega||^2 = g^{ij}\omega_i\omega_j
$$

is a scalar field. Then, locally, $\frac{d}{dt}||\omega||^2 = 0$, along any curve γ and for any parallel covector ω_i along γ, if and only if the equations (8.7.2) hold.

As usual, we can prove:

Theorem 8.7.1 *1) In a Hamilton space of order k, $H^{(k)n} = (M, H)$, there exists a unique N-linear connection D, with the coefficients $D\Gamma(N) = (H^i_{jh}, \underset{(1)}{C}{}^i_{jh}, ..., \underset{(k-1)}{C}{}^i_{jh}, C^{jh}_i)$ verifying the axioms:*

1⁰ N is the canonical nonlinear connection of $H^{(k)n}$.

2⁰ The fundamental tensor g^{ij} *is h-covariant constant:*

$$g^{ij}{}_{|h} = 0. \tag{8.7.3}$$

3⁰ g^{ij} *is* v_α*-covariant constant:*

$$g^{ij}\overset{(\alpha)}{|}_h = 0, \ (\alpha = 1, ..., k-1). \tag{8.7.4}$$

4⁰ g^{ij} *is* w_k*-covariant constant:*

$$g^{ij}|^h = 0. \tag{8.7.4'}$$

5⁰ $D\Gamma(N)$ *is h-torsion free:*

$$T^i_{jh} = H^i_{jh} - H^i_{hj} = 0. \tag{8.7.5}$$

6⁰ $D\Gamma(N)$ *is* v_α*-torsion free:*

$$\underset{(\alpha)}{S}{}^i_{jh} = \underset{(\alpha)}{C}{}^i_{jh} - \underset{(\alpha)}{C}{}^i_{hj} = 0, \ (\alpha = 1, ..., k-1). \tag{8.7.6}$$

7⁰ $D\Gamma(N)$ *is* w_k*-torsion free:*

$$S_i^{jh} = C_i^{jh} - C_i^{hj} = 0. \tag{8.7.7}$$

2) The connection $D\Gamma(N)$ *has the coefficients given by the generalized Christoffel symbols:*

$$H^i_{jh} = \frac{1}{2} g^{is}\left(\frac{\delta g_{sh}}{\delta x^j} + \frac{\delta g_{js}}{\delta x^h} - \frac{\delta g_{jh}}{\delta x^s}\right),$$

$$\underset{(\alpha)}{C}{}^i_{jh} = \frac{1}{2} g^{is}\left(\frac{\delta g_{sh}}{\delta y^{(\alpha)j}} + \frac{\delta g_{js}}{\delta y^{(\alpha)h}} - \frac{\delta g_{jh}}{\delta y^{(\alpha)s}}\right), \ (\alpha = 1, ..., k-1), \tag{8.7.8}$$

$$C_i^{jh} = -\frac{1}{2} g_{is}\left(\dot{\partial}^j g^{sh} + \dot{\partial}^h g^{js} - \dot{\partial}^s g^{jh}\right).$$

3) This connection depends only by the fundamental function H of the space $H^{(k)n}$ *and by the canonical nonlinear connection N.*

The connection $D\Gamma(N)$ with the coefficients (8.7.8) will be denoted by $C\Gamma(N)$ and called *canonical* for the space $H^{(k)n}$.

The canonical N-metrical connection $C\Gamma(N)$ has zero torsions T^i_{jh}, $\underset{(\alpha)}{S}{}^i_{jh}$, S_i^{jh}. Of course, we can determine all N-linear metrical connections of the space $H^{(k)n}$. Therefore we consider the Obata's operator, [115]:

$$\Omega^{ij}_{hm} = \frac{1}{2}\left(\delta^i_h \delta^j_m - g_{hm} g^{ij}\right), \ \Omega^{*ij}_{hm} = \frac{1}{2}\left(\delta^i_h \delta^j_m + g_{hm} g^{ij}\right) \tag{8.7.9}$$

and prove without difficulties:

Theorem 8.7.2 *The set of all N-linear metrical connections $D\overline{\Gamma}(N) = (\overline{H}^i_{jh}, \underset{(\alpha)}{\overline{C}}{}^{\ i}_{jh}, \overline{C}^{jh}_i)$ of the space $H^{(k)n}$ are expressed by*

$$(8.7.10)\qquad \begin{aligned} \overline{H}^i_{jh} &= H^i_{jh} + \Omega^{is}_{rj} X^r_{sh}, \\ \underset{(\alpha)}{\overline{C}}{}^{\ i}_{jh} &= \underset{(\alpha)}{C}{}^{\ i}_{jh} + \Omega^{is}_{rj} \underset{(\alpha)}{Y}{}^{\ r}_{sh}, \quad (\alpha = 1, ..., k-1), \\ \overline{C}^{jh}_i &= C^{jh}_i + \Omega^{js}_{rj} Z^{rh}_s. \end{aligned}$$

where $C\Gamma(N) = (H^i_{jh}, \underset{(\alpha)}{C}{}^{\ i}_{jh}, C^{jh}_i)$ is the canonical N-metrical connection and X^r_{sh}, $Y^{\ r}_{sh}$ and Z^{rh}_s are arbitrary d-tensors.

It is important to remark that a triple $\left(X^i_{jh}, \underset{(\alpha)}{Y}{}^{\ i}_{jh}, Z^{jh}_i\right)$ determines by means of (8.7.10) a transformation of the N-metrical connections $D\Gamma(N) \to D\overline{\Gamma}(N)$.

The set of transformations $\{D\Gamma(N) \to D\overline{\Gamma}(N)\}$ and the composition of these mappings is an Abelian group.

The last theorems lead to the following result:

Theorem 8.7.3 *There exists an unique N-linear connection $D\overline{\Gamma}(N) = = (\overline{H}^i_{jh}, \underset{(\alpha)}{\overline{C}}{}^{\ i}_{jh}, \overline{C}^{jh}_i)$, $(\alpha = 1, ..., k-1)$, metrical with respect to the fundamental tensor g^{ij} of the space $H^{(k)n}$ having as torsion the d -tensor fields $T^i{}_{jh}$ $(= -T^i{}_{hj})$, $\underset{(\alpha)}{S}{}^{\ i}_{jh}$ $(= -\underset{(\alpha)}{S}{}^{\ i}_{hj})$, S^{jh}_i $(= -S^{hj}_i)$, apriori given. The coefficients of $D\overline{\Gamma}(N)$ have the following expressions:*

(8.7.11)

$$\overline{H}^i_{jh} = \frac{1}{2} g^{is} \left(\frac{\delta g_{sh}}{\delta x^j} + \frac{\delta g_{js}}{\delta x^h} - \frac{\delta g_{jh}}{\delta x^s} \right) + \frac{1}{2} g^{is} \left(g_{sm} T^m{}_{jh} - g_{jm} T^m{}_{sh} + g_{hm} T^m{}_{sj} \right),$$

$$\underset{(\alpha)}{\overline{C}}{}^{\ i}_{jh} = \frac{1}{2} g^{is} \left(\frac{\delta g_{sh}}{\delta y^{(\alpha)j}} + \frac{\delta g_{js}}{\delta y^{(\alpha)h}} - \frac{\delta g_{jh}}{\delta y^{(\alpha)s}} \right) + \frac{1}{2} g^{is} (g_{sm} \underset{(\alpha)}{S}{}^{\ m}_{jh} - g_{jm} \underset{(\alpha)}{S}{}^{\ m}_{sh} +$$

$$+ g_{hm} \underset{(\alpha)}{S}{}^{\ m}_{sj}),$$

$$\overline{C}^{jh}_i = \frac{1}{2} g_{is} \left(\dot{\partial}^j g^{sh} + \dot{\partial}^h g^{js} - \dot{\partial}^s g^{jh} \right) - \frac{1}{2} g_{is} \left(g^{sm} S^{\ jh}_m - g^{jm} S^{\ sh}_m + g^{hm} S^{\ js}_m \right).$$

Now, consider the canonical N-metrical connection $C\Gamma(N)$ only. Then, we can write the Ricci identities, from Theorem 6.6.1, where $T^i{}_{jh} = 0$, $\underset{(\alpha)}{S}{}^{\ i}_{jh} = 0$, $(\alpha = 1, ..., k-1)$, $S^{jh}_i = 0$. Since the metricity conditions (6.6.7) are verified it follows:

Proposition 8.7.1 *The curvature d-tensor fields of $C\Gamma(N)$ satisfy the identities (6.6.8).*

Proposition 8.7.2 *The tensors of deflection of the canonical N-metrical connection $C\Gamma(N)$ satisfy the identities (6.6.9), with $T^i{}_{jh} = 0$, $\underset{(\alpha)}{S}{}^{\,i}_{\,jh} = 0$, $(\alpha = 1, ..., k-1)$, $S_i{}^{jh} = 0$.*

Let $\omega^i{}_j$ be the 1-forms connection of $C\Gamma(N)$

$$\omega^i{}_j = H^i_{js}dx^s + \underset{(1)}{C}{}^{\,i}_{\,js}\delta y^{(1)s} + \cdots + \underset{(k-1)}{C}{}^{\,i}_{\,js}\delta y^{(k-1)s} + C^{js}_i \delta p_s. \tag{8.7.12}$$

Theorem 6.7.1 is valid for $C\Gamma(N)$.
In particular, Theorem 6.7.2 for $C\Gamma(N)$ holds, too. Namely:

Theorem 8.7.4 *The manifold $T^{*k}M$ is with absolute parallelism of vectors with respect to the canonical N-metrical connection $C\Gamma(N)$ if and only if the curvature d-tensors of $C\Gamma(N)$ vanishes.*

Also, we have:

Theorem 8.7.5 *A smooth parametrized curve γ, given by $x^i = x^i(t)$, $y^{(\alpha)i} = y^{(\alpha)i}(t)$, $(\alpha = 1, ..., k-1)$, $p_i = p_i(t)$, $t \in I$ is an autoparallel curve with respect to $C\Gamma(N)$ if and only if the equations (6.7.13), are verified, $\omega^i{}_j$ being the 1-forms connection (8.7.12).*

The horizontal curves of the space $H^{(k)n}$ are characterized by the equations (6.7.14).

The horizontal paths of $H^{(k)n}$ are expressed by the equations (6.7.15) for the connection $C\Gamma(N)$.

The v_α-vertical paths of $H^{(k)n}$ are characterized by (6.7.16) and w_k-vertical paths are given by (6.7.17).

The canonical N-metrical connection $C\Gamma(N)$ of the Hamilton space of order k, $H^{(k)n} = (M, H)$ is h-, v_α- and w_k-torsion free if the d-tensors of torsion vanish:

$$\begin{aligned} &T^i_{jh} = H^i_{jh} - H^i_{hj} = 0, \ \underset{(\alpha)}{S}{}^{\,i}_{\,jh} = \underset{(\alpha)}{C}{}^{\,i}_{\,jh} - \underset{(\alpha)}{C}{}^{\,i}_{\,hj} = 0, \ (\alpha = 1, ..., k-1), \\ &S^{jh}_i = C^{jh}_i - C^{hj}_i. \end{aligned} \tag{8.7.13}$$

Therefore, we have:

Theorem 8.7.6 *The canonical N-metrical connection $C\Gamma(N)$, with the coefficients (8.7.8), of the Hamilton space of order k, $H^{(k)n}$, has the structure equations given by the equations (7.8.1), (7.8.2), (7.8.3) and (7.8.4) from Theorem 7.8.1, in the conditions (8.7.13) and with 1-forms $\omega^i{}_j$ from (8.7.12).*

The Bianchi identities of $C\Gamma(N)$ can be obtained from (7.6.11), (7.6.12), (7.6.13) by means of the conditions (8.7.13).

8.8 The Hamilton Space $H^{(k)n}$ of Electrodynamics

The classical Lagrangian of electrodynamics has been considered in section 5 of chapter 2:

$$\overset{\bullet}{L}(x, y^{(1)}) = mc\gamma_{ij}(x)y^{(1)i}y^{(1)j} + \frac{2e}{m}b_i(x)y^{(1)i}$$

in which m, c, e are known physical constants, $\gamma_{ij}(x)$ are gravitational potentials and $b_i(x)$ are electromagnetic potentials. This can be extended to the manifold of accelerations of order k, T^kM as follows:

$$L(x, y^{(1)}, ..., y^{(k)}) = mc\gamma_{ij}(x)z^{(k)i}z^{(k)j} + \frac{2e}{m}b_i(x)z^{(k)i}, \tag{8.8.1}$$

$z^{(k)i}$ being the Liouville d-vector field

$$kz^{(k)i} = ky^{(k)i} + (k-1)\underset{(1)}{M}{}^i_j y^{(k-1)j} + \cdots + \underset{(k-1)}{M}{}^i_j y^{(1)j}. \tag{8.8.2}$$

The dual coefficients $\underset{(1)}{M}{}^i_j$, ..., $\underset{(k-1)}{M}{}^i_j$ are given by the prolongation to T^kM of the Riemannian structure $\mathcal{R}^n = (M, \gamma_{ij}(x))$.

Consequently, $\underset{(1)}{M}{}^i_j$ depends on the variables $(x, y^{(1)})$; $\underset{(2)}{M}{}^i_j$ depends on the variables $(x, y^{(1)}, y^{(2)})$; ...; $\underset{(k-1)}{M}{}^i_j$ depends on the variables $(x, y^{(1)}, ..., y^{(k-1)})$. This is an important property, since $\left(\underset{(1)}{M}{}^i_j, ..., \underset{(k-1)}{M}{}^i_j\right)$ can be considered as *the dual coefficients of a nonlinear connection* $\overset{\circ}{N}$ on the manifold $T^{*k}M$, determined only by the tensor field $\gamma_{ij}(x)$.

It follows that:

1) The Lagrangian L, (8.8.1), is well determined only by $\gamma_{ij}(x)$ and $b_i(x)$;
2) L is differentiable on T^kM;
3) L is regular:

$$g_{ij} = \frac{1}{2}\frac{\partial^2 L}{\partial y^{(k)i}\partial y^{(k)j}} = mc\gamma_{ij}(x); \tag{8.8.3}$$

4) The pair $L^{(k)n} = (M, L)$ is a Lagrange space of order k.

It is called *the Lagrange space of order k of electrodynamics.* The geometry of the space $L^{(k)n}$ gives us a good geometrical model for the Analytical Mechanics based on the Lagrangian of electrodynamics. So, the integral of action $I(c) = \int_0^1 L(x, \frac{dx}{dt}, \cdots, \frac{1}{k!}\frac{d^kx}{dt^k})dt$ leads to the Euler-Lagrange equations $\overset{\circ}{E_i}(L) = 0$ (2.2.1). The energy of higher order can be determined. The energy of order k, $\mathcal{E}^k_c(L)$, satisfies the law of conservation. A Nöther theorem, holds.

In the same time the canonical k-semispray S is given by

$$S = y^{(1)}\frac{\partial}{\partial x^i} + 2y^{(2)i}\frac{\partial}{\partial y^{(1)i}} + \cdots + ky^{(k)i}\frac{\partial}{\partial y^{(k-1)i}} - (k+1)G^i\frac{\partial}{\partial y^{(k)i}} \tag{8.8.4}$$

whose coefficients G^i, calculated by means of (8.8.1) have the expressions

$$(k+1)G^i = \frac{1}{2mc}\gamma^{ij}\left\{\Gamma\left(\frac{\partial}{\partial y^{(k)j}}\right) - \frac{\partial}{\partial y^{(k-1)j}}\right\} \tag{8.8.4'}$$

with

$$\Gamma = y^{(1)}\frac{\partial}{\partial x^i} + 2y^{(2)i}\frac{\partial}{\partial y^{(1)i}} + \cdots + ky^{(k)i}\frac{\partial}{\partial y^{(k-1)i}} \tag{8.8.4''}$$

and

$$\frac{\partial L}{\partial y^{(k)j}} = 2mc\gamma_{ih}(x)z^{(k)h} + \frac{2e}{m}b_j(x). \tag{8.8.4'''}$$

The Legendre transformation $\xi : T^kM \to T^{*k}M$ determined by the Lagrange space of electrodynamics $L^{(k)n}$, is defined by (8.4.2), (8.4.2'). This is:

$$p_i = \frac{1}{2}\frac{\partial L}{\partial y^{(k)j}} = mc\gamma_{ih}(x)z^{(k)h} + \frac{e}{m}b_i(x). \tag{8.8.5}$$

The mapping φ is a local diffeomorphism. Its inverse $\varphi^{-1} = \xi$ is given by (8.4.3), where

(8.8.5')

$$\begin{aligned}\xi^i(x, y^{(1)}, ..., y^{(k-1)}, p) &= \frac{1}{mc}\gamma^{ih}(x)\left\{p_h - \frac{e}{m}b_h(x)\right\} - \\ &- \frac{1}{k}\left\{(k-1)\underset{(1)}{M}{}^i_h y^{(k-1)h} + \cdots + \underset{(k-1)}{M}{}^i_h y^{(1)h}\right\}.\end{aligned}$$

The canonical k-semispray S, (8.8.4) is transformed by Legendre mapping φ in the dual k-semispray S^*_ξ, from (8.4.8), with the coefficients η_i from (8.4.9).

Theorem 8.4.2 allows to determine, locally, the canonical nonlinear connection N^* on the manifold $T^{*k}M$.

As we remarked above the connection N of the prolongation of the Riemannian structure $(M, \gamma_{ij}(x))$, with the dual coefficients $\underset{(1)}{M}{}^i_j(x, y^{(1)})$, ..., $\underset{(k-1)}{M}{}^i_j(x, y^{(1)}, ..., y^{(k-1)})$ defines a nonlinear connection on the submanifold $T^{k-1}M$ in T^kM.

Thus, the Liouville vector field $z^{(k)i}$ from (8.8.2) is transformed by the Legendre mapping φ in the vector field $\overset{\vee}{z}{}^{(k)i}$:

$$k\overset{\vee}{z}{}^{(k)i} = k\xi^i + (k-1)\underset{(1)}{M}{}^i_m y^{(k-1)m} + \cdots + \underset{(k-1)}{M}{}^i_m y^{(1)m} \tag{8.8.6}$$

Let us consider the following Hamiltonian H, the dual of L:

(8.8.7)

$$H(x, y^{(1)}, ..., y^{(k-1)}, p) = 2p_i \overset{\vee}{z}{}^{(k)i} - L(x, y^{(1)}, ..., y^{(k-1)}, \xi(x, y^{(1)}, ..., y^{(k-1)}, p)).$$

We can state:

Theorem 8.8.1 *We have:*

1^0 The dual of the Lagrange space of electrodynamics $L^{(k)n} = (M, L)$, (8.8.1), is a Hamilton space $H^{(k)n} = (M, H)$, (8.8.7).

2^0 The fundamental function H of the space $H^{(k)n}$ is given by

$$H = \frac{1}{mc}\left\{\gamma^{ij}(x)p_i p_j - \frac{2e}{m}\gamma^{ij}p_i b_j + \frac{e}{m^2}\gamma^{ij}b_i b_j\right\}. \tag{8.8.8}$$

3^0 H depends only on the point x and the momenta p.

4^0 H is the fundamental function of a Hamilton space of order 1, $H^n = (M, H)$.

Proof: 1^0, 2^0. Applying theorem 7.4.3 and the formulas (8.8.5), (8.8.6) we obtain:

$$\overset{\vee}{z}{}^{(k)i} = \frac{1}{mc}\gamma^{ij}(x)\left[p_i - \frac{e}{m}b_i(x)\right]$$

and

$$L(x, y^{(1)}, ..., y^{(k-1)}, \xi) = mc\gamma_{ij}\frac{1}{mc}\gamma^{ij}\left[p_i - \frac{e}{m}b_i\right]\left[p_j - \frac{e}{m}b_j\right] +$$
$$+\frac{2e}{m^2 c}\gamma^{ij}b_j\left[p_i - \frac{e}{m}b_i\right].$$

Substituting in (8.8.7) we have the equality (8.8.8).

3^0 Looking at the function H one remarks that H does not depend on the variables $y^{(1)i}$, ..., $y^{(k-1)i}$.

4^0 H is a Hamilton function on the cotangent bundle T^*M and its Hessian, with respect to the momenta p_i, has the matrix $\left\|\frac{1}{mc}\gamma^{ij}(x)\right\|$. So, the pair $H^n = (M, H)$ is a Hamilton space. Q.E.D.

Remarks. 1^0 The Hamilton space $H^n = (M, H(x, p))$ is the dual of the Lagrange space of electrodynamics $L^n = (M, \overset{\circ}{L}(x, y^{(1)}))$.

2^0 $H^{(k)n}$ does not depend on the nonlinear connection N of the prolongation of the Riemann space $\mathcal{R}^n = (M, \gamma_{ij}(x))$.

For the Hamilton space $H^n = (M, H)$ with the fundamental function $H(x, p)$ from (8.8.8) we have a canonical nonlinear connection N, with the coefficients

$$N_{ij} = \gamma^h_{ij}(x)p_h + \frac{e}{c}\left(b_{i|j} + b_{j|i}\right), \tag{8.8.9}$$

where $b_{i|j} = \dfrac{\partial b_i}{\partial x^j} - b_s\gamma^s_{ij}$, $\gamma^i_{jh}(x)$ being the Christoffel symbols of the metric $\dfrac{1}{mc}\gamma_{ij}(x)$.

The canonical N-metrical connection $C\Gamma(N) = (H^i_{jh}, C_i^{jh})$ has the following coefficients

$$H^i_{jh} = \gamma^i_{jh}, \quad C_i^{jh} = 0. \tag{8.8.10}$$

The geometrical object fields H, $g_{ij} = \frac{1}{mc}\gamma_{ij}$, N_{ij} and H^i_{jh} allow to develop the geometry of the space $H^n = (M, H)$.

8.9 The Riemannian Almost Contact Structure Determined by the Hamilton Space $H^{(k)n}$

Consider a Hamilton space of order k, $H^{(k)n} = (M, H)$ having g^{ij} as fundamental tensor field. Let N be a nonlinear connection canonical associated to $H^{(k)n}$, according to Theorem 8.6.1. As usually, we consider the adapted basis $\left(\frac{\delta}{\delta x^i}, \frac{\delta}{\delta y^{(1)i}}, \cdots, \frac{\delta}{\delta y^{(k-1)i}}, \dot{\partial}^i\right)$ and its dual basis $(\delta x^i, \delta y^{(1)i}, ..., \delta y^{(k-1)i}, \delta p_i)$. Then, at every point $u^* = (x, y^{(1)}, ..., y^{(k-1)}, p) \in T^{*k}M$ we can define the tensor

(8.9.1)

$$\overset{\vee}{\mathbb{G}} = g_{ij} dx^i \otimes dx^j + g_{ij}\delta y^{(1)i} \otimes \delta y^{(1)j} + \cdots + g_{ij}\delta y^{(k-1)i} \otimes \delta y^{(k-1)j} + g^{ij}\delta p_i \otimes \delta p_j$$

$\overset{\vee}{\mathbb{G}}$ is the N-lift of the fundamental tensor g^{ij} of the Hamilton space of order k, $H^{(k)n} = (M, H)$, (cf. §6.7, ch. 6).

Theorem 8.9.1 *1^0 $\overset{\vee}{\mathbb{G}}$ is a (pseudo)-Riemannian structure on the manifold $\widetilde{T^{*k}M}$, determined only by the space $H^{(k)n}$ and the nonlinear connection N.*

2^0 The distributions N_0, N_1, ..., N_{k-2}, V_{k-1}, W_k are mutual orthogonal with respect to $\overset{\vee}{\mathbb{G}}$.

Indeed: 1^0. Every term of $\overset{\vee}{\mathbb{G}}$ is defined on $\widetilde{T^{*k}M}$ because g_{ij} is a d-tensor field and δx^i , ..., δp_i have geometrical meaning, i.e. they are transformed as in (6.3.4), $\det \| \overset{\vee}{\mathbb{G}} \| \neq 0$. 2^0. This property is obvious, taking into account the form of $\overset{\vee}{\mathbb{G}}$. Q.E.D.

The tensor $\overset{\vee}{\mathbb{G}}$ is of the form:

$$\begin{aligned} &\overset{\vee}{\mathbb{G}} = \mathbb{G}^H + \mathbb{G}^{V_1} + \cdots + \mathbb{G}^{V_{k-1}} + \mathbb{G}^{W_k}, \\ &\mathbb{G}^H = g_{ij} dx^i \otimes dx^j, \mathbb{G}^{V_\alpha} = g_{ij}\delta y^{(\alpha)i} \otimes \delta y^{(\alpha)j}, (\alpha = 1, ..., k-1), \\ &\mathbb{G}^{W_k} = g^{ij}\delta p_i \otimes \delta p_j. \end{aligned} \tag{8.9.2}$$

Here $\mathbb{G}^H$ is the restriction of $\overset{\vee}{\mathbb{G}}$ to the distribution $N_0 = N^*$, $\mathbb{G}^{V_\alpha}$ is the restriction of $\overset{\vee}{\mathbb{G}}$ to the distribution N_α , $\alpha = 1, ..., k-1$ $(N_{k-1} = V_{k-1})$ and

$\mathbb{G}^{W_k}$ is the restriction of $\overset{\vee}{\mathbb{G}}$ to the distribution W_k. Moreover, $\mathbb{G}^H$, $\mathbb{G}^{V_\alpha}$, $\mathbb{G}^{W_k}$ are d-tensor fields.

Theorem 8.9.2 *The tensor fields $\mathbb{G}^H$, $\mathbb{G}^{V_\alpha}$, $(\alpha = 1, ..., k-1)$, $\mathbb{G}^{W_k}$ are covariant constant with respect to any metrical N -connection $D\Gamma(N)$.*

Indeed, using the considerations of this section, it follows $D_X\mathbb{G}^H = 0$, $D_X\mathbb{G}^{V_\alpha} = 0$, $D_X\mathbb{G}^{W_k} = 0$ and, consequently $D_X \overset{\vee}{\mathbb{G}} = 0$. Q.E.D.

The geometry of the (pseudo)-Riemannian space $\left(\widetilde{T^{*k}M}, \overset{\vee}{\mathbb{G}}\right)$ can be studied by means of a metrical N-linear connection, presented in the section 7 of this chapter.

As we know (§6, Ch.6) on $\widetilde{T^{*k}M}$, endowed with the nonlinear connection N, there exists a natural almost $(k-1)n$-contact structure $\mathbb{F}$, defined by (6.6.3), having the properties from in Theorem 6.5.2. The condition of normality of $\mathbb{F}$ is (6.6.5). Exactly as in Theorem 6.7.2 we can deduce:

Theorem 8.9.3 *The pair $\left(\overset{\vee}{\mathbb{G}}, \mathbb{F}\right)$ is a Riemannian $(k-1)n$-almost contact structure determined by the space $H^{(k)n}$ and the nonlinear connection N.*

The considerations from section 6.8, ch. 8 show that the nonlinear connection N and the fundamental tensor g^{ij} of the Hamilton space of order k, $H^{(k)n}$, determine on the manifold $\widetilde{T^{*k}M}$, two important geometric object fields. One is the N-lift $\overset{\vee}{\mathbb{G}}$ from (8.9.1) and another one is $\overset{\vee}{\mathbb{F}}$ the almost $(k-1)n$-contact structure, given by (6.8.5).

Namely,

$$\overset{\vee}{\mathbb{F}}\left(\frac{\delta}{\delta x^i}\right) = -g_{ij}\overset{\cdot j}{\partial}, \ \overset{\vee}{\mathbb{F}}\left(\frac{\delta}{\delta y^{(\alpha)i}}\right) = 0, \ (\alpha = 1, ..., k-1), \ \overset{\vee}{\mathbb{F}}\left(\overset{\cdot i}{\partial}\right) = g^{ij}\frac{\delta}{\delta x^j}$$

By means of Theorem 7.9.1 it follows that $\overset{\vee}{\mathbb{F}}$ is a tensor field of type $(1,1)$ on the manifold $T^{*k}M$ given in (6.8.3), *rank* $\overset{\vee}{\mathbb{F}} = 2n$ and $\overset{\vee}{\mathbb{F}}^3 + \overset{\vee}{\mathbb{F}} = 0$.

The following version of the Theorem 6.8.2 holds:

Theorem 8.9.4 *For any Hamilton space of order k, $H^{(k)n} = (M, H)$, endowed with a canonical nonlinear connection N the following properties hold:*

1^0 The pair $\left(\overset{\vee}{\mathbb{G}}, \overset{\vee}{\mathbb{F}}\right)$ is a Riemannian almost $(k-1)n$-contact structure.

2^0 The associated 2-form is

$$\theta = \delta p_i \wedge dx^i.$$

3^0 If the coefficients N_{ij} of N are symmetric then

$$\theta = dp_i \wedge dx^i$$

and θ is the canonical presymplectic structure on $\widetilde{T^{*k}M}$.

4^0 The condition of normality of structure $\overset{\vee}{\mathbb{F}}$ is

$$\mathcal{N}_{\overset{\vee}{\mathbb{F}}}(X,Y) + \sum_{i=1}^{n}\left[\sum_{\alpha=1}^{k-1} d(\delta y^{(\alpha)i})(X,Y) + d\delta p_i(X,Y)\right] = 0,\ \forall X,Y \in \mathcal{X}(\widetilde{T^{*k}M}),$$

where $\mathcal{N}_{\overset{\vee}{\mathbb{F}}}$ is the Nijenhuis tensor of $\overset{\vee}{\mathbb{F}}$.

Now, taking the local expression of the tensor field $\overset{\vee}{\mathbb{F}}$:

$$\overset{\vee}{\mathbb{F}} = -g_{ij}\overset{\cdot j}{\partial} \otimes dx^i + g^{ij}\frac{\delta}{\delta x^i} \otimes \delta p_j$$

and applying the theory of h-, v_α- and w_k-covariant derivative from §7.5, ch. 7, it follows:

Theorem 8.9.5 *With respect to a metrical N-linear connection $D\Gamma(N)$, the Riemannian almost $(k-1)n$-contact structure $\left(\overset{\vee}{\mathbb{G}}, \overset{\vee}{\mathbb{F}}\right)$ is covariant constant, i.e.*

$$D_X \overset{\vee}{\mathbb{G}} = 0,\ D_X \overset{\vee}{\mathbb{F}} = 0.$$

It follows that the geometry of the Riemannian almost $(k-1)n$-contact space $\left(\widetilde{T^{*k}M}, \overset{\vee}{\mathbb{G}}, \overset{\vee}{\mathbb{F}}\right)$ can be studied by means of the metrical N-linear connection $D\Gamma(N)$. This space is called *the Riemannian almost $(k-1)n$ -contact model* of the Hamilton space of order k, $H^{(k)n}$.

Chapter 9

Subspaces in Hamilton Spaces of Order k

In this chapter we shall study the geometry of subspaces in a Hamilton space $H^{(k)n} = (M, H)$. A submanifold $\check{M}$ of the manifold M determines a dual manifold $T^{*k}\check{M}$. But the immersion $i\ \check{M} \to M$ does not automatically imply an immersion of $T^{*k}\check{M}$, into the dual manifold $T^{*k}M$, like in the case of manifold $T^k\check{M}$ into the total space of k tangent bundle T^kM, [94], since $T^{*k}\check{M} = T^{k-1}\check{M} \times_{\check{M}} T^*\check{M}$.

Therefore, by means of the immersion i and by an immersion of the cotangent manifold $T^*M\,\check{M}$ into T^*M we can define $T^{*k}\,\check{M}$ as an immersed submanifold in $T^{*k}M$. Thus, the main geometrical objects fields on $T^{*k}M$ induce the coresponding geometrical object fields on submanifold $T^{*k}\check{M}$. The Hamilton space $H^{(k)n} = (M, H)$ induces a Hamilton subspaces $\check{H}^{(k)m} = (\check{M}, \check{H})$. So we study the intrinsic geometrical object fields on $\check{H}^{(k)m}$ and the induced geometrical object fields, as well as the relations between them. These problems are approached by means of the method of moving frame, used in the case of subspaces $\check{L}^{(k)m} = (\check{M}, \check{L})$ in the Lagrange spaces of order k.

9.1 Submanifolds $T^{*k}\check{M}$ in the Manifold $T^{*k}M$

Let M be a $C^\infty-$ real, n-dimensional manifold and $\check{M}$ be a C^∞-real, m - dimensional manifold, $1 < m < n$, immersed in M through the immersion $i : \check{M} \to M$. Locally i can be given in the form

$$x^i = x^i(u^1, ..., u^m),\ rank\|\frac{\partial x^i}{\partial u^\alpha}\| = m, \forall (u^\alpha) \in \check{U} \subset \check{M} \tag{9.1.1}$$

The indices i, j, h, r, s, p, q run over the set $\{1, 2, ..., n\}$; the indices $\alpha, \beta, \gamma, ...$ run over the set $\{1, 2...m\}$ and $\overline{\alpha}, \overline{\beta}, \overline{\gamma}, ...$run over the set $\{1, ..n - m\}$.

If $i : \check{M} \to M$ is an embeding, then we identify $\check{M}$ with $i(\check{M}) \subset M$ and say that $\check{M}$ is a submanifold of M. In this case (9.1.1) are called the parametric equations of the submanifold $\check{M}$ in the manifold M.

Let us consider the manifold $T^{*k}\check{M}$ determined by $\check{M}$, i.e. $T^{*k}\check{M} = T^{k-1}\check{M} \times_{\check{M}} T^*\check{M}$. A point $\check{u} \in T^{*k}\check{M}$ will be denoted by $\check{u} = (u, v^{(1)}, ...v^{(k-1)}, \check{p})$ and its coordinates by $(u^\alpha, v^{(1)\alpha}, ..., v^{(k-1)\alpha}, \check{p}_\alpha)$. A change of local coordinates on the manifold $T^k\check{M}$ is given by

$$(9.1.2)\quad \begin{cases} \overline{u}^\alpha = \overline{u}^\alpha(u^1, ..., u^m), rank \left\| \dfrac{\partial \overline{u}^\alpha}{\partial u^\beta} \right\| = m, \\ \overline{v}^{(1)\alpha} = \dfrac{\partial \overline{u}^\alpha}{\partial u^\beta} v^{(1)\beta}, \\ \dots\dots\dots\dots\dots\dots\dots\dots, \\ (k-1)\overline{v}^{(k-1)\alpha} = \dfrac{\partial \overline{v}^{(k-2)\alpha}}{\partial u^\beta} v^{(1)\beta} + ... + (k-1)\dfrac{\partial \overline{v}^{(k-2)\alpha}}{\partial v^{(k-2)\beta}} v^{(k-1)\beta}, \\ \check{\overline{p}}_\alpha = \dfrac{\partial u^\beta}{\partial \overline{u}^\alpha} \check{p}_\beta, \end{cases}$$

with

$$\frac{\partial \overline{u}^\alpha}{\partial u^\beta} = \frac{\partial \overline{v}^{(1)\alpha}}{\partial v^{(1)\beta}} = ... = \frac{\partial \overline{v}^{(k-1)\alpha}}{\partial v^{(k-1)\beta}}.$$

Remarking that $T^{*k} : Man \to Man$ is a covariant functor from the category of differentiable manifolds Man to itself, it follows that the immersion $i : \check{M} \to M$ uniquelly determines the mapping $T^{*k}i : T^{*k}\check{M} \to T^{*k}M$ (see Ch.4) analytically given by the equations

$$(9.1.3)\quad \begin{cases} x^i = x^i(u^1, ..., u^m), rank \left\| \dfrac{\partial x^i}{\partial u^\alpha} \right\| = m, \\ y^{(1)i} = \dfrac{\partial x^i}{\partial u^\alpha} v^{(1)\alpha}, \\ \dots\dots\dots\dots\dots\dots\dots\dots \\ (k-1)\, y^{(k-1)i} = \dfrac{\partial y^{(k-2)i}}{\partial u^\alpha} v^{(1)\alpha} + 2\dfrac{\partial y^{(k-2)i}}{\partial v^{(1)\alpha}} v^{(2)\alpha} + ... + \\ \qquad + (k-1)\dfrac{\partial y^{(k-2)i}}{\partial v^{(k-2)\alpha}} v^{(k-1)\alpha}, \end{cases}$$

and

$$(9.1.3')\quad \frac{\partial x^i}{\partial u^\alpha} p_i = \check{p}_\alpha,$$

where

$$\text{(9.1.3")}\qquad \begin{aligned} &\frac{\partial x^i}{\partial u^\alpha}=\frac{\partial y^{(1)i}}{\partial v^{(1)\alpha}}=...=\frac{\partial y^{(k-1)i}}{\partial v^{(k-1)\alpha}},\\ &\frac{\partial y^{(a)i}}{\partial u^\alpha}=\frac{\partial y^{(a+1)i}}{\partial v^{(1)\alpha}}=...=\frac{\partial y^{(k-1)i}}{\partial v^{(k-1-a)\alpha}},(a=1,...,k-2).\end{aligned}$$

We shall denote

$$\text{(9.1.4)}\qquad B^i_\alpha(u)=\frac{\partial x^i}{\partial u^\alpha},B^i_{\alpha\beta}(u)=\frac{\partial^2 x^i}{\partial u^\alpha\partial u^\beta},...$$

In order to obtain an immersion i^* of $T^{*k}\check{M}=T^{k-1}\check{M}\times_{\check{M}}T^*\check{M}$ in $T^{*k}M=T^{k-1}M\times_M T^*M$, it is necessary that i^* to be of the form $i^*=T^{k-1}i\times f$, with $f:T^*\check{M}\to T^*M$. Analytically $T^{k-1}i$ is defined by the equations (9.1.3). Consequently, taking into account (9.1.3'), it follows that the mapping f can be given in the form

$$\text{(9.1.5)}\qquad p_i=A^\alpha_i(u)\,\check{p}_\alpha, rank\|A^\alpha_i(u)\|=m,\forall u=(u^1,...,u^m)\in\check{U}\subset\check{M}$$

We obtain

Theorem 9.1.1 *The equations (9.1.3), (9.1.5) define a local immersion*
$i^*:T^{*k}\check{M}\to T^{*k}M,$
$i^*:\check{u}=(u,v^{(1)},...,v^{(k-1)},\check{p})\to u^*=(x,y^{(1)},...,y^{(k-1)},p).$

Proof. The Jacobian matrix of the mapping i^* given by the equations (9.1.3), (9.1.5) at a point $\check{u}\in\check{U}$ is as follows:

$$\text{(9.1.6)}\qquad J(i^*)_{\check{u}}=\left\|\begin{matrix}\frac{\partial x^i}{\partial u^\alpha} & 0 & 0 & \cdots & 0 & 0\\ \frac{\partial y^{(1)j}}{\partial u^\alpha} & \frac{\partial x^i}{\partial u^\alpha} & 0 & \cdots & 0 & 0\\ \cdots & \cdots & \cdots & \cdots & \cdots & \cdots\\ \frac{\partial y^{(k-1)i}}{\partial u^\alpha} & \frac{\partial y^{(k-1)i}}{\partial v^{(1)\alpha}} & \frac{\partial y^{(k-1)i}}{\partial v^{(2)\alpha}} & \cdots & \frac{\partial x^i}{\partial u^\alpha} & 0\\ \frac{\partial A^\beta_i}{\partial u^\alpha}\check{p}_\beta & 0 & 0 & \cdots & 0 & A^\alpha_i\end{matrix}\right\|_{\check{u}}$$

Since the matrices $\left\|\frac{\partial x^i}{\partial u^\alpha}\right\|$ and $\|A^\alpha_i(u)\|$ have the rank m it follows that i^* defines a local immersion. Q.E.D.

The functions $A^\alpha_i(u)$ from (9.1.5) are not arbitrary. They must satisfy some conditions.

Proposition 9.1.1 *The following properties hold:*
1°

$$B^i_\alpha(u)A^\beta_i(u) = \delta^\beta_\alpha, \tag{9.1.7}$$

2° $A^\alpha_i(u)$ is a d-covector field with respect to index i,
$A^\alpha_i(u)$ is a d-vector field with respect to index α.

Indeed, (9.1.7) follows from (9.1.3') and (9.1.5'), 2° p_i being a d -covector it follows that $A^\alpha_i(u)$ has the same quality with respect to index i. The same remark is valid for the index α of $A^\alpha_i(u)$.

Of course $B^i_\alpha(u)$ is a d-vector with respect to index i and it is a covector with respect to index α.

The previous properties allow to call $B^i_\alpha(u)$ and $A^\alpha_i(u)$ the *mixed* d-tensor fields. Along $\check{M}$ a mixed d -tensor will be given by the components $T^{\alpha_1...\alpha_p i_1...i_r}_{\beta_1...\beta_q j_1...j_s}$ having the property that it a d-tensor of type (p,q) with respect

to indices $\alpha_1...\alpha_p$, and $\beta_1...\beta_q$ (to a change of coordinates on the submanifold $T^{*k}\check{M}$ and it is a d-tensor of type (r,s) with respect to indices $i_1...i_r$ and $j_1...j_s$, to a change of coordinates on the enveloping manifold $T^{*k}M$.

Remark. The notion of mixed d-tensor will be extended in the section 3.

As usual we set

$$\dot{\partial}^i = \frac{\partial}{\partial p_i}, \dot{\partial}^\alpha = \frac{\partial}{\partial \check{p}_\alpha}.$$

The differential di^* of the immersion i^* acts on the natural basis $(\frac{\partial}{\partial u^\alpha}, \frac{\partial}{\partial v^{(1)\alpha}}, ..., \frac{\partial}{\partial v^{(k-1)\alpha}}, \dot{\partial}^\alpha)$ by the rule:

$$di^* \left\| \frac{\partial}{\partial u^\alpha}, \frac{\partial}{\partial v^{(1)\alpha}}, ..., \frac{\partial}{\partial v^{(k-1)\alpha}}, \dot{\partial}^\alpha \right\| = \left\| \frac{\partial}{\partial x^i}, ..., \frac{\partial}{\partial y^{(k-1)\alpha}}, \dot{\partial}^i \right\| J(i^*) \tag{9.1.8}$$

and for the natural cobasis:

$$di^* \left\| dx^i, ..., dy^{(k-1)i}, dp_i \right\|^T = J(i^*) \left\| dx^\alpha, ..., dv^{(k-1)\alpha}, d\check{p}_\alpha \right\| \tag{9.1.8'}$$

Therefore, it is not difficult to see that:
1°. We have

$$\dot{\partial}^\alpha = A^\alpha_i \, \dot{\partial}^i ; dp_i = dA^\alpha_i \, \check{p}_\alpha + A^\alpha_i d\, \check{p}_\alpha \tag{9.1.9}$$

2° Along submanifold $T^{*k}\check{M}$ the vertical distributions $v_1, ..., v_{k-1}, w_k$ are subdistributions of the vertical distributions $V_1, ..., V_{k-1}, W_k$ from the manifold T^kM.

3° We have for the Liouville vector fields $\overset{1}{\gamma}, ..., \overset{k-1}{\gamma}$ from $T^{*k}M$ the relations

(9.1.10) $$di^{*}(\overset{1}{\gamma}) = \overset{1}{\Gamma}, ..., di^{*}(\overset{k-1}{\gamma}) = \overset{k-1}{\Gamma}$$

These equations will be applied in the theory of Hamilton subspaces.

9.2 Hamilton Subspaces $\check{H}^{(k)m}$ in $H^{(k)n}$. Darboux Frames

Let $H^{(k)n} = (M, H)$ be a Hamilton spaces of order k, its fundamental function H being defined on the manifold $T^{*k}M$ and an immersion $i^{*} : T^{*k}\check{M} \to T^{*k}M$ given locally by the equations (9.1.3), (9.1.5). For a point $u^{*} = (x, y^{(1)}, ..., y^{(k-1)}, p) \in T^{*k}M$ and a point $\check{u} = (u, v^{(1)}, ..., v^{(k-1)}, \check{p}) \in T^{*k}\check{M}$ with the property $u^{*} = i^{*}(\check{u})$ we obtain the restriction of the fundamental function H expressed by

(9.2.1) $$H \circ i^{*}(\check{u}) = \check{H}(\check{u}), \forall \check{u} \in T^{*k}\check{M}.$$

It follows that $\check{H}(\check{u})$ is a differentiable Hamiltonian on the submanifold $T^{*k}\check{M}$.

Consider the d-tensor field

(9.2.2) $$\check{g}^{\alpha\beta} = \frac{1}{2}\dot{\partial}^{\alpha}\dot{\partial}^{\beta}\check{H} \text{ on } \widetilde{T^{*k}\check{M}}.$$

¿From (9.2.1) we deduce that at the points $\check{u} \in \widetilde{T^{*k}\check{M}}$ the fundamental tensor field $\check{g}^{\alpha\beta}$ is given by

(9.2.3) $$\check{g}^{\alpha\beta}(\check{u}) = A_{i}^{\alpha}(u)A_{j}^{\beta}(u)g^{ij}(i^{*}\check{u}).$$

It follows from previous equality that

(9.2.3') $$rang\left\|\check{g}^{\alpha\beta}\right\| = m.$$

Therefore we have:

Theorem 9.2.1 *The pair* $\check{H}^{(k)m} = (\check{M}, \check{H})$ *is a Hamilton space of order k.*

Indeed, this property is a consequences of the equations (9.2.1), (9.2.2) and (9.2.3) and the fact that $A_{i}^{\alpha}(u^{1}, ..., u^{m})$ is a mixed d-tensor, with rank $\|A_{i}^{\alpha}\| = m$.

The space $\check{H}^{(k)m}$ will be called the Hamilton subspace of the Hamilton space $H^{(k)n} = (M, H)$.

In the point $\check{u}\in\check{U}\subset T^{*k}\check{M}$, the d-covector fields A_i^{α}, $(\alpha = 1,...,m)$ are defined by $A_i^{\alpha}(\check{u}) = A_i^{\alpha}(u)$. So $A_i^1(\check{u}),..., A_i^m(\check{u})$ are linearly independent. They are called tangent covectors to $T^{*k}\check{M}$. A d-covector $\omega_i(u^*)$, $u^* \in T^{*k}M$ is called normal to $T^{*k}\check{M}$ in $T^{*k}M$ if $u^* = i^*(\check{u})$ and

$$g^{ij}(i^*(\check{u}))A_i^{\alpha}(\check{u})\omega_j(i^*(\check{u})) = 0, \forall\, \check{u}\in\check{U}\subset T^{*k}\check{M}\ .$$

Then we can determine a Darboux coframe $\mathcal{R}^*$, at every point $\check{u}\in\check{U}$, of the form

$$\mathcal{R}^* = \{\check{u}, A_i^1(\check{u}), ..., A_i^m(\check{u}); A_i^{\overline{1}}(\check{u}), ..., A_i^{\overline{n-m}}(\check{u})\} \tag{9.2.4}$$

formed by m-tangent d-covectors $A_i^{\alpha}(\check{u})$ and by $n-m$ normal unit d-covectors $A_i^{\overline{\alpha}}(\check{u})$, $(\overline{\alpha} = \overline{1}, ..., \overline{n-m})$, which verify the conditions:

$$\begin{aligned} &g^{ij}(i^*\,\check{u})A_i^{\alpha}(\check{u})A_j^{\overline{\alpha}}(\check{u}) = 0,\\ &g^{ij}(i^*\,\check{u})A_i^{\overline{\alpha}}(\check{u})A_j^{\overline{\beta}}(\check{u}) = \delta^{\overline{\alpha}\overline{\beta}},\ \forall\,\check{u}\in\check{U} \end{aligned} \tag{9.2.4'}$$

Of course $\mathcal{R}^*$ exists and it has a geometrical meaning with respect to the change of local coordinates on the manifold $T^{*k}\check{M}$ and $T^{*k}M$ and with respect to the orthogonal (transformation of the normal d-covectors $A_i^{\overline{\alpha}}$) given by

$$A_i^{\overline{\alpha}'}(\check{u}) = C_{\overline{\beta}}^{\overline{\alpha}'}(\check{u})A_i^{\overline{\beta}}(\check{u}), \quad \left\|C_{\overline{\beta}}^{\overline{\alpha}'}(\check{u})\right\| \in O(n-m). \tag{9.2.5}$$

$\mathcal{R}^*$ is called a *moving coframe* on the submanifold $T^{*k}\check{M}$. The dual frame $\mathcal{R}$ of the coframe $\mathcal{R}^*$ is given by

$$\mathcal{R} = \{\check{u}, A_{\alpha}^i(\check{u}), A_{\overline{\alpha}}^i(\check{u})\}, \tag{9.2.4''}$$

where

$$\begin{cases} A_i^{\alpha}A_{\beta}^i = \delta_{\beta}^{\alpha}, A_i^{\alpha}A_{\overline{\beta}}^i = 0, A_i^{\overline{\alpha}}A_{\beta}^i = 0, A_i^{\overline{\alpha}}A_{\overline{\beta}}^i = \delta_{\overline{\beta}}^{\overline{\alpha}},\\ A_i^{\alpha}A_{\alpha}^j + A_i^{\overline{\alpha}}A_{\overline{\alpha}}^j = \delta_i^j, \end{cases} \tag{9.2.6}$$

at every point $\check{u}\in T^{*k}\check{M}$.

So, $\mathcal{R}$ will be called the *moving frame or the Darboux frame* along the Hamilton subspaces $\check{H}^{(k)m}$ in the Hamilton space $H^{(k)n}$.

The conditions (9.2.4'), (9.2,6) imply

$$\begin{aligned} \check{g}^{\alpha\beta}\, A_{\beta}^i &= g^{ij}\, A_j^{\alpha}\\ \delta^{\overline{\alpha}\overline{\beta}}\, A_{\overline{\beta}}^i &= g^{ij}\, A_j^{\overline{\alpha}} \end{aligned} \tag{9.2.7}$$

Indeed,

$\check{g}^{\alpha\beta} A^i_\beta = g^{rs} A^\alpha_r A^\beta_s A^i_\beta = g^{rs} A^\alpha_r (\delta^i_s - A^{\overline{\beta}}_s A^i_{\overline{\beta}}) = g^{ij} A^\alpha_j.$

In the following investigation we use the Darboux frame $\mathcal{R}$ in order to represent in $\mathcal{R}$ the d-tensor from the Hamilton spaces $H^{(k)n}$.

We get:

Proposition 9.2.1 *The fundamental tensor g^{ij} of the Hamilton space $H^{(k)n}$ and its covariant g_{ij} are represented in the Darboux frame $\mathcal{R}$ by*

$$(9.2.8)\qquad \begin{cases} g^{ij} = \check{g}^{\alpha\beta} A^i_\alpha A^j_\beta + \delta^{\overline{\alpha}\overline{\beta}} A^i_{\overline{\alpha}} A^j_{\overline{\beta}} \\ \\ g_{ij} = \check{g}_{\alpha\beta} A^\alpha_i A^\beta_j + \delta_{\overline{\alpha}\overline{\beta}} A^{\overline{\alpha}}_i A^{\overline{\beta}}_j \end{cases}$$

Indeed, the formulas (9.2.8) are consequence of (9.2.7)

The moving frame $\mathcal{R}$ will be used in the next section in order to derive the Gauss-Weingarden formulae and Gauss-Codazzi equations of the Hamilton subspaces $\check{H}^{(k)m}$ in $H^{(k)n}$.

9.3 Induced Nonlinear Connection

Now, let us consider the canonical nonlinear connection N of the Hamilton space of order k, $H^{(k)n} = (M, H)$. N has the dual coefficients $(\underset{(1)}{M^i_j}, ..., \underset{(k-1)}{M^i_j}, N_{ij})$. We shall prove that the restriction of N to the Hamilton subspaces $\check{H}^{(k)m} = (\check{M}, \check{H})$ determines an induced nonlinear connection $\check{N}$ on the manifold $T^{*k}\check{M}$.

The nonlinear connection $\check{N}$ is well determined by its dual coefficients or by means of its adapted cobasis $(du^\alpha, \delta v^{(1)\alpha}, ..., \delta v^{(k-1)\alpha}, \delta \check{p}_\alpha)$.

Definition 9.3.1 *A nonlinear connection $\check{N}$ is called* **induced** *by the canonical nonlinear connection N if the following conditions hold:*

$$(9.3.1)\qquad \begin{aligned} &\delta v^{(1)\alpha} = A^\alpha_i \delta y^{(1)i}, ..., \delta v^{(k-1)\alpha} = A^\alpha_i \delta y^{(k-1)i} \\ &\delta \check{p}_\alpha = B^i_\alpha \delta p_i, \end{aligned}$$

The previous conditions have a geometrical meaning. In fact, to a change of coordinates on $T^{*k}\check{M}$ the previous formulae are preserved. Tha same happens if we change the coordinates on the manifold $T^{*k}M$.

Theorem 9.3.1 *The dual coefficients of the induced nonlinear connection $\check{N}$ are given by the following formulae*

$$(9.3.2)\qquad \underset{(1)}{\check{M}^\alpha_\beta} = A^\alpha_i \underset{(1)}{\check{M}^i_\beta}, ..., \underset{(k-1)}{\check{M}^\alpha_\beta} = A^\alpha_i \underset{(k-1)}{\check{M}^i_\beta}, \quad \check{N}_{\alpha\beta} = \check{N}_{\alpha i} B^i_\beta,$$

where

(9.3.3)

$$\overset{\vee}{\underset{(1)}{M}}{}^i_\beta = \underset{(1)}{M}{}^i_j \frac{\partial y^{(1)j}}{\partial v^{(1)\beta}} + \frac{\partial y^{(1)i}}{\partial u^\beta},$$

$$\overset{\vee}{\underset{(2)}{M}}{}^i_\beta = \underset{(2)}{M}{}^i_j \frac{\partial y^{(2)j}}{\partial v^{(2)\beta}} + \underset{(1)}{M}{}^i_j \frac{\partial y^{(2)j}}{\partial v^{(1)\beta}} + \frac{\partial y^{(2)i}}{\partial u^\beta},$$

....

$$\overset{\vee}{\underset{(k-1)}{M}}{}^i_\beta = \underset{(k-1)}{M}{}^i_j \frac{\partial y^{(k-1)j}}{\partial v^{(k-1)\beta}} + \underset{(k-2)}{M}{}^i_j \frac{\partial y^{(k-1)j}}{\partial v^{(k-2)\beta}} + \dots + \underset{(1)}{M}{}^i_j \frac{\partial y^{(k-1)j}}{\partial v^{(1)\beta}} + \frac{\partial y^{(k-1)i}}{\partial u^\alpha}$$

and

$$\overset{\vee}{N}_{\alpha i} = N_{ji} B^j_\alpha - \frac{\partial p_i}{\partial u^\alpha}. \tag{9.3.4}$$

Proof. The first equations (9.3.1) lead to $dv^{(1)\alpha} + \overset{\vee}{\underset{(1)}{M}}{}^\alpha_\beta du^\beta = A^\alpha_i (dy^{(1)i} + \underset{(1)}{M}{}^i_j dx^j) = A^\alpha_i (\frac{\partial y^{(1)i}}{\partial u^\beta} + \frac{\partial y^{(1)i}}{\partial v^{(1)\beta}} dv^{(1)\beta} + \underset{(1)}{M}{}^i_j B^\alpha_\beta du^\beta)$.

Because of $\frac{\partial y^{(1)i}}{\partial v^{(1)\alpha}} = B^i_\alpha$ and $A^\alpha_i B^i_\beta = \delta^\alpha_\beta$ the last equality leads to (9.3.2): $\overset{\vee}{\underset{(1)}{M}}{}^\alpha_\beta = A^\alpha_i \overset{\vee}{\underset{(1)}{M}}{}^i_\beta$, where $\overset{\vee}{\underset{(1)}{M}}{}^i_\beta$ is given in (9.3.3), etc.

Remark that

$$\delta y^{(a)i} = dy^{(a)i} + \underset{(1)}{M}{}^i_j dy^{(a-1)i} + \dots + \underset{(a)}{M}{}^i_j dx^j, (a = 1, \dots, k-1) \tag{9.3.5}$$

and

$$dy^{(a)i} = \frac{\partial y^{(a)i}}{\partial u^\alpha} du^\alpha + \dots + \frac{\partial y^{(a)i}}{\partial v^{(a-1)\alpha}} dv^{(a-1)\alpha} + B^i_\alpha dv^{(a)\alpha}. \tag{9.3.5'}$$

We can prove:

Proposition 9.3.1 *The following formulae hold:*

(9.3.6)

$$dx^i = B^i_\beta du^\beta,$$

$$\delta y^{(1)i} = B^i_\beta \delta v^{(1)\beta} + A^i_{\overline{\alpha}} \underset{(1)}{H}{}^{\overline{\alpha}}_\beta du^\beta,$$

....

$$\delta y^{(k-1)i} = B^i_\beta \delta v^{(k-1)\beta} + A^i_{\overline{\alpha}} (\underset{(1)}{H}{}^{\overline{\alpha}}_\beta dv^{(k-2)\beta} + \dots + \underset{(k-1)}{H}{}^{\overline{\alpha}}_\beta du^\beta),$$

$$\delta p_i = A^\beta_i \delta \overset{\vee}{p}_\beta - A^{\overline{\alpha}}_i H_{\overline{\alpha}\beta} du^\beta,$$

where

$$(9.3.7)\qquad A^i_{\overline{\alpha}} \underset{(1)}{H}{}^{\overline{\alpha}}_{\beta} = \underset{(1)}{\check{M}}{}^i_\beta - B^i_\alpha \underset{(1)}{\check{M}}{}^\alpha_\beta ; \ldots ; A^i_{\overline{\alpha}} \underset{(k-1)}{H}{}^{\overline{\alpha}}_{\beta} = \underset{(k-1)}{\check{M}}{}^i_\beta - B^i_\alpha \underset{(k-1)}{\check{M}}{}^\alpha_\beta ,$$

$$(9.3.7')\qquad A^{\overline{\alpha}}_i H_{\overline{\alpha}\beta} = \check{N}_{\beta i} .$$

Proof. The formulae (9.3.5), (9.3.5') and (9.3.3) have as consequence the formulae (9.3.6). Also, we have $A^{\overline{\alpha}}_i H_{\overline{\alpha}\beta} = -A^i_\beta \delta p_i = \check{N}_{\beta i}$. q.e.d.

The previous expressions (9.3.6) is not convenient for us, because in the right hand sides we have the natural cobasis $du^\alpha, dv^{(1)i}, \ldots, dv^{(k-2)i}$. This means that the corresponding coefficients $A^i_{\overline{\alpha}} \underset{(a)}{H}{}^{\overline{\alpha}}_{\beta}, (a = 1, \ldots, k-1)$ do not have a geometrical meaning.

So, we prove:

Theorem 9.3.2 *In every point* $\check{u} \in T^{*k}\check{M}$ *the following formulas hold:*

(9.3.8)

$$dx^i = B^i_\beta du^\beta,$$

$$\delta y^{(1)i} = B^i_\beta \delta y^{(1)\beta} + A^i_{\overline{\alpha}} \underset{(1)}{K}{}^{\overline{\alpha}}_{\beta} du^\beta,$$

$$\delta y^{(2)i} = B^i_\beta \delta v^{(2)\beta} + A^i_{\overline{\alpha}} (\underset{(1)}{K}{}^{\overline{\alpha}}_{\beta} \delta v^{(1)\beta} + \underset{(2)}{K}{}^{\overline{\alpha}}_{\beta} du^\beta),$$

....

$$\delta y^{(k-1)i} = B^i_\beta \delta v^{(k-1)\beta} + A^i_{\overline{\alpha}} (\underset{(1)}{K}{}^{\overline{\alpha}}_{\beta} \delta v^{(k-2)\beta} + \underset{(2)}{K}{}^{\overline{\alpha}}_{\beta} \delta v^{(k-3)\beta} + \ldots + \underset{(k-1)}{K}{}^{\overline{\alpha}}_{\beta} du^\beta)$$

and

$$\delta p_i = A^\beta_i \delta \check{p}_\beta - A^{\overline{\alpha}}_i H_{\overline{\alpha}\beta} du^\beta,$$

where $A^{\overline{\alpha}}_i H_{\overline{\alpha}\beta}$ *is given by (9.3.7') and*

$$\underset{(1)}{K}{}^{\overline{\alpha}}_{\beta} = \underset{(1)}{H}{}^{\overline{\alpha}}_{\beta},$$

$$(9.3.9)\qquad \underset{(2)}{K}{}^{\overline{\alpha}}_{\beta} = \underset{(2)}{H}{}^{\overline{\alpha}}_{\beta} - \underset{(1)}{H}{}^{\overline{\alpha}}_{\gamma} \underset{(1)}{\check{N}}{}^{\alpha}_{\beta},$$

....

$$\underset{(k-1)}{K}{}^{\overline{\alpha}}_{\beta} = \underset{(k-1)}{H}{}^{\overline{\alpha}}_{\beta} - \underset{(1)}{H}{}^{\overline{\alpha}}_{\gamma} \underset{(k-2)}{\check{N}}{}^{\gamma}_{\beta} - \ldots - \underset{(k-2)}{H}{}^{\overline{\alpha}}_{\gamma} \underset{(1)}{\check{N}}{}^{\gamma}_{\beta} .$$

Of course, $\underset{(1)}{\check{N}}{}^{\alpha}_{\beta}, \ldots, \underset{(k-1)}{\check{N}}{}^{\alpha}_{\beta}$ are the coefficients of the induced connection $\check{N}$.

Proof. Taking into account the formulas (6.3.5) we have

$$dv^{(a)\alpha} = \delta v^{(a)\alpha} - \overset{\vee}{\underset{(1)}{N}}{}^{\alpha}_{\beta}\, \delta v^{(a-1)\beta} - \dots - \overset{\vee}{\underset{(a)}{N}}{}^{\alpha}_{\beta}\, \delta u^{\beta}, \ (a = 1, \dots, k-1).$$

So, $\delta y^{(a)i}$ from (9.3.6) is as follows

$$\delta y^{(a)i} = B^{i}_{\beta}\delta v^{(a)\beta} + A^{i}_{\overline{\alpha}}[\underset{(1)}{H}{}^{\alpha}_{\beta}(\delta v^{(a-1)\gamma} - \overset{\vee}{\underset{(1)}{N}}{}^{\beta}_{\gamma}\, \delta v^{(a-2)\gamma} - \dots - \overset{\vee}{\underset{(a-1)}{N}}{}^{\beta}_{\gamma}\, \delta u^{\gamma})$$

$$+\underset{(2)}{H}{}^{\overline{\alpha}}_{\beta}(\delta v^{(a-2)\beta} - \overset{\vee}{\underset{(1)}{N}}{}^{\beta}_{\gamma}\, \delta v^{(a-3)\gamma} - \dots - \overset{\vee}{\underset{(a-2)}{N}}{}^{\beta}_{\gamma}\, \delta u^{\gamma}) + \dots +$$

$$+\underset{(a-1)}{H}{}^{\overline{\alpha}}_{\beta}(\delta v^{(1)\beta} - \overset{\vee}{\underset{(1)}{N}}{}^{\beta}_{\gamma}\, \delta u^{\gamma}) + \overset{\vee}{\underset{(a)}{N}}{}^{\overline{\alpha}}_{\beta}\, \delta u^{\beta}).$$

Identifying with $\delta y^{(a)i}$ from (9.3.6):

$$\delta y^{(a)i} = B^{i}_{\beta}\delta v^{(a)\beta} + A^{i}_{\overline{\alpha}}[\underset{(1)}{K}{}^{\overline{\alpha}}_{\beta}\delta v^{(a-1)\beta} + \underset{(2)}{K}{}^{\overline{\alpha}}_{\beta}\delta v^{(a-2)\beta} + \dots + \underset{(a)}{K}{}^{\overline{\alpha}}_{\beta} du^{\beta}]$$

we obtain the formulae (9.3.9). q.e.d

The previous theorem has an important consequence.

Corollary 9.3.1 *With respect to the transformations of coordinates on the submanifold $T^{*k}\overset{\vee}{M}$ and to the transformation (9.2.5), $\underset{(1)}{K}{}^{\overline{\alpha}}_{\beta}, \dots, \underset{(k-1)}{K}{}^{\overline{\alpha}}_{\beta}$ and $H_{\overline{\alpha}\beta}$ are the mixed d-tensor fields.*

Generally a set of functions $T^{i\dots\alpha\dots\overline{\alpha}}_{j\dots\beta\dots\overline{\beta}}(\overset{\vee}{u})$ which are the components of a d-tensor in the indices $i, j, \dots$and d-tensor in the indices $\alpha, \beta, \dots$ and tensor with respect to the transformation (9.2.5) in the indices $\overline{\alpha}, \overline{\beta}, \dots$ is called a *mixed* d-tensor field on the submanifolds $T^{*k}\overset{\vee}{M}$.

For instance $B^{i}_{\alpha}, A^{i}_{\alpha}, A^{i}_{\overline{\alpha}}, A^{\alpha}_{i}, A^{\overline{\alpha}}_{i}, \overset{\vee}{g}_{\alpha\beta}, g_{ij}, \delta_{\overline{\alpha}\overline{\beta}}, \underset{(1)}{K}{}^{\overline{\alpha}}_{\beta}, \dots, \underset{(k-1)}{K}{}^{\overline{\alpha}}_{\beta}$ and $H_{\overline{\alpha}\beta}$ are mixed d-tensor fields. The previous definition of mixed d-tensor fields can be extended to any geometrical d -object fields.

9.4 The Relative Covariant Derivative

In this section we shall construct an operator ∇ of relative covariant derivation in the algebra of mixed d-tensor fields. It is clear that ∇ will be well determined if we know its action on the mixed d-vector field:

$$X^{i}(\overset{\vee}{u}), X^{\alpha}(\overset{\vee}{u}), X^{\overline{\alpha}}(\overset{\vee}{u}), \forall\, \overset{\vee}{u} = (u, v^{(1)}, \dots, v^{(k-1)}, \overset{\vee}{p}) \in T^{*k}\overset{\vee}{M}\ . \tag{9.4.1}$$

Definition 9.4.1 *We call a coupling of the canonical metrical N-connection D of the Hamilton spaces of order k, $H^{(k)n} = (M, H)$ to the induced nonlinear* *connection* $\check{N}$, *of the submanifold* $\check{M}$, *in the manifold* M *the operator* $\check{D}$ *with the property*

$$\check{D}\, X^i = DX^i \text{ modulo } \{(9.3.8),\ (9.3.8')\}, \tag{9.4.2}$$

where

$$DX^i = dx^i + X^i \omega^i_j \tag{9.4.2'}$$

and where ω^i_j *are 1-forms connection:*

$$\omega^i_j = H^i_{jh} dx^h + \underset{(1)}{C^i_{jh}} \delta y^{(1)h} + ... + \underset{(k-1)}{C^i_{jh}}\, \delta y^{(k-1)h} + C^{ih}_j \delta p_h. \tag{9.4.2"}$$

Thus, in every point $\check{u} \in T^{*k}\, \check{M}$, $\check{D}\, X^i$ has the form

$$\check{D}\, X^i = dx^i + X^j\, \check{\omega}^i_j, \tag{9.4.3}$$

where

$$\check{\omega}^i_j = \check{H}^i_{j\beta}\, du^\beta + \underset{(1)}{\check{C}^i_{j\beta}}\, \delta v^{(1)\beta} + ... + \underset{(k-1)}{\check{C}^i_{j\beta}}\, \delta v^{(k-1)\beta} + \check{C}^{i\beta}_j\, \delta\, \check{p}_\beta\, . \tag{9.4.4}$$

Theorem 9.4.1 *The connection one forms* $\check{\omega}^i_j$ *of* $\check{D}$ *are given by (9.4.4), where:*

$$\check{H}^i_{j\beta} = H^i_{j\beta} B^h_\beta + (\underset{(1)}{C^i_{jh}} \underset{(1)}{K^{\overline{\alpha}}_\beta} + ... + \underset{(k-1)}{C^i_{jh}}\ \underset{(k-1)}{K^{\overline{\alpha}}_\beta}) A^h_{\overline{\alpha}} - C^{jh}_i A^{\overline{\alpha}}_h H_{\overline{\alpha}\beta},$$

$$\begin{aligned}
&\underset{(1)}{\check{C}^i_{j\beta}} = \underset{(1)}{C^i_{jh}} B^h_\beta + (\underset{(2)}{C^i_{jh}} \underset{(1)}{K^{\overline{\alpha}}_\beta} + ... + \underset{(k-1)}{C^i_{jh}}\ \underset{(k-2)}{K^{\overline{\alpha}}_\beta}) A^h_{\overline{\alpha}},\\
&\cdots\cdots\cdots\cdots\cdots\cdots\cdots\cdots\cdots\cdots\\
&\underset{(k-1)}{\check{C}^i_{j\beta}} = \underset{(k-1)}{C^i_{jh}}\, B^h_\beta + \underset{(k-2)}{C^i_{jh}}\, \underset{(1)}{K^{\overline{\alpha}}_\beta} A^h_{\overline{\alpha}},
\end{aligned} \tag{9.4.5}$$

$$\check{C}^{j\beta}_i = C^{jh}_i A^\beta_h.$$

Proof. Using the formulae (9.3.8), (9.3.8'), ω^i_j modulo (9.3.8), (9.3.8') from (9.4.2") leads to

(9.4.4')

$$\begin{aligned}
\check{\omega}^i_j = H^i_{jh} B^h_\beta du^\beta + \ & \underset{(1)}{C^i_{jh}} (B^h_\beta \delta v^{(1)\beta} + A^h_{\overline{\alpha}} \underset{(1)}{K^{\overline{\alpha}}_\beta} du^\beta) + ... + \\
& \underset{(k-1)}{C^i_{jh}}\, [B^h_\beta \delta v^{(k-1)\beta} + A^h_{\overline{\alpha}} (\underset{(1)}{K^{\overline{\alpha}}_\beta} \delta v^{(k-2)\beta} + ... + \underset{(k-1)}{K^{\overline{\alpha}}_\beta}\ du^\beta)] + \\
& C^{jh}_i [A^h_\beta \delta\, \check{p}_\beta - A^{\overline{\alpha}}_h H_{\overline{\alpha}\beta} du^\beta].
\end{aligned}$$

Identifying with $\overset{\vee}{\omega}{}^i_j$ from (9.4.4) we obtain the formulae (9.4.5). Evidently the coupling connection $\overset{\vee}{D}$ of the canonical metrical N-connection D, depend only on the fundamental function H of the space $H^{(k)n}$ and by the immersion $i^* : T^{*k}\overset{\vee}{M}$ in $T^{*k}M$.

Of course, we can write $\overset{\vee}{D}\, X^i$ in the form

$$\text{(9.4.6)} \quad \overset{\vee}{D}\, X^i = X^i_{|\alpha} du^\alpha + X^i \overset{(1)}{|}_\alpha \delta v^{(1)\alpha} + \ldots + X^i \overset{(k-1)}{|}_\alpha \delta v^{(k-1)\alpha} + X^i|^\alpha \delta \overset{\vee}{p}_\alpha,$$

where

$$X^i_{|\alpha} = \frac{\delta X^i}{\delta u^\alpha} + X^j \overset{\vee}{H}{}^i_{j\alpha},$$

$$\text{(9.4.7)} \qquad X^i \overset{(a)}{|}_\alpha = \frac{\delta X^i}{\delta v^{(a)\alpha}} + X^j \underset{(a)}{\overset{\vee}{C}}{}^i_{j\alpha}, (a = 1, ..., k-1),$$

$$X^i|^\alpha = \dot{\partial}^\alpha X^i + X^j \overset{\vee}{C}{}^{i\alpha}_j.$$

We can extend, without difficulties the action of the linear connection $\overset{\vee}{D}$ to the d-tensors $T^{i_1 \ldots i_r}_{j_1 \ldots j_s}(\overset{\vee}{u})$, $\forall\, \overset{\vee}{u} \in T^{*k}\overset{\vee}{M}$.

Definition 9.4.2 *We call the induced tangent connection on $T^{*k}\overset{\vee}{M}$ by the canonical metrical N-connection D the operator D^T given by*

$$\text{(9.4.8)} \qquad D^T X^\alpha = A^\alpha_i \overset{\vee}{D}\, X^i, \quad for \quad X^i = A^i_\alpha X^\alpha$$

We can see that:

$$\text{(9.4.9)} \qquad D^T X^\alpha = dX^\alpha + X^\beta \omega^\alpha_\beta,$$

where ω^α_β are the connection one forms of D^T :

$$\text{(9.4.10)} \qquad \omega^\alpha_\beta = H^\alpha_{\beta\gamma} du^\gamma + \underset{(1)}{C}{}^\alpha_{\beta\gamma} \delta v^{(1)\gamma} + \ldots + \underset{(k-1)}{C}{}^\alpha_{\beta\gamma} \delta v^{(k-1)\gamma} + C^{\alpha\gamma}_\beta \delta \overset{\vee}{p}_\gamma$$

Theorem 9.4.2 *1° The coefficients $(H^\alpha_{\beta\gamma}, \underset{(1)}{C}{}^\alpha_{\beta\gamma}, ..., \underset{(k-1)}{C}{}^\alpha_{\beta\gamma}, C^{\alpha\gamma}_\beta)$ of the one forms connection ω^α_β have the following expressions*

$$H^\alpha_{\beta\gamma} = A^\alpha_i\Big(\frac{\delta A^i_\beta}{\delta u^\gamma} + A^j_\beta \overset{\vee}{H}{}^i_{j\gamma}\Big),$$

$$\text{(9.4.11)} \qquad \underset{(a)}{C}{}^\alpha_{\beta\gamma} = A^\alpha_i\Big(\frac{\delta A^i_\beta}{\delta v^{(a)\gamma}} + A^h_\beta \underset{(a)}{\overset{\vee}{C}}{}^\alpha_{h\gamma}\Big), (a = 1, ..., k-1),$$

$$C^{\alpha\gamma}_\beta = A^\alpha_i(\dot{\partial}^\gamma A^i_\beta + A^h_\beta \overset{\vee}{C}{}^{i\gamma}_h).$$

2° With respect of the change of local coordinates on the manifold $T^{*k}\overset{\vee}{M}$, the coefficients

$$H^{\alpha}_{\beta\gamma},\ \underset{(1)}{C^{\alpha}_{\beta\gamma}},...,\ \underset{(k-1)}{C^{\alpha}_{\beta\gamma}},...,C^{\alpha\gamma}_{\beta}$$

have the rule of transformation as the coefficients of $\overset{\vee}{N}$ -linear connection on $T^{*k}\overset{\vee}{M}$.

Proof. 1° The equalities (9.4.8) and (9.4.9) imply

$$dX^{\alpha}+X^{\beta}\omega^{\alpha}_{\beta}=A^{\alpha}_{i}(dX^{i}+X^{j}\ \overset{\vee}{\omega}{}^{i}_{j})=A^{\alpha}_{i}[(dA^{i}_{\beta}+A^{h}_{\beta}\ \overset{\vee}{\omega}{}^{i}_{h})X^{\beta}+A^{i}_{\beta}dx^{\beta}]$$

Consequently,

$$\omega^{\alpha}_{\beta}=A^{\alpha}_{i}(dA^{i}_{\beta}+\overset{\vee}{\omega}{}^{i}_{j}\ A^{j}_{\beta}). \tag{9.4.12}$$

Thus, the equalities (9.4.4), (9.4.10) and (9.4.12) lead to the form (9.4.11) of the coefficients of induced tangent connection D^{T}.

2° Using the rule of transformations of the coefficients of $\overset{\vee}{\omega}{}^{i}_{j}$ it follows the sentence 2° of theorem . q.e.d.

As in the case of $\overset{\vee}{D}$ we can write:

(9.4.13)

$$D^{T}X^{\alpha}=X^{\alpha}_{|\beta}du^{\beta}+X^{\alpha}\overset{(1)}{|}_{\beta}\delta v^{(1)\beta}+...+X^{\alpha}\overset{(k-1)}{|}_{\beta}\delta v^{(k-1)\beta}+X^{\alpha}|^{\beta}\delta\overset{\vee}{p}_{\beta},$$

where

$$X^{\alpha}_{|\gamma}=\frac{\delta X^{\alpha}}{\delta u^{\gamma}}+X^{\beta}H^{\alpha}_{\beta\gamma},$$

$$X^{\alpha}\overset{(a)}{|}_{\gamma}=\frac{\delta X^{\alpha}}{\delta v^{(a)\gamma}}+X^{\beta}\underset{(a)}{C^{\alpha}_{\beta\gamma}},\quad (a=1,...k-1), \tag{9.4.14}$$

$$X^{\alpha}|^{\gamma}=\dot{\partial}^{\gamma}\ X^{\alpha}+X^{\beta}C^{\alpha\gamma}_{\beta}.$$

Evidently, the induced tangent connection D^{T} depends on the canonical $\overset{\vee}{N}$-metrical connection D and depends on the immersion $i^{*}:T^{*k}\overset{\vee}{M}\rightarrow T^{*k}M$.

This operator D^{T}, as well the operators „$|_{\gamma}$" , „ $\overset{(a)}{|}_{\gamma}$ " and „$|^{\gamma}$" can be extended to the tensor fields $T^{\alpha_{1}...\alpha_{a}}_{\beta_{1}...\beta_{q}}(\overset{\vee}{u})$.

Definition 9.4.3 *We call the induced normal connection on $T^{*k}\overset{\vee}{M}$ by the canonical metrical N-connection D, the operator $D^{\perp}$ given by*

$$D^{\perp}X^{\overline{\alpha}}=A^{\overline{\alpha}}_{i}\ \overset{\vee}{D}\ X^{i},\ for\ X^{i}=A^{i}_{\overline{\alpha}}X^{\overline{\alpha}}. \tag{9.4.15}$$

As before we set

$$D^{\perp} X^{\overline{\alpha}} = dX^{\overline{\alpha}} + X^{\overline{\beta}}\omega^{\overline{\alpha}}_{\overline{\beta}}, \tag{9.4.16}$$

with the connection 1-forms of $D^{\perp}$.

$$\omega^{\overline{\alpha}}_{\overline{\beta}} = H^{\overline{\alpha}}_{\overline{\beta}\gamma}du^{\gamma} + \underset{(1)}{C^{\overline{\alpha}}_{\overline{\beta}\gamma}}\delta v^{(1)\gamma} + ... + \underset{(k-1)}{C^{\overline{\alpha}}_{\overline{\beta}\gamma}}\delta v^{(k-1)\gamma} + C^{\overline{\alpha}\gamma}_{\overline{\beta}}\delta \overset{\vee}{p}_{\gamma}\,. \tag{9.4.17}$$

Applying the same method as in the case of $\overset{\vee}{D}$ and D^T we can calculate the coefficients of the 1-forms connection $\omega^{\overline{\alpha}}_{\overline{\beta}}$.

Theorem 9.4.3 *The coefficients* $(H^{\overline{\alpha}}_{\overline{\beta}\gamma}, \underset{(1)}{C^{\overline{\alpha}}_{\overline{\beta}\gamma}}, ..., \underset{(k-1)}{C^{\overline{\alpha}}_{\overline{\beta}\gamma}}, C^{\overline{\alpha}\gamma}_{\overline{\beta}})$ *of the induced normal connection* $D^{\perp}$ *are given by*

$$\begin{aligned} H^{\overline{\alpha}}_{\overline{\beta}\gamma} &= A^{\overline{\alpha}}_i\Big(\frac{\delta A^i_{\overline{\beta}}}{\delta u^{\gamma}} + A^j_{\overline{\beta}}\overset{\vee}{H}{}^i_{j\gamma}\Big), \\ \underset{(a)}{C^{\overline{\alpha}}_{\overline{\beta}\gamma}} &= A^{\overline{\alpha}}_i\Big(\frac{\delta A^i_{\overline{\beta}}}{\delta v^{(a)\gamma}} + A^j_{\overline{\beta}}\underset{(a)}{\overset{\vee}{C}{}^i_{j\gamma}}\Big), (a = 1, ..., k-1), \\ C^{\overline{\alpha}\gamma}_{\overline{\beta}} &= A^{\overline{\alpha}}_i(\dot{\partial}^{\gamma} A^i_{\overline{\beta}} + A^i_{\gamma}\overset{\vee}{C}{}^{\overline{\alpha}\gamma}_{\overline{\beta}}). \end{aligned} \tag{9.4.18}$$

Proof. Indeed, (9.4.15) and (9.4.16) have as consequence:

$$\omega^{\overline{\alpha}}_{\overline{\beta}} = A^{\overline{\alpha}}_i(dA^i_{\overline{\beta}} + A^j_{\overline{\beta}}\overset{\vee}{\omega}{}^i_j). \tag{9.4.19}$$

Therefore, $\overset{\vee}{\omega}{}^i_j$ from (9.4.4) and $\omega^{\overline{\alpha}}_{\overline{\beta}}$ expressed in (9.4.17) substituted in the equality (9.4.19) leads to (9.4.18). q.e.d.

As usual, we may set for the induced normal connection

$$D^{\perp}X^{\overline{\alpha}} = X^{\overline{\alpha}}_{|\beta}du^{\beta} + X^{\overline{\alpha}}\overset{(1)}{|}_{\beta}\delta v^{(1)\beta} + ... + X^{\overline{\alpha}}\overset{(k-1)}{|}_{\beta}\delta v^{(k-1)\beta} + X^{\overline{\alpha}}|^{\beta}\delta \overset{\vee}{p}_{\beta} \tag{9.4.20}$$

with

$$\begin{aligned} X^{\overline{\alpha}}_{|\beta} &= \frac{\delta X^{\overline{\alpha}}}{\delta u^{\beta}} + X^{\overline{\gamma}}H^{\overline{\alpha}}_{\overline{\gamma}\beta}, \\ X^{\overline{\alpha}}\overset{(a)}{|}_{\beta} &= \frac{\delta X^{\overline{\alpha}}}{\delta v^{(a)\beta}} + X^{\gamma}\underset{(a)}{C^{\overline{\alpha}}_{\overline{\gamma}\beta}}, (a = 1, ...k-1), \\ X^{\overline{\alpha}}|^{\beta} &= \dot{\partial}^{\beta} X^{\overline{\alpha}} + X^{\gamma}C^{\overline{\alpha}\beta}_{\overline{\gamma}}. \end{aligned} \tag{9.4.21}$$

These are the induced normal covariant derivations.

Now, we can define the relative (or mixed) derivation ∇ introduced at the begining of this section.

Definition 9.4.4 *A relative (mixed) covariant derivation in the algebra of mixed d-tensor field is an operator ∇ which has the following properties.*

$$\nabla f = df, \forall f \in \mathcal{F}(T^{*k}\check{M})$$
(9.4.22)
$$\nabla X^i = \check{D} X^i, \nabla X^\alpha = D^T X^\alpha, \nabla X^{\overline{\alpha}} = D^\perp X^{\overline{\alpha}}$$

for any mixed vector fields $X^i(\check{u}), X^\alpha(\check{u})$ and $X^{\overline{\alpha}}(\check{u})$.

The connection 1-forms $\check{\omega}^i_j, \omega^\alpha_\beta$ and $\omega^{\overline{\alpha}}_{\overline{\beta}}$ will be called the *connections* 1-forms of the relative covariant derivation ∇. Of course, the operator ∇ can be extended to any mixed d-tensor $T^{i\ldots\alpha\ldots\overline{\alpha}}_{j\ldots\beta\ldots\overline{\beta}}(\check{u})$, $\check{u} \in T^{*k}\check{M}$.

We can write the Ricci identities for the relative covariant derivative ∇, and its torsion and curvature tensor, taking into account every components $\check{D}, D^T$,and $D^\perp$ of ∇.

Then, it is easy to prove:

Theorem 9.4.4 *The structure equations of the mixed covariant derivation ∇ are as follows:*

$$d(du^\alpha) - du^\beta \wedge \omega^\alpha_\beta = -\overset{(0)}{\Omega^\alpha},$$
(9.4.23)
$$d(dv^{(a)\alpha}) - \delta v^{(a)\beta} \wedge \omega^\alpha_\beta = -\overset{(a)}{\Omega^\alpha}, (\alpha = 1, ..., k-1),$$
$$d(\delta \check{p}_\alpha) - \delta \check{p}_\beta \wedge \omega^\beta_\alpha = -\Omega_\alpha$$

and

$$d(\check{\omega}^i_j) - \check{\omega}^h_j \wedge \check{\omega}^j_h = -\check{\Omega}^i_j,$$
(9.4.24)
$$d(\omega^\alpha_\beta) - \omega^\gamma_\beta \wedge \omega^\alpha_\gamma = -\Omega^\alpha_\beta,$$
$$d(\omega^{\overline{\alpha}}_{\overline{\beta}}) - \omega^{\overline{\gamma}}_{\overline{\beta}} \wedge \omega^{\overline{\alpha}}_{\overline{\gamma}} = -\Omega^{\overline{\alpha}}_{\overline{\beta}},$$

in which $\Omega^\alpha, \overset{(a)}{\Omega^\alpha}$ and Ω_α are the 2-forms of torsion:

(9.4.25)
$$\overset{(0)}{\Omega^\alpha} = du^\beta \wedge \{\frac{1}{2}T^\alpha_{\beta\gamma}du^\gamma + \underset{(1)}{C^\alpha_{\beta\gamma}}\delta v^{(1)\gamma} + \ldots + \underset{(k-1)}{C^\alpha_{\beta\gamma}}\delta v^{(k-1)\gamma} + C^{\alpha\gamma}_\beta \delta \check{p}_\gamma\},$$

$$\overset{(a)}{\Omega^\alpha} = du^\beta \wedge \underset{(a0)}{P^\alpha_\beta} + \sum_{b=1}^{k-1} \delta v^{(b)\beta} \wedge \underset{(ab)}{P^\alpha_\beta} + \delta v^{(a)\beta} \wedge \{H^\alpha_{\beta\gamma}du^\gamma +$$

$$+ \sum_{b=1}^{k-1} \underset{(b)}{C^\alpha_{\beta\gamma}}\delta v^{(b)\gamma} + C^{\alpha\gamma}_\beta \delta \check{p}_\gamma\}, (a = 1, ..., k-1),$$

with $\underset{(ab)}{P^{\alpha}_{\beta}}$ from (5.4.8) and where the 2-forms of curvature are given by

$$\overset{\vee}{\Omega}{}^{i}_{j}=\frac{1}{2}\overset{\vee}{R}{}_{j}{}^{i}{}_{\alpha\beta}du^{\alpha}\wedge du^{\beta}+\sum_{a=1}^{k-1}\underset{(a)}{\overset{\vee}{P}{}_{j}{}^{i}{}_{\alpha\beta}}\,du^{\alpha}\wedge\delta v^{(a)\beta}+\overset{\vee}{P}{}_{j}{}^{i}{}_{\alpha}{}^{\beta}du^{\alpha}\wedge\delta\overset{\vee}{p}_{\beta}+$$

$$+\sum_{a,b=1}^{k-1}\underset{(ab)}{\overset{\vee}{S}{}_{j}{}^{i}{}_{\alpha\beta}}\,dv^{(a)\alpha}\wedge\delta v^{(b)\beta}+\sum_{a=1}^{k-1}\overset{\vee}{S}{}_{j}{}^{i}{}_{\alpha}{}^{\beta}\,dv^{(a)\alpha}\wedge\delta\overset{\vee}{p}_{\beta}+\frac{1}{2}\overset{\vee}{S}{}_{j}{}^{i\alpha\beta}\delta\overset{\vee}{p}_{\alpha}\wedge\overset{\vee}{p}_{\beta}\ ....,$$

(9.4.26)

$$\Omega^{\alpha}_{\beta}=\frac{1}{2}R^{\alpha}_{\beta}{}_{\gamma\varphi}du^{\gamma}\wedge du^{\varphi}+\sum_{a=1}^{k-1}\underset{(a)}{P^{\alpha}_{\beta}{}_{\gamma\varphi}}du^{\gamma}\wedge\delta v^{(a)\varphi}+P^{\alpha\varphi}_{\beta\gamma}du^{\gamma}\wedge\delta\overset{\vee}{p}_{\varphi}+$$

$$+\sum_{a,b=1}^{k-1}S^{\alpha}_{\beta}{}_{\gamma\varphi}\delta v^{(a)\gamma}\wedge\delta v^{(b)\varphi}+\sum_{a=1}^{k-1}S_{\beta}{}^{\alpha}{}_{\gamma}{}^{\varphi}dv^{(a)\gamma}\wedge\delta\overset{\vee}{p}_{\varphi}+\frac{1}{2}S_{\beta}{}^{\alpha\gamma\varphi}\delta\overset{\vee}{p}_{\gamma}\wedge\delta\overset{\vee}{p}_{\varphi},$$

$$\Omega^{\overline{\alpha}}_{\overline{\beta}}=\frac{1}{2}R^{\overline{\alpha}}_{\overline{\beta}}{}_{\gamma\varphi}du^{\gamma}\wedge du^{\varphi}+\sum_{a=1}^{k-1}\underset{(a)}{P^{\overline{\alpha}}_{\overline{\beta}}{}_{\gamma\varphi}}du^{\gamma}\wedge\delta v^{(a)\varphi}+P_{\overline{\beta}}{}^{\overline{\alpha}}{}_{\gamma}{}^{\varphi}du^{\gamma}\wedge\delta\overset{\vee}{p}_{\varphi}+$$

$$+\sum_{a,b=1}^{k-1}S^{\overline{\alpha}}_{\overline{\beta}}{}_{\gamma\varphi}\delta v^{(a)\gamma}\wedge\delta v^{(b)\varphi}+\sum_{a=1}^{k-1}S^{\overline{\alpha}}_{\overline{\beta}}{}_{\gamma\varphi}dv^{(a)\gamma}\wedge\delta\overset{\vee}{p}_{\varphi}+\frac{1}{2}S^{\overline{\alpha}\gamma\varphi}_{\overline{\beta}}\delta\overset{\vee}{p}_{\gamma}\wedge\delta\overset{\vee}{p}_{\varphi}\,.$$

The previous equations can be particularized in the case when the induced tangent connection D^T has vanishing tensors of torsion $T^{\alpha}_{\beta\gamma}$, $\underset{(a)}{S}{}^{\alpha}{}_{\beta\gamma}$ and $S_{\alpha}{}^{\beta\gamma}$.

In the following we shall adopt the notations

$$\overset{\vee}{\Omega}_{ij}=\overset{\vee}{\Omega}{}^{h}_{i}\,g_{hj},\Omega_{\alpha\beta}=\Omega^{\gamma}_{\alpha}\,\overset{\vee}{g}_{\gamma\beta},\Omega_{\overline{\alpha}\overline{\beta}}=\Omega^{\overline{\gamma}}_{\overline{\alpha}}\delta_{\overline{\gamma}\overline{\beta}}. \tag{9.4.27}$$

These covariant 2-forms of curvature of the mixed covariant connection $\overset{\vee}{\nabla}$ will be useful to write the fundamental equations of the immersion $i^{*}:T^{*k}\overset{\vee}{M}\to T^{k}M$.

As a direct consequence of the Theorems from the present section we get:

Theorem 9.4.5 *The mixed covariant derivation* ∇ *is a metrical one, i.e.:*

$$\nabla g^{ij}=0,\nabla\delta^{\overline{\alpha}\overline{\beta}}=0,\nabla\overset{\vee}{g}{}^{\alpha\beta}=0. \tag{9.4.28}$$

Proof. The first formula is immediate, because of $\nabla g^{ij}=\overset{\vee}{D}g^{ij}=0$ while $\nabla\delta^{\overline{\alpha}\overline{\beta}}=0,\nabla\overset{\vee}{g}{}^{\alpha\beta}=0$ can be proved by means of Gauss-Weingarten formulas, given in next section.

9.5 The Gauss-Weingarten Formula

We need to study the moving equations of the Darboux frame:

$$\mathcal{R} = \{\check{u}, A^i_\alpha(\check{u}), A^i_{\overline{\alpha}}(\check{u})\}, \forall\, \check{u} \in T^{*k}\check{M}\ . \tag{9.5.1}$$

So, we obtain

Theorem 9.5.1 *The following Gauss-Weingarten formulae hold good:*

$$\nabla A^i_\alpha = A^i_{\overline{\alpha}}\Pi^{\overline{\alpha}}_\alpha, \quad \nabla A^i_{\overline{\alpha}} = -A^i_\alpha \Pi^\alpha_{\overline{\alpha}}, \tag{9.5.2}$$

where

$$\begin{aligned} \Pi^{\overline{\beta}}_\alpha &= \underset{(0)}{H}{}^{\overline{\beta}}_{\alpha\gamma} du^\gamma + \underset{(1)}{H}{}^{\overline{\beta}}_{\alpha\gamma} \delta v^{(1)\gamma} + \dots + \underset{(k-1)}{H}{}^{\overline{\beta}}_{\alpha\gamma}\ \delta v^{(k-1)\gamma} + H^{\overline{\beta}\gamma}_\alpha \delta \check{p}_\gamma, \\ \Pi^\alpha_{\overline{\beta}} &= \check{g}^{\alpha\gamma}\ \delta_{\overline{\beta}\overline{\gamma}} \Pi^{\overline{\gamma}}_\gamma \end{aligned} \tag{9.5.3}$$

and where

$$\begin{aligned} \underset{(0)}{H}{}^{\overline{\beta}}_{\alpha\gamma} &= A^{\overline{\beta}}_h\Big(\frac{\delta A^h_\alpha}{\delta u^\gamma} + A^j_\alpha\ \check{H}^h_{j\gamma}\Big), \\ \underset{(a)}{H}{}^{\overline{\beta}}_{\alpha\gamma} &= A^{\overline{\beta}}_h\Big(\frac{\delta A^h_\alpha}{\delta v^{(a)\gamma}} + A^j_\alpha\ \underset{(a)}{\check{C}}{}^h_{j\gamma}\ \Big), (a = 1, \dots, k-1), \\ H^{\overline{\beta}\gamma}_\alpha &= A^{\overline{\beta}}_h(\dot{\partial}^\gamma\ A^h_\alpha + A^j_\alpha\ \check{C}^{h\gamma}_j), \end{aligned} \tag{9.5.4}$$

with $\check{H}^h_{j\gamma}$, $\underset{(a)}{\check{C}}{}^h_{j\gamma}$ *and* $\check{C}^{h\gamma}_j$ *are from (9.4.5).*

Proof. Taking into account that A^i_α is a mixed d-tensor we have for its relative covariant derivation:

$$\begin{aligned} \nabla A^i_\alpha = \ & dA^i_\alpha + A^j_\alpha\ \check{\omega}^i_j\ - A^i_\beta\omega^\beta_\alpha = dA^i_\alpha + A^j_\alpha\ \check{\omega}^i_j\ - \\ & -A^i_\gamma A^\gamma_h(dA^h_\alpha + \check{\omega}^h_j\ A^j_\alpha) = dA^i_\alpha + A^j_\alpha\ \check{\omega}^i_j\ - \\ & -(\delta^i_h - A^i_{\overline{\gamma}}A^{\overline{\gamma}}_h)(dA^h_\alpha + \check{\omega}^h_j\ A^j_\alpha) = A^i_{\overline{\alpha}}A^{\overline{\alpha}}_h(dA^h_\alpha + \check{\omega}^h_j\ A^j_\alpha). \end{aligned}$$

So, we get

$$\Pi^{\overline{\alpha}}_\alpha = A^{\overline{\alpha}}_h(dA^h_\alpha + \check{\omega}^h_j A^j_\alpha) \tag{9.5.5}$$

Remarking that

$$dA^h_\alpha = \frac{\delta A^h_\alpha}{\delta u^\beta}du^\beta + \frac{\delta A^h_\alpha}{\delta v^{(1)\beta}}\delta v^{(1)\beta} + \dots + \frac{\delta A^h_\alpha}{\delta v^{(k-1)\beta}}\delta v^{(k-1)\beta} + \dot{\partial}^\beta\ A^h_\alpha \delta\, \check{p}_\beta$$

and $\overset{\vee}{\omega}{}^i_j$ is from (9.4.4) we obtain

$$\Pi^{\overline{\alpha}}_{\alpha} = A^{\overline{\alpha}}_h[(\frac{\delta A^h_\alpha}{\delta u^\beta} + A^j_\alpha \overset{\vee}{H}{}^h_{j\beta})du^\beta + (\frac{\delta A^h_\alpha}{\delta v^{(1)\beta}} + A^j_\alpha \underset{(1)}{\overset{\vee}{C}}{}^h_{j\beta})\delta v^{(1)\beta} + ...$$

$$... + (\frac{\delta A^h_\alpha}{\delta v^{(k-1)\beta}} + A^j_\alpha \underset{(k-1)}{\overset{\vee}{C}}{}^h_{j\beta})\delta v^{(k-1)\beta} + (\dot{\partial}^\beta A^h_\alpha + A^j_\alpha \underset{(1)}{\overset{\vee}{C}}{}^{h\beta}_j)\delta \overset{\vee}{p}_\beta .$$

The last expression of the 1-forms $\prod^{\overline{\alpha}}_{\alpha}$ can be written as in (9.5.3) with the coefficients (9.5.4).

In order to prove the second formula (9.5.2) we remark, using the same method, the formula

$$\nabla A^i_{\overline{\alpha}} = A^i_\alpha[A^\alpha_j(dA^j_{\overline{\alpha}} + A^r_{\overline{\alpha}}\overset{\vee}{\omega}{}^j_r)] = -A^i_\alpha \Pi^\alpha_{\overline{\alpha}}. \tag{9.5.6}$$

So, we have

$$\Pi^\alpha_{\overline{\alpha}} = -A^\alpha_j(dA^j_{\overline{\alpha}} + A^r_{\overline{\alpha}}\overset{\vee}{\omega}{}^j_r). \tag{9.5.6'}$$

The second relation between the 1-forms $\Pi^{\overline{\alpha}}_{\alpha}$ and $\Pi^\alpha_{\overline{\alpha}}$ is proved without difficulties by means of the equations $\nabla g_{ij} = 0, \nabla \overset{\vee}{g}_{\alpha\beta} = 0$ and $\nabla\delta_{\overline{\alpha}\overline{\beta}} = 0$, which will be proved in the next Lemma. q.e.d

Lemma 9.5.1 *The following properties of the relative covariant derivation ∇ hold:*

$$\nabla g_{ij} = 0,\ \nabla g_{\alpha\beta} = 0,\ \nabla\delta_{\overline{\alpha}\overline{\beta}} = 0. \tag{9.5.7}$$

Evidently, the equation $\nabla g_{ij} = 0$ is immediate. After the formula (9.2.8') we have

$$\overset{\vee}{g}_{\alpha\beta} = g_{ij}A^i_\alpha A^j_\beta,\ \delta_{\overline{\alpha}\overline{\beta}} = g_{ij}A^i_{\overline{\alpha}}A^j_{\overline{\beta}} \tag{9.5.8}$$

So, using (9.5.2) and $\nabla g_{ij} = 0$, we obtain $\nabla \overset{\vee}{g}_{\alpha\beta} = 0, \nabla\delta_{\overline{\alpha}\overline{\beta}} = 0$. q.e.d.

Remark 9.5.1 *1° Theorem 8.4.5 is a consequences of the previous Lemma.*

2° Because of $g_{ij}A^i_\alpha A^j_{\overline{\beta}} = 0$, applying the operator ∇ we get by means of (9.5.2) that $g_{ij}(\Pi^{\overline{\alpha}}_\alpha A^i_{\overline{\alpha}}A^j_{\overline{\beta}} - A^i_\alpha A^j_\beta \Pi^\beta_{\overline{\alpha}}) = 0$. But, this is $\delta_{\overline{\alpha}\overline{\beta}}\Pi^{\overline{\alpha}}_\alpha = \overset{\vee}{g}_{\alpha\beta}\,\Pi^\beta_{\overline{\alpha}}$, which imply the second formula (9.5.3).

The formulae (9.5.4) allows to prove that the coefficients

$$\underset{(0)}{H}{}^{\overline{\beta}}_{\alpha\gamma}, \underset{(1)}{H}{}^{\overline{\beta}}_{\alpha\gamma}, ..., \underset{(k-1)}{H}{}^{\overline{\beta}}_{\alpha\gamma}, H^{\overline{\beta}\gamma}_\alpha \tag{9.5.9}$$

are mixed d-tensor. They will be called the *second fundamental tensors* of the Hamilton subspace $\overset{\vee}{H}{}^{(k)m} = (\overset{\vee}{M}, \overset{\vee}{H})$ of the Hamilton space $H^{(k)n} = (M, H)$.

As an application of the previous considerations we get:

Proposition 9.5.1 *We have*

$$\overset{\vee}{\Omega}_{ij} = -\overset{\vee}{\Omega}_{ji}, \Omega_{\alpha\beta} = -\Omega_{\beta\alpha}, \Omega_{\overline{\alpha}\overline{\beta}} = -\Omega_{\overline{\beta}\overline{\alpha}} \tag{9.5.10}$$

The proof follows the same way as for the covariant curvature 2-forms Ω_{ij} of a metrical connection.

Let us consider a parametrized smooth curve $\overset{\vee}{c}$ on the Hamilton subspace $\overset{\vee}{H}{}^{(k)m} = (\overset{\vee}{M}, \overset{\vee}{H})$. Locally $\overset{\vee}{c}$ can be given by

$$u^\alpha = u^\alpha(t), v^{(1)\alpha} = v^{(1)\alpha}(t), ..., v^{(k-1)\alpha} = v^{(k-1)\alpha}(t), \overset{\vee}{p}_\alpha = \overset{\vee}{p}_\alpha(t)$$

A tangent vector X^i along $\overset{\vee}{c}$ is given by $X^i = A^i_\alpha X^\alpha$.

But, along curve $\overset{\vee}{c}$,we have

$$\frac{\nabla X^i}{dt} = A^i_{\overline{\alpha}} \Pi^{\overline{\alpha}}_\alpha X^\alpha + A^i_\alpha \frac{\nabla X^\alpha}{dt} \tag{9.5.11}$$

$\dfrac{\nabla X^\alpha}{dt} = 0$ along $\overset{\vee}{c}$, implies $\dfrac{\nabla X^i}{dt} = A^i_{\overline{\alpha}} \Pi^{\overline{\alpha}}_\alpha X^\alpha$. This means that $\dfrac{\nabla X^i}{dt}$ is normal to the subspace $\overset{\vee}{H}{}^{(k)m}$.

We say that the subspace $\overset{\vee}{H}{}^{(k)n}$ is *totally geodesic* in the space $H^{(k)n}$ if along any curve $\overset{\vee}{c}$, $\dfrac{\nabla X^\alpha}{dt} = 0$ implies $\dfrac{\nabla X^i}{dt} = 0$. The geometrical meaning of the condition is evident.

Theorem 9.5.2 *The Hamilton subspaces of order k, $\overset{\vee}{H}{}^{(k)n} = (\overset{\vee}{H}, \overset{\vee}{M})$ is totally geodesic in the Hamilton space of order k, $H^{(k)n} = (H, M)$ if and only if, the second fundamental tensors $\underset{(0)}{H}{}^{\overline{\beta}}_{\alpha\gamma}, \underset{(1)}{H}{}^{\overline{\beta}}_{\alpha\gamma}, ..., \underset{(k-1)}{H}{}^{\overline{\beta}}_{\alpha\gamma}, H^{\overline{\beta}\gamma}_\alpha$ vanish.*

Proof. If the second fundamental tensors vanish then $\Pi^{\overline{\alpha}}_\beta$ identically vanish and by (9.5.11) it follows that $\dfrac{\nabla X^\alpha}{dt} = 0$ imply $\dfrac{\nabla X^i}{dt} = 0$, for any $X^i = A^i_\alpha X^\alpha$.

Conversely, the condition $\dfrac{\nabla X^\alpha}{dt} = 0 \Rightarrow \dfrac{\nabla X^i}{dt} = 0$ for any $X^i = A^i_\alpha X^\alpha$ and the formula (9.5.11) leads to $\Pi^{\overline{\alpha}}_\alpha = 0$. Taking into account (9.5.3) and the fact that $\overset{\vee}{c}$ is arbitrary, it follows that all second fundamental tensors vanish.

9.6 The Gauss-Codazzi Equations

The Gauss-Codazzi equations of a Hamilton subspaces of order k, $\overset{\vee}{H}{}^{(k)m} = (\overset{\vee}{M}, \overset{\vee}{H})$ in the Hamilton spaces of order k, $H^{(k)n} = (H, M)$ endowed with a canonical metrical N-connection D, are obtained from the integrability conditions of the system of equalities (9.5.2). We can deduce these equations using the structure equations (9.4.23) and (9.4.24).

Theorem 9.6.1 *The Gauss-Codazzi equations of the Hamilton subspaces* $\check{H}^{(k)m}$ *in the Hamilton space* $H^{(k)n}$ *are as follows:*

$$A^i_\alpha A^j_\beta \check{\Omega}_{ij} - \Omega_{\alpha\beta} = \Pi_{\beta\overline{\gamma}} \wedge \Pi^{\overline{\gamma}}_{\alpha},$$

$$(9.6.1) \qquad A^i_{\overline{\alpha}} A^j_{\overline{\beta}} \check{\Omega}_{ij} - \Omega_{\overline{\alpha}\overline{\beta}} = \Pi_{\gamma\overline{\beta}} \wedge \Pi^{\gamma}_{\overline{\alpha}},$$

$$-A^i_\alpha A^{\overline{\alpha}}_\beta \check{\Omega}_{ij} = \delta_{\overline{\alpha}\overline{\gamma}}(d\Pi^{\overline{\gamma}}_\alpha + \Pi^{\overline{\gamma}}_\beta \wedge \omega^\beta_\alpha - \Pi^{\overline{\alpha}}_\alpha \wedge \omega^{\overline{\gamma}}_{\overline{\alpha}}),$$

where $\Pi_{\alpha\overline{\beta}} = \check{g}_{\alpha\gamma}\, \Pi^{\gamma}_{\overline{\beta}}$.

Proof. Consider the first equations (9.5.2), written in the form

$$(*) \qquad dA^i_\alpha + A^j_\alpha \check{\omega}^i_j - A^i_\beta \omega^\beta_\alpha = A^i_{\overline{\alpha}} \Pi^{\overline{\alpha}}_\alpha.$$

By exterior differentiation of the both side of this equation we get:

$$(**) \qquad dA^j_\alpha \wedge \check{\omega}^i_j + A^j_\alpha d\check{\omega}^i_j - dA^i_\beta \wedge \omega^\beta_\alpha - A^i_\beta d\omega^\beta_\alpha = dA^i_{\overline{\alpha}} \wedge \Pi^{\overline{\alpha}}_\alpha + A^i_{\overline{\alpha}} d\Pi^{\overline{\alpha}}_\alpha.$$

Looking to the equality (*) and at:

$$(***) \qquad dA^i_{\overline{\alpha}} + A^j_{\overline{\alpha}} \check{\omega}^i_j - A^i_{\overline{\beta}} \omega^{\overline{\beta}}_{\overline{\alpha}} = -A^i_\beta \Pi^\beta_{\overline{\alpha}},$$

the equality (**) becomes

$$(9.6.2) \qquad -A^j_{\overline{\alpha}} \check{\Omega}^i_j + A^i_\beta(\Omega^\beta_\alpha + \Pi^\beta_{\overline{\alpha}} \wedge \Pi^{\overline{\alpha}}_\alpha) = A^i_{\overline{\alpha}}(d\Pi^{\overline{\alpha}}_\alpha + \Pi^{\overline{\alpha}}_\beta \wedge \omega^\beta_\alpha - \Pi^{\overline{\beta}}_\alpha \wedge \omega^{\overline{\alpha}}_{\overline{\beta}}).$$

Now, multiplying (9.6.2) by $g_{ih}A^h_\beta = \check{g}_{\beta\gamma}\, A^\gamma_i$ we obtain the first equations (9.6.1). The same operation, taking the factor $g_{ij}A^j_{\overline{\beta}} = \delta_{\overline{\beta}\overline{\alpha}} A^{\overline{\alpha}}_i$ leads to third equation (9.6.1). Of course, to deduce the second equations (9.6.1) we will take the exterior differential of both sides of equations (***) and apply the same method. q.e.d.

In order to obtain the system of all fundamental equations of the subspaces $\check{H}^{(k)m}$ in $H^{(k)n}$, we must find the relations between the torsion 2-forms $\overset{(0)}{\Omega}{}^i, \overset{(a)}{\Omega}{}^i$, $(a = 1, ..., k-1), \Omega_i$ of the canonical metrical N-connection D of $H^{(k)n}$ and the torsion 2-forms $\overset{(0)}{\Omega}{}^\alpha, \overset{(a)}{\Omega}{}^\alpha, (a = 1, ..., k-1), \Omega_\alpha$ of the relative connection ∇.

Therefore, we obtain:

Theorem 9.6.2 *The fundamental equation of the Hamilton subspaces of order* k, $\check{H}^{(k)m} = (\check{M}, \check{H})$ *in a Hamilton space of order* k, $H^{(k)n} = (M, H)$ *endowed with the canonical metrical* N*-connection* D *are: the Gauss-Codazzi equations (9.6.1), as well as the following equations:*

$$d(dx^i) - dx^j \wedge \omega^i_j = \overset{(0)}{\Omega^i}, \text{ modulo (9.3.8)},$$

$$d(\delta y^{(a)i}) - \delta y^{(a)j} \wedge \omega^i_j = - \overset{(a)}{\Omega^i} \text{ modulo (9.3.8)}, \ (a = 1, ..., k-1),$$

$$d(\delta p_i) + \delta p_j \wedge \omega^j_i = -\Omega_i, \text{ modulo (9.3.8')}.$$

We end here the theory of Hamilton subspaces of order k in a Hamilton space of the same order.

We underline the importance of this theory for applications in the Higher Order Hamiltonian Mechanics.

The particular case $m = n - 1$ of the hypersurfaces $\check{M}$ in M can be obtained from the previous study without difficulties. If $k = 1$ we have the theory of Hamilton subspaces $\check{H}^n = (\check{M}, \check{H}(x, \check{p}))$ in a Hamilton space $H^n = (M, H(x, p))$.

Chapter 10

The Cartan Spaces of Order k as Dual of Finsler Spaces of Order k

The geometry of Finsler spaces of order k, $F^{(k)n} = (M, F(x, y^{(1)}, ..., y^{(k)}))$, studied in chapter 3, is a particular case of the geometry of Lagrange spaces of order k, $L^{(k)n} = (M, L(x, y^{(1)}, ..., y^{(k)}))$. The fundamental function F being a k-homogeneous regular Lagrangian on the fibres of the bundle T^kM.

As we know, the 'dual' of the space $L^{(k)n}$, via Legendre transformation *Leg*, is a Hamilton space of order k, $H^{(k)n} = (M, H(x, y^{(1)}, ..., y^{(k-1)}, p))$. In this chapter we will prove that the restriction of the mapping *Leg* to the Finsler spaces of order k determines a new class of Hamilton spaces of order k, that are called the Cartan spaces of order k and are denoted by $\mathcal{C}^{(k)n} = (M, K(x, y^{(1)}, ..., y^{(k-1)}, p))$.

Also, the spaces $\mathcal{C}^{(k)n}$ are the Hamilton spaces $H^{(k)n} = (M, H)$ in which the fundamental function $K(x, y^{(1)}, ..., y^{(k-1)}, p)$ is regular and k-homogeneous on the fibres of the dual bundle $T^{*k}M$.

For the spaces $\mathcal{C}^{(k)n}$ it is important to determine the fundamental geometrical object fields which are important in their differential geometry.

10.1 $\mathcal{C}^{(k)n}$-Spaces

Definition 10.1.1 *A Cartan space of order $k \geq 1$ is a pair $\mathcal{C}^{(k)n} = (M, K(x, y^{(1)}, ..., y^{(k-1)}, p))$ for which the following axioms hold:*

1⁰ K is a real function on the manifold $T^{*k}M$, differentiable on $\widetilde{T^{*k}M}$ and continuous on the null section of the projection $\pi^{*k} : T^{*k}M \to M$.

2⁰ $K > 0$ on $\widetilde{T^{*k}M}$.

3⁰ K is positively k-homogeneous on the fibres of bundle $T^{*k}M$, i.e.

$$K(x, ay^{(1)}, ..., a^{k-1}y^{(k-1)}, a^k p) = a^k K(x, y^{(1)}, ..., y^{(k-1)}, p), \ \forall a \in \mathbf{R}^+. \tag{10.1.1}$$

4^0 The Hessian of K^2, with respect to the momenta p_i, having the elements

$$g^{ij}(x, y^{(1)}, ..., y^{(k-1)}, p) = \frac{1}{2} \dot{\partial}^i \dot{\partial}^j K^2 \tag{10.1.2}$$

is positively defined.

It follows that the k-homogeneity of the function K is not k -homogeneity of K with respect to p_i. It is considered as in section 1, ch. 3. In the book [115], ch. 13 is considered only the case of homogeneity of K with respect to p_i.

We deduce from (10.1.2) that g^{ij} is contravariant of order two, symmetric and nondegenerate, i.e.

$$rank \left\| g^{ij} \right\| = n, \quad on \quad \widetilde{T^{*k}M}. \tag{10.1.3}$$

The function K is called the *fundamental* (or *metric*) function of $\mathcal{C}^{(k)n}$ and g^{ij} is *the fundamental tensor* of this space.

The first problem is if exist the Cartan spaces $\mathcal{C}^{(k)n}$.

Theorem 10.1.1 *If the base manifold M is paracompact, then on $T^{*k}M$ there exist functions $K(x, y^{(1)}, ..., y^{(k-1)}, p)$ such that the pair (M, K) is a Cartan space of order k.*

Proof: The manifold M being paracompact there exists a Finsler space of order $k-1$, $F^{(k-1)n} = (M, F(x, y^{(1)}, ..., y^{(k-1)}))$. Its fundamental tensor a_{ij}:

$$a_{ij}(x, y^{(1)}, ..., y^{(k-1)}) = \frac{1}{2} \frac{\partial^2 F^2}{\partial y^{(k-1)i} \partial y^{(k-1)j}}$$

is positively defined on $T^{k-1}M$ and 0-homogeneous on the fibres of bundle $(T^{k-1}M, \pi^{k-1}, M)$.

Now, we can construct on the dual bundle $T^{*k}M = T^{k-1}M \times_M T^*M$ the following Hamiltonian

$$K(x, y^{(1)}, ..., y^{(k-1)}, p) = \left\{ a^{ij}(x, y^{(1)}, ..., y^{(k-1)}) p_i p_j \right\}^{1/2}, \tag{10.1.4}$$

where a^{ij} is the contravariant tensor of a_{ij}.

It follows, without difficulties, that K is a scalar function (i.e. it does not depend of the transformations of coordinates on $T^{*k}M$) which satisfies the axioms 1^0-4^0 from Definition 1.1.

The fact that K is positively k-homogeneous on the fibres of the bundle $T^{*k}M$ follows directly. As a^{ij} are 0-homogeneous we have that

$$K(x, ay^{(1)}, ..., a^{k-1}y^{(k-1)}, a^k p) = a^k K(x, y^{(1)}, ..., y^{(k-1)}, p).$$

The fundamental tensor $g^{ij}(x, y^{(1)}, ..., y^{(k-1)}, p)$ coincides with $a^{ij}(x, y^{(1)}, ..., y^{(k-1)})$. The conclusion is that the pair (M, K), with K from (10.1.4) is a Cartan space of order k.

Remark 10.1.1 The Cartan space $\mathcal{C}^{(k)n}$ with the fundamental function K from (10.1.4) is a particular one, but it is an important and useful example of Cartan space of order k.

Remark 10.1.2 If $\mathcal{C}^{(k)n} = (M, K)$ is a Cartan space of order k, then $H^{(k)n} = (M, K^2)$ is a Hamilton space of order k having the same fundamental tensor g^{ij} as the space $\mathcal{C}^{(k)n}$. $H^{(k)n}$ will be called the Hamilton space associated to the Cartan space of order k , $\mathcal{C}^{(k)n}$. All geometrical properties of $H^{(k)n} = (M, K^2)$ are geometrical properties of $\mathcal{C}^{(k)n} = (M, K)$.

10.2 Geometrical Properties of the Cartan Spaces of Order k

First of all we shall study those properties of the spaces $\mathcal{C}^{(k)n} = (M, K)$ which result from the k-homogeneity of the fundamental function K expressed in the identity (10.1.1).

Consider the vector field on $T^{*k}M$:

$$\overset{k-1}{\Gamma} + kC^* = y^{(1)i}\frac{\partial}{\partial y^{(1)i}} + \cdots + (k-1)y^{(k-1)i}\frac{\partial}{\partial y^{(k-1)i}} + kp_i \dot{\partial}^i \tag{10.2.1}$$

and the Lie derivation with respect to this vector field $\mathcal{L}_{\overset{k-1}{\Gamma} + kC^*}$.

The functions K, K^2, $\dot{\partial}^j K^{2k}$, g^{ij} $(i, j = 1, ..., n)$ and C^{ijh}:

$$C^{ijh} = \frac{1}{2}\dot{\partial}^h g^{ij} = -\frac{1}{4}\dot{\partial}^i\dot{\partial}^j\dot{\partial}^h K^2 \tag{10.2.2}$$

are k, $2k$, k, 0 and $-k$ homogeneous, respectively on the fibres of the bundle $T^{*k}M$. Taking into account the Theorems 4.5.7 and 4.5.8 we have:

Proposition 10.2.1 *The following identities hold:*

$$\mathcal{L}_{\overset{k-1}{\Gamma} + kC^*} K = kK, \tag{10.2.3}$$

$$\mathcal{L}_{\overset{k-1}{\Gamma} + kC^*} K^2 = 2kK, \tag{10.2.4}$$

$$\mathcal{L}_{\overset{k-1}{\Gamma} + kC^*} \dot{\partial}^h K^2 = k\,\dot{\partial}^h K^2, \tag{10.2.5}$$

$$\mathcal{L}_{\overset{k-1}{\Gamma} + kC^*} g^{ij} = 0, \tag{10.2.6}$$

$$\mathcal{L}_{\overset{k-1}{\Gamma} + kC^*} C^{ijh} = -kC^{ijh}. \tag{10.2.7}$$

We remark, especially, 0-homogeneity of the functions $g^{ij}(x, y^{(1)}, ..., y^{(k-1)}, p)$. By virtue of (10.2.1) and (10.2.6) we have the identity:

$$y^{(1)i}\frac{\partial g^{jh}}{\partial y^{(1)i}} + \cdots + (k-1)y^{(k-1)i}\frac{\partial g^{jh}}{\partial y^{(k-1)i}} + kp_i \overset{\cdot i}{\partial} g^{jh} = 0. \tag{10.2.6'}$$

Of course, the covariant tensor field g_{ij} of the fundamental tensor g^{ij} of the Cartan space $\mathcal{C}^{(k)n}$ is 0-homogeneous on the fibres of $T^{*k}M$. Indeed, the equality (10.2.6) implies $\mathcal{L}_{\overset{k-1}{\Gamma}+kC^*} g_{ij} = 0$. Therefore we have the identity:

$$y^{(1)i}\frac{\partial g_{jh}}{\partial y^{(1)i}} + \cdots + (k-1)y^{(k-1)i}\frac{\partial g_{jh}}{\partial y^{(k-1)i}} + kp_i \overset{\cdot i}{\partial} g_{jh} = 0. \tag{10.2.6''}$$

In any Cartan space of order k, $\mathcal{C}^{(k)n} = (M, K)$, there exists two important tensors $\underset{(k-1)}{C}{}^{i}_{jh}$, C_i^{jh} which are the v_{k-1}- and w_k-coefficients of a canonical metrical connection.

We have:

Theorem 10.2.1 *For any Cartan space of order k, $\mathcal{C}^{(k)n} = (M, K)$, $\underset{(k-1)}{C}{}^{i}_{jh}$, C_i^{jh} given by*

$$\underset{(k-1)}{C}{}^{i}_{jh} = \frac{1}{2}g^{is}\left(\frac{\partial g_{sh}}{\partial y^{(k-1)j}} + \frac{\partial g_{js}}{\partial y^{(k-1)h}} - \frac{\partial g_{jh}}{\partial y^{(k-1)s}}\right), \tag{10.2.8}$$

$$C_i^{jh} = -\frac{1}{2}g_{is}\left(\frac{\partial g^{sh}}{\partial p_j} + \frac{\partial g^{js}}{\partial p_h} - \frac{\partial g^{jh}}{\partial p_s}\right) \tag{10.2.9}$$

have the following properties:

1⁰ They are d-tensor fields of type (1,2) and (2,1), respectively.

2⁰ $\underset{(k-1)}{C}{}^{i}_{jh}$ are $1-k$ homogeneous and C_i^{jh} are $-k$ homogeneous on the fibres of $T^{*k}M$.

3⁰ They are the v_{k-1}- and w_k-coefficients of a canonical metrical connection D, i.e.:

$$g^{ij} \overset{(k-1)}{|}_h = 0, \quad g^{ij}|^h = 0. \tag{10.2.10}$$

4⁰ We have

$$C_i^{jh} = g_{is}C^{sjh}. \tag{10.2.11}$$

5⁰ $\underset{(k-1)}{S}{}^{i}_{jh} = \underset{(k-1)}{C}{}^{i}_{jh} - \underset{(k-1)}{C}{}^{i}_{hj} = 0$, $S_i^{jh} = C_i^{jh} - C_i^{hj} = 0$.

The proofs of these affirmations are not difficult.

10.3 Canonical Presymplectic Structures, Variational Problem of the Space $\mathcal{C}^{(k)n}$

Consider a Cartan space of order k, $\mathcal{C}^{(k)n} = (M, K(x, y^{(1)}, ..., y^{(k-1)}, p))$, and the canonical presymplectic structure

$$\theta = dp_i \wedge dx^i, \tag{10.3.1}$$

where

$$\theta = d\omega, \omega = p_i dx^i. \tag{10.3.1'}$$

Of course the structure θ can be directly studied using the property of integrability, given by $d\theta = 0$. The relations between the structure θ and the Poisson structure $\{f, g\}_0$:

$$\{f, g\}_0 = \frac{\partial f}{\partial x^i}\frac{\partial g}{\partial p_i} - \frac{\partial f}{\partial p_i}\frac{\partial g}{\partial x^i}, \forall f, g, \in \mathcal{F}(\Sigma_0), \tag{10.3.2}$$

and

$$\Sigma_0 = \left\{(x, y^{(1)}, ..., y^{(k-1)}, p) | y^{(1)i} = \cdots = y^{(k-1)i} = 0\right\}, \tag{10.3.3}$$

lead to some important geometrical results.

Recall that the submanifold Σ_0 has been introduced in the section 3 of chapter 8 as a section of the canonical projection of the differentiable bundle $(T^{*k}M, \overline{\pi}^*, T^*M)$.

Let us consider the restriction K_0 of the fundamental function K of space $\mathcal{C}^{(k)n}$ to Σ_0:

$$K_0 = K_{|\Sigma_0}. \tag{10.3.4}$$

Thus we have

$$K_0(x, p) = K(x, 0, ..., 0, p).$$

It follows that K_0 is 1-homogeneous with respect to p_i.

Therefore the pair $\mathcal{C}_0^{(1)n} = (M, K_0)$ is a classical Hamilton space, [115], with the fundamental tensor field

$$g_0^{ij}(x, p) = g^{ij}(x, 0, ..., 0, p) = \frac{1}{2}\dot{\partial}^i\dot{\partial}^j K_0^2. \tag{10.3.5}$$

Theorem 7.3.1 is valid in the particular case of Cartan space $\mathcal{C}_0^{(1)n}$:

Theorem 10.3.1 *The pair (Σ_0, θ_0), with $\theta_0 = \theta_{|\Sigma_0}$, is a symplectic manifold.*

Since $\theta_0 = dp_i \wedge dx^i$ in every point $(x,p) \in \Sigma_0$, it is a closed 2-form of rank $2n = \dim \Sigma_0$.

The tangent space $T_u\Sigma_0$ at a point $u \in \Sigma_0$ has the natural basis $\left(\frac{\partial}{\partial x^i}, \frac{\partial}{\partial p_i}\right)_u$, $(i = 1, ..., n)$ and the natural cobasis (dx^i, dp_i), $(i = 1, ..., n)$.

Consider the $\mathcal{F}(\Sigma_0)$-module $\mathcal{X}(\Sigma_0)$ of vector fields and the $\mathcal{F}(\Sigma_0)$-module $\mathcal{X}^*(\Sigma_0)$ of 1-form fields on the submanifold Σ_0.

Taking into account the theory from ch. 8, section 3 we have:

1^0 The $\mathcal{F}(\Sigma_0)$-linear mapping $S_{\theta_0} : \mathcal{X}(\Sigma_0) \to \mathcal{X}^*(\Sigma_0)$ given by

$$S_{\theta_0}(X) = i_X\theta_0, \ \forall X \in \mathcal{X}(\Sigma_0)$$

is an isomorphism.

2^0 There exists an unique vector field $X_{K_0^2} \in \mathcal{X}(\Sigma_0)$ such that

$$S_{\theta_0}(X_{K_0^2}) = i_{X_{K_0^2}}\theta_0 = -dK_0^2.$$

3^0 The Hamiltonian vector field $X_{K_0^2}$ is given by

$$X_{K_0^2} = \frac{\partial K_0^2}{\partial p_i}\frac{\partial}{\partial x^i} - \frac{\partial K_0^2}{\partial x^i}\frac{\partial}{\partial p_i}.$$

Consequently, we can formulate:

Theorem 10.3.2 *The integral curves of the Hamiltonian vector field $X_{K_0^2}$ are given by the Σ_0-canonical equations:*

$$\frac{dx^i}{dt} = \frac{\partial K_0^2}{\partial p_i}, \ \frac{dp_i}{dt} = -\frac{\partial K_0^2}{\partial x^i}, \ y^{(\alpha)i} = 0, \ (\alpha = 1, ..., k-1). \tag{10.3.6}$$

Now, let us consider two functions $f, g \in \mathcal{F}(\Sigma_0)$ and the vector fields X_f, X_g given by

$$i_{X_f}\theta_0 = -df, \ i_{X_g}\theta_0 = -dg.$$

Proposition 10.3.1 *The following relations between the structures θ_0 and $\{,\}_0$ on the submanifold Σ_0, hold:*

$$\{f,g\}_0 = \theta_0(X_f, X_g), \ \forall f, g \in \mathcal{F}(\Sigma_0). \tag{10.3.7}$$

Remark 10.3.1 The triple $(T^{*k}M, K^2(x, y^{(1)}, ..., y^{(k-1)}, p), \theta)$ is a particular Hamiltonian system. It can be studied directly applying the method of Gotay, [115].

But, the equations (10.3.6) are extremely particular. For Cartan spaces $\mathcal{C}^{(k)n} = (M, K(x, y^{(1)}, ..., y^{(k-1)}, p))$ the integral of action (see Ch.5)

$$I(c) = \int_0^1 [p_i\frac{dx^i}{dt} - K(x, \frac{dx}{dt}, ..., \frac{1}{(k-1)!}\frac{d^{k-1}}{dt^{k-1}}, p)]dt \tag{10.3.8}$$

leads to the fundamental equations of the space:

$$(10.3.9)\qquad \begin{aligned} \frac{dx^i}{dt} &= \frac{1}{2}\frac{\partial K}{\partial p_i}, \\ \frac{dp_i}{dt} &= -\frac{1}{2}\Big[\frac{\partial K}{\partial x^i} - \frac{d}{dt}\frac{\partial K}{\partial y^{(1)i}} + \cdots + (-1)^{k-1}\frac{d^{k-1}}{dt^{k-1}}\frac{\partial K}{\partial y^{(k-1)i}}\Big] \end{aligned}$$

The integrand of the integral of action is k-homogeneous on the fibres of $T^{*k}M$. The Hamilton-Jacobi equations (10.3.9) are homogeneous on the fibres of $T^{*k}M$, too.

The energy of order $k-1$, $\mathcal{E}^{k-1}$, of the Cartan space $\mathcal{C}^{(k)n} = (M, K)$, by means of formula (5.3.1), is given by:

(10.3.10)

$$\mathcal{E}^{k-1}(K) = I^{k-1}(K) - \frac{1}{2!}\frac{d}{dt}I^{k-2}(K) + \cdots + (-1)^{k-2}\frac{1}{(k-1)!}\frac{d^{k-2}}{dt^{k-2}}I^1(K) - K.$$

We have:

Theorem 10.3.3 *For a Cartan space $\mathcal{C}^{(k)n} = (M, K)$, the energy of order $k-1$, $\mathcal{E}^{k-1}(K)$ is constant along every solution curve of the Hamilton-Jacobi equations.*

10.4 The Cartan Spaces $\mathcal{C}^{(k)n}$ as Dual of Finsler Spaces $F^{(k)n}$

Let $F^{(k)n} = (M, F(x, y^{(1)}, ..., y^{(k)}))$ be a Finsler space of order k having $F(x, y^{(1)}, ..., y^{(k)})$ as fundamental function and

$$(10.4.1)\qquad a_{ij}(x, y^{(1)}, ..., y^{(k)}) = \frac{1}{2}\frac{\partial^2 F^2}{\partial y^{(k)i}\partial y^{(k)j}}$$

as fundamental tensor.

Remembering the Definition 3.1.1, F is a function from T^kM to $\mathbf{R}$, differentiable on the manifold $\widetilde{T^kM} = T^kM \setminus \{0\}$ and continuous on the null section. F is a positive function, k-homogeneous:

$$F(x, ay^{(1)}, ..., a^ky^{(k)}) = a^kF(x, y^{(1)}, ..., y^{(k)}),\ \forall a \in \mathbf{R}^+$$

and the tensor field a_{ij} is positively defined on $\widetilde{T^kM}$.

The Legendre mapping $\varphi : \widetilde{T^kM} \to \widetilde{T^{*k}M}$ defined in section 4, ch.8, by

$$(10.4.2)\qquad \begin{cases} x^i = y^{(0)i}, y^{(1)i} = y^{(1)i}, ..., y^{(k-1)i} = y^{(k-1)i}, \\ p_i = \dfrac{1}{2}\dfrac{\partial F^2}{\partial y^{(k)i}} \end{cases}$$

is a local diffeomorphism.

We denote

$$p_i = \varphi_i(y^{(0)}, y^{(1)}, ..., y^{(k)}). \tag{10.4.2'}$$

Evidently $y^{(0)i} = x^i$ is the point of the base manifold M, $x = \pi^k(x, y^{(1)}, ..., y^{(k)})$.

The local inverse $\xi = \varphi^{-1} : \widetilde{T^{*k}M} \to \widetilde{T^kM}$ is expressed by (8.4.3):

$$\begin{cases} y^{(0)i} = x^i, y^{(1)i} = y^{(1)i}, ..., y^{(k-1)i} = y^{(k-1)i}, \\ y^{(k)i} = \xi^i(x, y^{(1)}, ..., y^{(k-1)}, p) \end{cases} \tag{10.4.3}$$

Remarking that the Legendre transformation φ, from (10.4.2) is k-homogeneous, it follows that its local inverse ξ^{-1} is k -homogeneous on the fibres of bundle $T^{*k}M$. So, we have:

$$\xi^i(x, ay^{(1)}, ..., a^{k-1}y^{(k-1)}, a^kp) = a^k\xi^i(x, y^{(1)}, ..., y^{(k-1)}, p), \ \forall a \in \mathbf{R}^+. \tag{10.4.4}$$

Applying the Theorem 8.4.1 we have:

Theorem 10.4.1 *The Legendre mapping φ transforms the canonical k-spray of the Finsler space $F^{(k)n}$:*

$$S = y^{(1)i}\frac{\partial}{\partial x^i} + \cdots + ky^{(k)i}\frac{\partial}{\partial y^{(k-1)i}} - (k+1)G^i\frac{\partial}{\partial y^{(k)i}} \tag{10.4.5}$$

with the coefficients

$$(k+1)G^i(x, y^{(1)}, ..., y^{(k)}) = \frac{1}{2}a^{ij}\left\{\Gamma\left(\frac{\partial F^2}{\partial y^{(k)j}}\right) - \frac{\partial F^2}{\partial y^{(k-1)j}}\right\}$$

in the ξ-dual k-spray

$$\begin{aligned} S^*_\xi = {} & y^{(1)i}\frac{\partial}{\partial x^i} + \cdots + (k-1)y^{(k-1)i}\frac{\partial}{\partial y^{(k-2)i}} + \\ & + k\xi^i(x, y^{(1)}, ..., y^{(k-1)}, p)\frac{\partial}{\partial y^{(k-1)i}} + \eta_i(x, y^{(1)}, ..., y^{(k-1)}, p)\frac{\partial}{\partial p_i} \end{aligned} \tag{10.4.6}$$

with ξ^i from (10.4.3) and

$$\begin{aligned} \eta_i = {} & -a_{is}\Big[\frac{\partial \xi^s}{\partial x^r}y^{(1)r} + \cdots + \\ & + (k-1)\frac{\partial \xi^s}{\partial y^{(k-1)r}}\xi^r + (k+1)G^s(x, y^{(1)}, ..., y^{(k-1)}, \xi)\Big] \end{aligned} \tag{10.4.7}$$

Proof: By means of theorem 8.4.1, the k-spray S, which is 2 -homogeneous on the fibres of T^kM, is transformed in the dual k -semispray S^*_ξ from (10.4.6), (10.4.7). We must prove that S^*_ξ is 2-homogeneous on the fibres of $T^{*k}M$. But every term of S^*_ξ is 2-homogeneous on the fibres of $T^{*k}M$. Thus, S^*_ξ is a dual k-spray.

Theorem 10.4.2 *The dual k-spray S^*_{ξ} from (10.4.6) determines a local nonlinear connection N^* on the manifold $T^{*k}M$, which depends on the Finsler space of order k, $F^{(k)n}$, and N^* has the following dual coefficients*

$$(10.4.8)\qquad \begin{aligned} &\underset{(1)}{M^*}{}^i_j = -\frac{\partial \xi^i}{\partial y^{(1)j}}, \ldots, \underset{(k-1)}{M}{}^{*}{}^i_j = -\frac{\partial \xi^i}{\partial y^{(k-1)j}},\\ &N^*_{ij} = \frac{\delta \eta_i}{\delta y^{(1)j}}, \end{aligned}$$

where the operators $\dfrac{\delta}{\delta y^{(1)j}}$ are constructed by means of the coefficients $\underset{(1)}{N^*}{}^i_j$, ..., $\underset{(k-1)}{N}{}^{*}{}^i_j$ from the dual coefficients (10.4.8).

Indeed, this theorem is just the Theorem 8.4.2 applied to Finsler spaces of order k.

We remark the following property of homogeneity of the coefficients M^*, N^*:

Proposition 10.4.1 *The coefficients $\underset{(1)}{M^*}{}^i_j$, ..., $\underset{(k-1)}{M}{}^{*}{}^i_j$, N^*_{ij} are homogeneous on the fibres of $T^{*k}M$ of degree $k-1$, ..., 1, k, respectively.*

Indeed, ξ^i being k-homogeneous and η_i from (10.4.7) being $k+1$ homogeneous, by means of (10.4.8) the property follows.

Now, let us consider $\overset{\circ}{N}$ a fixed nonlinear connection on the submanifold $T^{k-1}M$ of $T^{*k}M = T^{k-1}M \times_M T^*M$. We assume that the coefficients $\underset{(1)}{\overset{\circ}{M}}{}^i_j$, ..., $\underset{(k-1)}{\overset{\circ}{M}}{}^i_j$ are homogeneous of degree 1, ..., $k-1$ on the fibres of $T^{k-1}M$.

Thus, the Legendre mapping $\varphi : u = (x, y^{(1)}, ..., y^{(k-1)}, y^{(k)}) \in T^kM \rightarrow u^* = (x, y^{(1)}, ..., y^{(k-1)}, p) \in T^{*k}M$ transforms the Liouville d-vector field $z^{(k)i}$ at the point u:

$$(10.4.9)\qquad kz^{(k)i} = ky^{(k)i} + (k-1)\underset{(1)}{\overset{\circ}{M}}{}^i_m y^{(k-1)m} + \cdots + \underset{(k-1)}{\overset{\circ}{M}}{}^i_m y^{(1)m}$$

in the d-vector field $\overset{\vee}{z}{}^{(k)i}$ at the point u^*. The vector field $\overset{\vee}{z}{}^{(k)i}$ is:

$$(10.4.9')\qquad k\overset{\vee}{z}{}^{(k)i} = k\xi^i + (k-1)\underset{(1)}{\overset{\circ}{M}}{}^i_m y^{(k-1)m} + \cdots + \underset{(k-1)}{\overset{\circ}{M}}{}^i_m y^{(1)m}.$$

Of course, $\overset{\vee}{z}{}^{(k)i}$ is k-homogeneous on the fibres of $T^{*k}M$. Consider the function (10.4.10)

$$K^2(x, y^{(1)}, ..., y^{(k-1)}, p) = 2p_i\overset{\vee}{z}{}^{(k)i} - F^2(x, y^{(1)}, ..., y^{(k-1)}, \xi(x, y^{(1)}, ..., y^{(k-1)}, p)).$$

Theorem 10.4.3 *We have:*

1⁰ The pair $(M, K^2(x, y^{(1)}, ..., y^{(k-1)}, p))$ is a Hamilton space. Its fundamental function K^2 is $2k$-homogeneous on the fibres of $T^{*k}M$.

2⁰ The fundamental tensor of the space (M, K^2) is positively defined and it does not depend on the apriori given nonlinear connection $\overset{\circ}{N}$. It is given by:

$$g^{ij}(x, y^{(1)}, ..., y^{(k-1)}, p) = a^{ij}(x, y^{(1)}, ..., y^{(k-1)}, \xi^i(x, y^{(1)}, ..., y^{(k-1)}, p)).$$

The proof follows from Theorem 8.4.3 in which $H = K^2$.

We shall say that (M, K^2) is a Cartan spaces of order k, dual (via Legendre transformation) to the Lagrange space $L^{(k)n} = (M, F^2)$ associated to the Finsler space $F^{(k)n} = (M, F)$.

The inverse problem: being given a Cartan space of order k, $\mathcal{C}^{(k)n} = (M, K(x, y^{(1)}, ..., y^{(k-1)}, p))$ let us determine its dual as a Finsler space of order k, $F^{(k)n} = (M, F(x, y^{(1)}, ..., y^{(k-1)}, y^{(k)}))$. We follow the theory that has been done in the section 5, ch. 8.

Let $\overset{\circ}{N}$ be an apriori given nonlinear connection on $T^{k-1}M$, with the dual coefficients $\left(\underset{(1)}{\overset{\circ}{M}}{}^{\,i}_{\,j}, ..., \underset{(k-1)}{\overset{\circ}{M}}{}^{\,i}_{\,j}\right)$ and the mapping

$$\xi^* : u^* = (x, y^{(1)}, ..., y^{(k-1)}, p) \in T^{*k}M \to u = (x, y^{(1)}, ..., y^{(k-1)}, y^{(k)}) \in T^kM$$

defined by

$$\text{(10.4.11)} \qquad \begin{cases} y^{(0)i} = x^i, y^{(1)i} = y^{(1)i}, ..., y^{(k-1)i} = y^{(k-1)i}, \\ y^{(k)i} = \xi^{*i}(x, y^{(1)}, ..., y^{(k-1)}, p), \end{cases}$$

where ξ^{*i} is expressed from the formula

$$\text{(10.4.12)} \qquad k\xi^{*i} + (k-1)\underset{(1)}{\overset{\circ}{M}}{}^{\,i}_{\,j} y^{(k-1)j} + \cdots + \underset{(k-1)}{\overset{\circ}{M}}{}^{\,i}_{\,j} y^{(1)j} = \frac{k}{2}\dot{\partial}^i K^2.$$

Theorem 10.4.4 *The mapping ξ^* is a local diffeomorphism which preserves the fibres of $T^{*k}M$ and T^kM.*

Indeed, by means of theorem 7.5.1, the Jacobian of ξ^* is det $||g^{ij}||$, g^{ij} being the fundamental tensor of the Cartan space $\mathcal{C}^{(k)n}$. q.e.d.

The formulas (10.4.11) and (10.4.12) imply:

$$\text{(10.4.13)} \qquad \begin{aligned} &\dot{\partial}^i \xi^{*j} = g^{ij}, \\ &z^{(k)i}(x, y^{(1)}, ..., y^{(k-1)}, \xi^*(u^*)) = \frac{1}{2}\dot{\partial}^i K^2(x, y^{(1)}, ..., y^{(k-1)}, p), \end{aligned}$$

where $z^{(k)i}$ is the Liouville vector field (10.4.9).

Let $F^2 = \xi^*(K^2)$ be the Lagrangian

(10.4.14)
$$F^2(x, y^{(1)}, ..., y^{(k)}) = 2p_i z^{(k)i}(x, y^{(1)}, ..., y^{(k-1)}, \xi^*) - K^2(x, y^{(1)}, ..., y^{(k-1)}, \varphi^*).$$

where φ^* is the local inverse of ξ^*.

Theorem 8.5.2 allows to state:

Theorem 10.4.5 *The pair (M, F), with F from (10.4.14), has the properties:*

1^0 It is a Finsler space, having the fundamental function F^2, $2k$ -homogeneous on the fibres of T^kM.

2^0 Its fundamental tensor field is given by

$$a_{ij}(u) = g_{ij}(x, y^{(1)}, ..., y^{(k-1)}, \varphi^*(u)).$$

So, the space (M, F) is called *the Finsler space of order k dual to the Cartan space of order k, $\mathcal{C}^{(k)n} = (M, K)$.*

¿From (10.4.14) we deduce

$$(10.4.15) \qquad p_i = \frac{1}{2}\frac{\partial F^2}{\partial y^{(k)i}}, \; p_i = \varphi_i^*(x, y^{(1)}, ..., y^{(k)}).$$

Therefore, by means of the Theorem 8.5.3, we have:

Theorem 10.4.6 *The Legendre transformation determined by the Lagrange space $L^{(k)n} = (M, F^2)$, with F^2 from (10.4.14) is defined by the local diffeomorphism φ^*, the inverse of the local diffeomorphism ξ^* and $K^2 = \varphi^*(F^2)$ is given by*

$$K^2(x, y^{(1)}, ..., y^{(k-1)}, p) = 2p_i z^{(k)i}(x, y^{(1)}, ..., y^{(k-1)}, \xi^*) - F^2(x, y^{(1)}, ..., y^{(k-1)}, \xi^*).$$

Remark 10.4.1 *It follows $\varphi = \varphi^*$ and $\varphi^{-1} = \xi^*$. Therefore, locally, we get $K^2 = \varphi^*(\xi^*(K^2))$ and $F^2 = \xi^*(\varphi^*(F^2))$.*

10.5 Canonical Nonlinear Connection. N-Linear Connections

As we know, from section 6, Ch. 8 and from Theorem 10.4.6 we can determine a nonlinear connection N^* of the Cartan space $\mathcal{C}^{(k)n} = (M, K)$ by means of an apriori given nonlinear connection $\overset{\circ}{N}$ on the manifold $T^{k-1}M$. We construct a bundle morphism and determine the space $L^{(k)n} = (M, F^2)$, with $F^2 = \xi^*(K^2)$.

The Legendre transformation $\varphi^* = \xi^{*-1}$ (with $\varphi_i^* = \frac{1}{2}\frac{\partial F^2}{\partial y^{(k)i}}$) transforms the k-spray S of $L^{(k)n}$ in the dual k-spray $S^*_{\xi^*}$:

$$(10.5.1) \; S^*_{\xi^*} = y^{(1)i}\frac{\partial}{\partial x^i} + \cdots + (k-1)y^{(k-1)i}\frac{\partial}{\partial y^{(k-2)i}} + k\xi^{*i}\frac{\partial}{\partial y^{(k-1)i}} + \eta_i\frac{\partial}{\partial p_i}$$

Theorem 10.5.1 *The following systems of functions*

$$(10.5.2)\qquad \underset{(1)}{M}{}^{*i}_{j} = -\frac{\partial \xi^{*i}}{\partial y^{(1)j}}, \;\ldots,\; \underset{(k-1)}{M}{}^{*i}_{j} = -\frac{\partial \xi^{*i}}{\partial y^{(k-1)j}},$$

and

$$(10.5.2')\qquad N^{*}_{ij} = \frac{\delta \eta_i}{\delta y^{(1)j}},$$

where the operators $\dfrac{\delta}{\delta y^{(1)i}}$ *are constructed by means of (10.5.2), give the dual coefficients of a nonlinear connection* N^* *which depends on the fundamental function* K *of Cartan space* $\mathcal{C}^{(k)n} = (M, K)$ *and on the apriori given nonlinear connection* $\overset{\circ}{N}$ *of the manifold* $T^{k-1}M$.

Indeed, this is a particular case of the Theorem 8.6.1.

The nonlinear connection N^* is called *canonical* for the Cartan space $\mathcal{C}^{(k)n} = (M, K)$.

In the following we denote N^* by N and consider the adapted basis and adapted cobasis determined by N and by the vertical distribution W_k:

$$(10.5.3)\qquad \left\{ \frac{\delta}{\delta x^i}, \frac{\delta}{\delta y^{(1)i}}, \ldots, \frac{\delta}{\delta y^{(k-1)i}}, \frac{\partial}{\partial p_i} = \dot{\partial}^{i} \right\}$$

and

$$(10.5.3')\qquad \left\{ dx^i, \delta y^{(1)i}, \ldots, \delta y^{(k-1)i}, \delta p_i \right\}.$$

Let D be a N-linear connection with the coefficients $D\Gamma(N) = (H^i_{jh}, \underset{(\alpha)}{C}{}^{i}_{jh}, C_i^{jh})$, $(\alpha = 1, \ldots, k-1)$.

The fundamental tensor field g^{ij} of $\mathcal{C}^{(k)n}$ is absolute parallel with respect to D if

$$(10.5.4)\qquad g^{ij}{}_{|h} = 0,\; g^{ij} \overset{(\alpha)}{|}_{h} = 0,\; (\alpha = 1, \ldots, k-1),\; g^{ij}|^{h} = 0.$$

Assume that the equations (10.5.4) hold. Then D is called the *metrical* N*-linear connection.*

In the case when h-, v_α-, and w_k-torsions of D vanish, D is called *canonical metrical* N*-linear connection* of the Cartan space $\mathcal{C}^{(k)n}$.

Theorem 8.7.1 leads to the following important result:

Theorem 10.5.2 *1⁰ There exists an unique canonical metrical* N*-linear connection* D *of the space* $\mathcal{C}^{(k)n} = (M, K)$*. Its coefficients are given by the gener-*

alized Christoffel symbols:

$$H^i_{jh} = \frac{1}{2} g^{is} \left(\frac{\delta g_{sh}}{\delta x^j} + \frac{\delta g_{js}}{\delta x^h} - \frac{\delta g_{jh}}{\delta x^s} \right),$$

(10.5.5) $$\underset{(\alpha)}{C}{}^i_{jh} = \frac{1}{2} g^{is} \left(\frac{\delta g_{sh}}{\delta y^{(\alpha)j}} + \frac{\delta g_{js}}{\delta y^{(\alpha)h}} - \frac{\delta g_{jh}}{\delta y^{(\alpha)s}} \right), \ (\alpha = 1, ..., k-1),$$

$$C^{jh}_i = -\frac{1}{2} g_{is} \left(\frac{\partial g^{sh}}{\partial p_j} + \frac{\partial g^{js}}{\partial p_h} - \frac{\partial g^{jh}}{\partial p_s} \right).$$

2⁰ This connection depends only on the canonical nonlinear connection N and on the fundamental function K of the Cartan space of order k, $\mathcal{C}^{(k)n}$.

3⁰ The coefficients $(H^i_{jh}, \underset{(1)}{C}{}^i_{jh}, ..., \underset{(k-1)}{C}{}^i_{jh}, C^{jh}_i)$ *are homogeneous on the fibres of bundle* $T^{*k}M$ *of degree:* $0, -1, ..., 1-k, -k$, *respectively.*

Corollary 10.5.1 *The following identities hold*

$$\mathcal{L}_{\overset{k-1}{\Gamma}+kC^*} H^i_{jh} = 0,$$

(10.5.6) $$\mathcal{L}_{\overset{k-1}{\Gamma}+kC^*} \underset{(\alpha)}{C}{}^i_{jh} = -\alpha \underset{(\alpha)}{C}{}^i_{jh}, \ (\alpha = 1, ..., k-1),$$

$$\mathcal{L}_{\overset{k-1}{\Gamma}+kC^*} C^{jh}_i = -k C^{jh}_i.$$

Corollary 10.5.2 *We have*

(10.5.7) $$C^{jh}_i = g_{is} C^{sjh}.$$

Consider the d-Liouville vector fields of the space $\mathcal{C}^{(k)n}$: $z^{(1)i}$, ..., $z^{(k-1)i}$, (cf. (6.2.7), (6.2.7')). They are homogeneous on the fibres of $T^{*k}M$ of degree 1, 2, ..., $k-1$, respectively.

The deflection tensor fields of the canonical metrical N-linear connection D are given by (see ch. 7, section 5):

(10.5.8)
$$\overset{(\alpha)}{D}{}^i_j = \overset{(\alpha)}{z}{}^i{}_{|j} = \frac{\delta \overset{(\alpha)}{z}{}^i}{\delta x^j} + \overset{(\alpha)}{z}{}^m H^i_{mj},$$
$$\overset{(\alpha\beta)}{D}{}^i_j = \overset{(\alpha)}{z}{}^i \overset{(\beta)}{|}_j = \frac{\delta \overset{(\alpha)}{z}{}^i}{\delta y^{(\beta)j}} + \overset{(\alpha)}{z}{}^m \underset{(\beta)}{C}{}^i_{mj},$$
$$\overset{(\alpha)}{D}{}^{ij} = \overset{(\alpha)}{z}{}^i{}^{|j} = \frac{\delta \overset{(\alpha)}{z}{}^i}{\delta p_j} + \overset{(\alpha)}{z}{}^m C^{ij}_m, \ (\alpha, \beta = 1, ..., k-1)$$

and

(10.5.8')
$$\Delta_{ij} = p_{i|j} = -p_m H^m_{ij},$$
$$\overset{(\alpha)}{\delta}_{ij} = p_i \overset{(\alpha)}{|}_j = -p_m \underset{(\alpha)}{C}{}^m_{ij}, \ (\alpha = 1, ..., k-1),$$
$$\delta^j_i = p_i|^j = \delta^j_i - p_m C^{mi}_i.$$

The degrees of homogeneity on the fibres of $T^{*k}M$ for these tensor fields are easily determined.

The deflection tensors (10.5.8) and (10.5.8') lead to important identities for the canonical metrical N-linear connection D, derived from the Ricci identities applied to the Liouville d-vector fields $z^{(1)i}, \dots, z^{(k-1)i}$.

Indeed, Theorem 7.6.2 and 7.6.3 give us:

Theorem 10.5.3 *The canonical metrical N-linear connection of Cartan space $\mathcal{C}^{(k)n}$ satisfies the following identities:*

$$\overset{(\alpha)}{D}{}^{i}{}_{j|h} - \overset{(\alpha)}{D}{}^{i}{}_{h|j} = z^{(\alpha)s} R_{s}{}^{i}{}_{jh} - \sum_{\beta=1}^{k-1}\left\{\overset{(\alpha\beta)}{D}{}^{i}{}_{s} \underset{(0\beta)}{R}{}^{s}{}_{jh}\right\} + \overset{(\alpha)}{D}{}^{is} \underset{(0)}{R}_{sjh},$$

$$\overset{(\alpha)}{D}{}^{i}{}_{j}\overset{(\beta)}{|}_{h} - \overset{(\alpha\beta)}{D}{}^{i}{}_{h|j} = z^{(\alpha)s} \underset{(\beta)}{P}{}^{i}{}_{jh} - \overset{(\alpha)}{D}{}^{i}{}_{s} \underset{(\beta)}{C}{}^{s}{}_{jh} - \overset{(\alpha\beta)}{D}{}^{i}{}_{s} H^{s}_{jh} -$$
$$- \sum_{\gamma=1}^{k-1}\left\{\overset{(\alpha\gamma)}{D}{}^{i}{}_{s} \underset{(\beta\gamma)}{B}{}^{s}{}_{jh}\right\} - \overset{(\alpha)}{D}{}^{is} \underset{(\beta)}{B}_{sjh},$$

$$\overset{(\alpha)}{D}{}^{i}{}_{j}|^{h} - \overset{(\alpha)}{D}{}^{ih}{}_{|j} = z^{(\alpha)s} P_{s}{}^{i}{}_{j}{}^{h} - \overset{(\alpha)}{D}{}^{i}_{s} C^{sh}{}_{j} - \overset{(\alpha)}{D}{}^{is} H_{s}{}^{h}{}_{j} -$$
$$- \sum_{\gamma=1}^{k-1}\left\{\overset{(\alpha\gamma)}{D}{}^{i}{}_{s} \underset{(\gamma)}{B}{}^{s}{}^{h}{}_{j}\right\} - \overset{(\alpha)}{D}{}^{is} \underset{(0)}{B}{}_{sj}{}^{h},$$

$$\overset{(\alpha\beta)}{D}{}^{i}_{j}\overset{(\gamma)}{|}_{h} - \overset{(\alpha\gamma)}{D}{}^{i}_{h}\overset{(\beta)}{|}_{j} = z^{(\alpha)s} \underset{(\beta\gamma)}{S}{}_{s}{}^{i}{}_{jh} - \overset{(\alpha\beta)}{D}{}^{is} \underset{(\gamma)}{C}{}^{s}_{jh} + \overset{(\alpha\gamma)}{D}{}^{i}_{s} \underset{(\beta)}{C}{}^{s}_{hj} -$$
$$(10.5.9)\qquad - \sum_{\sigma=1}^{k-1}\left\{\overset{(\alpha\sigma)}{D}{}^{i}{}_{s} \underset{(\beta\gamma)}{\overset{(\sigma)}{C}}{}^{s}{}_{jh}\right\} - \overset{(\alpha)}{D}{}^{is} \underset{(\beta\gamma)}{B}{}_{sjh},$$

$$\overset{(\alpha\beta)}{D}{}^{i}{}_{j}|^{h} - \overset{(\alpha)}{D}{}^{ih}\overset{(\beta)}{|}_{j} = z^{(\alpha)s} \underset{(\beta)}{S}{}_{s}{}^{i}{}_{j}{}^{h} - \overset{(\alpha)}{D}{}^{is} \underset{(\beta)}{C}{}_{s}{}^{h}{}_{j} - \overset{(\alpha\beta)}{D}{}^{i}_{s} C^{sh}_{j} -$$
$$- \sum_{\sigma=1}^{k-1}\left\{\overset{(\alpha\sigma)}{D}{}^{i}{}_{s} \underset{(\beta\sigma)}{B}{}^{s}{}^{h}{}_{j}\right\} - \overset{(\alpha)}{D}{}^{is} \underset{(\beta)}{C}{}_{sj}{}^{h},$$

$$\overset{(\alpha)}{D}{}^{ij}|^{h} - \overset{(\alpha)}{D}{}^{ih}|^{j} = z^{(\alpha)s} S^{ijh}_{s}$$

and

$$\Delta_{ij|h} - \Delta_{ih|j} = -p_s R_{i}{}^{s}{}_{jh} - \sum_{\gamma=1}^{k-1}\left(\overset{(\gamma)}{\delta}_{is} \underset{(0\gamma)}{R}{}^{s}_{jh}\right) - \delta^{s}_{i} \underset{(0)}{R}_{sjh},$$

$$\Delta_{ij}\overset{(\beta)}{|}_{h} - \delta_{ih|j} = -p_s \underset{(\beta)}{P}{}_{i}{}^{s}{}_{jh} - \Delta_{is} \underset{(\beta)}{C}{}^{s}_{jh} - \overset{(\beta)}{\delta}_{is} H^{s}_{jh} - \sum_{\gamma=1}^{k-1}\left(\overset{(\gamma)}{\delta}_{is} \underset{(\alpha\gamma)}{B}{}^{s}{}_{jh}\right) - \delta^{s}_{i} \underset{(\beta)}{B}_{sjh},$$

$$\Delta_{ij}|^{h} - \delta_{i}{}^{h}{}_{|j} = -p_s P_{i}{}^{s}{}_{j}{}^{h} - \Delta_{is} C^{sh}_{j} - \delta^{s}_{i} H_{s}{}^{h}{}_{j} - \sum_{\gamma=1}^{k-1}\left(\overset{(\gamma)}{\delta}_{is} \underset{(\gamma)}{B}{}^{sh}_{j}\right) - \delta^{s}_{i} \underset{(0)}{B}{}_{sj}{}^{h},$$

$$
\begin{aligned}
\overset{(\beta)}{\delta}_{ij}\,\overset{(\gamma)}{|}_{h} - \overset{(\gamma)}{\delta}_{ih}\,\overset{(\beta)}{|}_{j} = & -p_s \underset{(\beta\gamma)}{S}{}_{i}{}^{s}{}_{jh} - \overset{(\beta)}{\delta}_{is} \underset{(\gamma)}{C}{}^{s}{}_{jh} - \\
& - \overset{(\gamma)}{\delta}_{is} \underset{(\beta)}{C}{}^{s}_{hj} - \\
& - \sum_{\sigma=1}^{k-1} \left(\overset{(\sigma)}{\delta}_{is} \underset{(\beta\gamma)}{\overset{(\sigma)}{C}}{}^{s}{}_{jh} \right) - \delta_i^s \underset{(\beta\gamma)}{B}{}_{sjh},
\end{aligned}
\tag{10.5.9'}
$$

$$
\overset{(\beta)}{\delta}_{ij}\,|^{h} - \delta_i^h\,\overset{(\beta)}{|}_{j} = -p_s \underset{(\beta)}{S}{}_{i}{}^{s}{}_{j}{}^{h} - \delta_i^s \underset{(\beta)}{C}{}_{s}{}^{h}{}_{j} - \overset{(\beta)}{\delta}_{is} C_j^{sh} - \sum_{\sigma=1}^{k-1} \left(\overset{(\sigma)}{\delta}_{is} \underset{(\beta\sigma)}{C}{}^{sh}{}_{j} \right) - \delta_i^s \underset{(\beta)}{C}{}_{sj}^{h},
$$

$$
\delta_i{}^{j}|^{h} - \delta_i{}^{h}|^{j} = -p_s S_i{}^{sjh}.
$$

The Ricci identities and Bianchi identities of the canonical metrical N -linear connection of the Cartan space $\mathcal{C}^{(k)n}$ can be written using the corresponding indentities of the N-linear connections of $T^{*k}M$, given in the chapter 6.

Applying the Ricci identities to the fundamental tensor g^{ij} of the space $\mathcal{C}^{(k)n}$, with respect to the canonical metrical N -linear connection, we obtain the identities (7.6.8).

So, we have:

Theorem 10.5.4 *In a Cartan space $\mathcal{C}^{(k)n}$, with respect to the canonical metrical N-linear connection, the following identities hold:*

$$
g^{sj} R_s{}^{i}{}_{hm} + g^{is} R_s{}^{j}{}_{hm} = 0,
$$

$$
g^{sj} \underset{(\alpha)}{P}{}_{s}{}^{i}{}_{hm} + g^{is} \underset{(\alpha)}{P}{}_{s}{}^{j}{}_{hm} = 0, \ (\alpha = 1, ..., k-1),
$$

$$
g^{sj} S_s{}^{ihm} + g^{is} S_s{}^{jhm} = 0.
$$

10.6 Parallelism of Vector Fields in Cartan Space $\mathcal{C}^{(k)n}$

Consider a Cartan space of order k, $\mathcal{C}^{(k)n} = (M, K)$, endowed with the canonical metrical N-linear connection $C\Gamma(N) = (H^i_{jm}, \underset{(\alpha)}{C}{}^{i}{}_{jm}, C_i^{jm})$, $(\alpha = 1, ..., k-1)$, the coefficients being given by (10.5.5). The local vector fields $\left(\dfrac{\delta}{\delta x^i}, \dfrac{\delta}{\delta y^{(\alpha)i}}, \dfrac{\delta}{\delta p_i} \right)$, $(\alpha = 1, .., k-1)$ determine an adapted basis and $(dx^i, \delta y^{(\alpha)i}, \delta p_i)$, $(\alpha = 1, ..., k-1)$ is its dual basis.

Along a smooth parametrized curve $\gamma : I \to T^{*k}M$, having the image in a domain of a local chart:

$$
x^i = x^i(t),\ y^{(\alpha)i} = y^{(\alpha)i}(t),\ p_i = p_i(t),\ t \in I,\ (\alpha = 1, .., k-1)
\tag{10.6.1}
$$

the tangent vector field $\dot{\gamma}$ is given by (7.7.2), i.e.

$$\dot{\gamma} = \frac{dx^i}{dt}\frac{\delta}{\delta x^i} + \sum_{\alpha=1}^{k-1}\frac{\delta y^{(\alpha)i}}{dt}\frac{\delta}{\delta y^{(\alpha)i}} + \frac{\delta p_i}{dt}\dot{\partial}^i, \tag{10.6.2}$$

where

$$\begin{aligned} \frac{\delta y^{(\alpha)i}}{dt} &= \frac{dy^{(\alpha)i}}{dt} + \underset{(1)}{M}{}^i_j\frac{dy^{(\alpha-1)j}}{dt} + \dots + \underset{(\alpha-1)}{M}{}^i_j\frac{dy^{(1)j}}{dt} + \underset{(\alpha)}{M}{}^i_j\frac{dx^j}{dt}, \\ \frac{\delta p_i}{dt} &= \frac{dp_i}{dt} - N_{ji}\frac{dx^j}{dt}, \qquad (\alpha = 1, ..., k-1). \end{aligned} \tag{10.6.3}$$

Consider the 1-forms of metrical canonical N-linear connection $C\Gamma(N)$:

$$\omega^i{}_j = H^i_{js}dx^s + \sum_{\alpha=1}^{k-1}\underset{(\alpha)}{C}{}^i_{js}\delta y^{(\alpha)s} + C_j^{is}\delta p_s. \tag{10.6.4}$$

Then the vector field $X \in \mathcal{X}(T^{*k}M)$:

$$X = \overset{(0)}{X}{}^i\frac{\delta}{\delta x^i} + \sum_{\alpha=1}^{k-1}\overset{(\alpha)}{X}{}^i\frac{\delta}{\delta y^{(\alpha)i}} + X_i\dot{\partial}^i \tag{10.6.5}$$

has the covariant differential along curve γ:

$$\begin{aligned} \frac{DX}{dt} = &\left(\frac{d\overset{(0)}{X}{}^i}{dt} + \overset{(0)}{X}{}^s\frac{\omega^i_s}{dt}\right)\frac{\delta}{\delta x^i} + \sum_{\alpha=1}^{k-1}\left(\frac{d\overset{(\alpha)}{X}{}^i}{dt} + \overset{(\alpha)}{X}{}^s\frac{\omega^i_s}{dt}\right)\frac{\delta}{\delta y^{(\alpha)i}} + \\ &+\left(\frac{dX_i}{dt} - X_s\frac{\omega^s_i}{dt}\right)\dot{\partial}^i. \end{aligned} \tag{10.6.6}$$

Theorem 7.7.1 takes the form:

Theorem 10.6.1 *The vector X from (10.6.5) is parallel along curve γ if and only if $(\overset{(0)}{X}{}^i, \overset{(\alpha)}{X}{}^i, X_i)$ are the solutions of the system of differential equations:*

$$\begin{cases} \dfrac{d\overset{(0)}{X}{}^i}{dt} + \overset{(0)}{X}{}^s\dfrac{\omega^i_s}{dt} = 0, \\ \dfrac{d\overset{(\alpha)}{X}{}^i}{dt} + \overset{(\alpha)}{X}{}^s\dfrac{\omega^i_s}{dt} = 0, \ (\alpha = 1, ..., k-1), \\ \dfrac{dX_i}{dt} - X_s\dfrac{\omega^s_i}{dt} = 0. \end{cases} \tag{10.6.7}$$

We obtain, also

Theorem 10.6.2 *The Cartan spaces $\mathcal{C}^{(k)n}$ is with absolute parallelism of vectors, with respect to $C\Gamma(N)$, if and only if all curvature d -tensors of $C\Gamma(N)$ vanish.*

In the case $X = \dot{\gamma}$, the equation $\dfrac{D\dot{\gamma}}{dt} = 0$ says that γ is an autoparallel curve of $\mathcal{C}^{(k)n}$ with respect to $C\Gamma(N)$.

¿From Theorem 7.7.3 it follows:

Theorem 10.6.3 *A smooth parametrized curve γ, (10.6.1), is an autoparallel curve of the Cartan space $\mathcal{C}^{(k)n}$, endowed with metrical canonical N -linear connection $C\Gamma(N)$, if and only if the following system of differential equations is verified:*

$$\text{(10.6.8)}\qquad \begin{cases} \dfrac{d^2x^i}{dt^2} + \dfrac{dx^s}{dt}\dfrac{\omega^i_s}{dt} = 0, \\[2mm] \dfrac{d}{dt}\dfrac{\delta y^{(\alpha)i}}{dt} + \dfrac{\delta y^{(\alpha)s}}{dt}\dfrac{\omega^i_s}{dt} = 0, \\ \qquad\qquad (\alpha = 1, ..., k-1), \\[2mm] \dfrac{d}{dt}\dfrac{\delta p_s}{dt} - \dfrac{\delta p_s}{dt}\dfrac{\omega^s_i}{dt} = 0. \end{cases}$$

As we know from section 7, ch. 7, a curve γ is horizontal if $\dot{\gamma} = \overset{\cdot}{\gamma}{}^H$, i.e.

$$\text{(10.6.9)}\qquad x^i = x^i(t),\ \frac{\delta y^{(\alpha)i}}{dt} = 0,\ (\alpha = 1, ..., k-1),\ \frac{\delta p_i}{dt} = 0,\ t \in I.$$

Therefore, taking into account the definitions of horizontal paths, v_α-paths and w_k-paths (§7, ch. 6), we obtain:

Theorem 10.6.4 *The Cartan space $\mathcal{C}^{(k)n}$ endowed with the metrical canonical N -linear connection $C\Gamma(N)$ has the following properties:*

a. The horizontal paths are characterized by the system of differential equations:

$$\text{(10.6.10)}\qquad \frac{d^2x^i}{dt^2} + H^i_{jh}\frac{dx^j}{dt}\frac{dx^h}{dt} = 0,\ \frac{\delta y^{(\alpha)i}}{dt} = 0,\ \frac{\delta p_i}{dt} = 0.$$

b. The v_α-paths at the point $x = x_0$ are characterized by the system of differential equations:

$$\text{(10.6.11)}\qquad \begin{aligned} &\frac{dx^i}{dt} = 0,\ \frac{dy^{(\beta)i}}{dt} = 0,\ (\beta \neq \alpha),\ \frac{dp_i}{dt} = 0, \\ &\frac{d}{dt}\frac{dy^{(\alpha)i}}{dt} + \underset{(\alpha)}{C}{}^{\,i}_{sj}\frac{dy^{(\alpha)s}}{dt}\frac{dy^{(\alpha)j}}{dt} = 0. \end{aligned}$$

c. The w_k-paths at the point $x = x_0$ are characterized by the system of differential equations:

$$\frac{dx^i}{dt} = 0, \quad \frac{dy^{(\alpha)i}}{dt} = 0, \quad (\alpha = 1, ..., k-1),$$
$$\frac{d^2 p_i}{dt^2} - C_i^{js}(x, 0, ..., 0, p)\frac{dp_j}{dt}\frac{dp_s}{dt} = 0. \tag{10.6.12}$$

10.7 Structure Equations of Metrical Canonical N-Connection

In a Cartan space of order k, $\mathcal{C}^{(k)n} = (M, K)$, endowed with the metrical canonical N-linear connection $C\Gamma(N)$, lemma 7.8.1 (ch. 7) holds:

The following object fields

$$d(dx^i) - dx^m \wedge \omega^i_m, \quad d(\delta y^{(\alpha)i}) - dy^{(\alpha)m} \wedge \omega^i_m, \quad (\alpha = 1, ..., k-1),$$
$$d(\delta p_i) + \delta p_m \wedge \omega^i_m$$

are d-vector fields and

$$d\omega^i_j - \omega^m_j \wedge \omega^i_m$$

is a d-tensor field of type $(1,1)$.

Consequently, we have:

Theorem 10.7.1 *A Cartan space of order k, $\mathcal{C}^{(k)n} = (M, K)$, has the following structure equations of the metrical canonical N-linear connection $C\Gamma(N)$:*

$$d(dx^i) - dx^m \wedge \omega^i_m = -\overset{(0)}{\Omega}{}_i,$$
$$d(\delta y^{(\alpha)i}) - dy^{(\alpha)m} \wedge \omega^i_m = -\overset{(\alpha)}{\Omega}{}_i, \quad (\alpha = 1, ..., k-1),$$
$$d(\delta p_i) + \delta p_m \wedge \omega^i_m = -\Omega_i. \tag{10.7.1}$$

and

$$d\omega^i_j - \omega^m_j \wedge \omega^i_m = -\Omega^i_j, \tag{10.7.2}$$

where the 2-forms of torsion are:
(10.7.3)

$$\overset{(0)}{\Omega}{}_i = dx^j \wedge \left(\sum_{\alpha=1}^{k-1} \underset{(\alpha)}{C}{}^{i}_{jm} \delta y^{(\alpha)m} + C^{im}_j \delta p_m \right),$$

$$\overset{(\alpha)}{\Omega}{}_i = dx^j \wedge \underset{(\alpha 0)}{P}{}^{i}_{j} + \sum_{\gamma=1}^{k-1} \delta y^{(\gamma)j} \wedge \underset{(\alpha\gamma)}{P}{}^{i}_{j} +$$
$$+\delta y^{(\alpha)j} \wedge \left\{ H^i_{jm} dx^m + \sum_{\gamma=1}^{k-1} \underset{(\gamma)}{C}{}^{i}_{jm} \delta y^{(\gamma)m} + C^{im}_j \delta p_m \right\}, \ (\alpha = 1, ..., k-1),$$

$$\Omega_i = dx^j \wedge \left\{ \frac{1}{2} \underset{(0)}{R}{}_{ijm} dx^m + \sum_{\gamma=1}^{k-1} \underset{(\gamma)}{B}{}_{ijm} \delta y^{(\gamma)m} + \underset{(0)}{B}{}^{m}_{ij} \delta p_m \right\} -$$
$$-\delta p_j \wedge \left\{ H^j_{im} dx^m + \sum_{\gamma=1}^{k-1} \underset{(\gamma)}{C}{}^{j}_{im} \delta y^{(\gamma)m} + C^{jm}_i \delta p_m \right\}$$

and where the 2-forms of curvature are

$$\Omega^i_j = \frac{1}{2} R^i_{jhm} dx^h \wedge dx^m +$$
$$+ \sum_{\gamma=1}^{k-1} \underset{(\gamma)}{P}{}_j{}^i{}_{hm} dx^h \wedge \delta y^{(\gamma)m} + P_j{}^i{}_h{}^m dx^h \wedge \delta p_m +$$
$$+ \sum_{\alpha \le \beta, \alpha=1}^{k-1} \sum_{\beta=1}^{k-1} \underset{(\alpha\beta)}{S}{}_j{}^i{}_{hm} \delta y^{(\alpha)h} \wedge \delta y^{(\beta)m} + \tag{10.7.4}$$
$$+ \sum_{\gamma=1}^{k-1} \underset{(\gamma)}{S}{}_j{}^i{}_{hm} \delta y^{(\gamma)h} \wedge \delta p_m + \frac{1}{2} S_j{}^{ihm} \delta p_h \wedge \delta p_m,$$

where, according to §5, ch. 6, $\underset{(\alpha,0)}{P}{}^{i}_{j}, \ldots, \underset{(\alpha,\alpha-1)}{P}{}^{i}_{j}$ *are as follows:*

$$\underset{(\alpha,0)}{P}{}^{i}_{j} = d\underset{(\alpha)}{M}{}^{i}_{j} - \underset{(1)}{N}{}^{m}_{j} d \underset{(\alpha-1)}{M}{}^{i}_{m} - \cdots - \underset{(\alpha-1)}{N}{}^{m}_{j} d\underset{(1)}{M}{}^{i}_{m},$$

$$\underset{(\alpha,1)}{P}{}^{i}_{j} = d \underset{(\alpha-1)}{M}{}^{i}_{j} - \underset{(1)}{N}{}^{m}_{j} d \underset{(\alpha-2)}{M}{}^{i}_{m} - \cdots - \underset{(\alpha-2)}{N}{}^{m}_{j} d\underset{(1)}{M}{}^{i}_{m}, \tag{10.7.5}$$

....

$$\underset{(\alpha,\alpha-1)}{P}{}^{i}_{j} = d\underset{(1)}{M}{}^{i}_{j}.$$

Of course this theorem is important for the theory of metrical canonical connection. Also, it allows to determine the Bianchi identities of the spaces $\mathcal{C}^{(k)n}$.

We can use the previous results in a theory of Cartan subspaces of order k in the Cartan space $\mathcal{C}^{(k)n}$, cf. Ch.9.

10.8 Riemannian Almost Contact Structure of the Space $\mathcal{C}^{(k)n}$

Consider a Cartan space of order k, $\mathcal{C}^{(k)n} = (M, K(x, y^{(1)}, ..., y^{(k-1)}, p))$ and its canonical nonlinear connection N.

The adapted basis $(\frac{\delta}{\delta x^i}, \frac{\delta}{\delta y^{(1)i}}, ..., \frac{\delta}{\delta y^{(k-1)i}}, \frac{\delta}{\delta p_i}), \frac{\delta}{\delta p_i} = \frac{\partial}{\partial p_i} = \dot{\partial}^i$ and its dual basis $(\delta x^i, \delta y^{(1)i}, ..., \delta y^{(k-1)i}, \delta p_i)$, where $\delta x^i = dx^i$ are well determined by N. As we know from section 1, the fundamental tensor of $\mathcal{C}^{(k)n}$ is:

$$g^{ij} = \frac{1}{2} \dot{\partial}^i \dot{\partial}^j K^2. \tag{10.8.1}$$

Taking into account the associated Hamilton space $H^{(k)n} = (M, K^2)$ of the space $\mathcal{C}^{(k)n}$, we can study the Riemannian almost contact structure of the Cartan space of order k by means of the corresponding structure of the space $H^{(k)n} = (M, K^2)$. Such that, using the covariant tensor g_{ij} of the fundamental tensor g^{ij} we define the tensor

$$\check{\mathbb{G}} = g_{ij} dx^i \otimes dx^j + \sum_{\alpha=1}^{k-1} g_{ij} \delta y^{(\alpha)i} \otimes \delta y^{(\alpha)j} + g^{ij} \delta p_i \otimes \delta p_j, \tag{10.8.2}$$

$\check{\mathbb{G}}$ is the N-lift of the fundamental tensor g^{ij} of the space $\mathcal{C}^{(k)n}$. Since g^{ij} is positively defined on $\widetilde{T^{*k}M}$ and N is given on $T^{*k}M$, it follows:

Theorem 10.8.1 *1⁰* $\check{\mathbb{G}}$ *is a Riemannian structure on the manifold* $\widetilde{T^{*k}M}$ *determined only by the fundamental tensor* g^{ij} *of the Cartan space* $\mathcal{C}^{(k)n}$ *and by the canonical nonlinear connection* N.

2⁰ The distributions $N_0, N_1, ..., N_{k-2}, V_{k-1}, W_k$ *are mutual orthogonal with respect to* $\check{\mathbb{G}}$.

Proposition 10.8.1 *The tensor* $\check{\mathbb{G}}$ *is not homogeneous on the fibres of the bundle* $T^{*k}M$.

Indeed, the first term in $\check{\mathbb{G}}$ is 0-homogeneous, the second term is 2-homogeneous, ..., the last term is $2k$-homogeneous. So, the whole $\check{\mathbb{G}}$ is not homogeneous.

Let us consider the following invariants:

(10.8.3)
$$K_1^2 = g_{ij} z^{(1)i} z^{(1)j},\ K_2^2 = g_{ij} z^{(2)i} z^{(2)j},\ ...,\ K_{k-1}^2 = g_{ij} z^{(k-1)i} z^{(k-1)j},\ K_0^2 = g^{ij} p_i p_j,$$

where $z^{(1)i}$, ..., $z^{(k)i}$ are the Liouville vector fields determined by the canonical nonlinear connection of space $\mathcal{C}^{(k)n}$.

Of course, all invariants K_0^2, K_1^2, ..., K_{k-1}^2 are positive.

Thus we can construct a new Riemannian structure on $\widetilde{T^{*k}M}$:

$$\overset{\circ}{\mathbb{G}} = g_{ij}dx^i \otimes dx^j + \sum_{\alpha=1}^{k-1} \frac{1}{K_\alpha^2} g_{ij}\delta y^{(\alpha)i} \otimes \delta y^{(\alpha)j} + \frac{1}{K_0^2} g^{ij}\delta p_i \otimes \delta p_j. \tag{10.8.4}$$

Theorem 10.8.2 *1⁰ $\overset{\circ}{\mathbb{G}}$ is a Riemannian structure on the manifold $\widetilde{T^{*k}M}$ determined only by g^{ij} and N.*

2⁰ $\overset{\circ}{\mathbb{G}}$ is 0-homogeneous on the fibres of the bundle $T^{*k}M$.

3⁰ The distributions N_0, N_1, ..., N_{k-2}, V_{k-1}, W_k are mutual orthogonal with respect to $\overset{\circ}{\mathbb{G}}$.

The proof follows without difficulties.

The Riemannian structure $\overset{\vee}{\mathbb{G}}$ is of the form

$$\overset{\vee}{\mathbb{G}} = \overset{H}{\mathbb{G}} + \overset{V_1}{\mathbb{G}} + \cdots + \overset{V_{k-1}}{\mathbb{G}} + \overset{W_k}{\mathbb{G}} \tag{10.8.5}$$

with

$$\begin{aligned} &\overset{H}{\mathbb{G}} = g_{ij}dx^i \otimes dx^j,\ \overset{V_1}{\mathbb{G}} = g_{ij}\delta y^{(1)i} \otimes \delta y^{(1)j},\ \ldots, \\ &\overset{V_{k-1}}{\mathbb{G}} = g_{ij}\delta y^{(k-1)i} \otimes \delta y^{(k-1)j},\ \overset{W_k}{\mathbb{G}} = g^{ij}\delta p_i \otimes \delta p_j. \end{aligned} \tag{10.8.6}$$

The tensors $\overset{H}{\mathbb{G}}$, $\overset{V_\alpha}{\mathbb{G}}$, $(\alpha = 1, ..., k-1)$ and $\overset{W_k}{\mathbb{G}}$ are d-tensor fields on the manifold $\widetilde{T^{*k}M}$.

Exactly as in the chapter 8 we can prove:

Theorem 10.8.3 *The d-tensor fields $\overset{H}{\mathbb{G}}$, $\overset{V_\alpha}{\mathbb{G}}$, $(\alpha = 1, ..., k-1)$ and $\overset{W_k}{\mathbb{G}}$ are covariant constant with respect to the canonical metrical N-connection of the Cartan space of order k, $\mathcal{C}^{(k)n}$.*

Now, let us consider the natural almost $(k-1)n$-contact structure $\mathbb{F}$ determined by the canonical nonlinear connection N. It is defined by (6.6.3):

(10.8.7)

$$\mathbb{F}\left(\frac{\delta}{\delta x^i}\right) = -\frac{\partial}{\partial y^{(k)i}},\ \mathbb{F}\left(\frac{\partial}{\partial y^{(k-1)i}}\right) = \frac{\delta}{\delta x^i},\mathbb{F}\left(\frac{\delta}{\delta y^{(\alpha)i}}\right) = 0,\ (\alpha = \overline{1, k-1}),$$

$$\mathbb{F}\left(\frac{\delta}{\delta p_i}\right) = 0.$$

$\mathbb{F}$ is a tensor field of type $(1,1)$ and in adapted basis it is expressed by (3.5.4).

The condition of normality of $\mathbb{F}$ is given by the equation (6.6.5).

But the pair of structures $(\overset{\vee}{\mathbb{G}}, \mathbb{F})$ is a Riemannian almost $(k-1)n$ contact structure on $\widetilde{T^{*k}M}$ determined only by N and by the fundamental function K of the Cartan space $\mathcal{C}^{(k)n}$. So, we have:

Theorem 10.8.4 *On the manifold $T^{*k}M$ there exists a natural Riemannian almost $(k-1)n$ -contact structure $(\overset{\vee}{\mathbb{G}}, \mathbb{F})$, determined only by the canonical nonlinear connection N and the fundamental function K of the Cartan space $\mathcal{C}^{(k)n} = (M, K)$.*

The canonical nonlinear connection N and the fundamental tensor g^{ij} of the Cartan space $\mathcal{C}^{(k)n}$ determine the almost $(k-1)n$ -contact structure $\overset{\vee}{\mathbb{F}}$:

(10.8.8)
$$\overset{\vee}{\mathbb{F}}\left(\frac{\delta}{\delta x^i}\right) = -g_{ij}\frac{\partial}{\partial p_j},\ \overset{\vee}{\mathbb{F}}\left(\frac{\delta}{\delta y^{(\alpha)i}}\right) = 0,\ (\alpha = \overline{1, k-1}),\ \overset{\vee}{\mathbb{F}}\left(\frac{\delta}{\delta p_i}\right) = g^{ij}\frac{\delta}{\delta x^j}.$$

Theorem 6.7.1, reads:

Theorem 10.8.5 *The structure $\overset{\vee}{\mathbb{F}}$ of a Cartan space $\mathcal{C}^{(k)n}$ has the following properties:*

$$\overset{\vee}{\mathbb{F}} = -g_{ij}\frac{\delta}{\delta p_j} \otimes dx^i + g^{ij}\frac{\delta}{\delta x^i} \otimes \delta p_j,$$

(10.8.9)
$$Ker\ \overset{\vee}{\mathbb{F}} = N_1 \oplus \cdots \oplus N_{k-1},\ Im\ \overset{\vee}{\mathbb{F}} = N_0 \oplus W_k,$$

$$rank\ \overset{\vee}{\mathbb{F}} = 2n,$$

(10.8.10)
$$\overset{\vee}{\mathbb{F}}{}^3 + \overset{\vee}{\mathbb{F}} = 0.$$

Consequently, $\overset{\vee}{\mathbb{F}}$ is an almost $(k-1)n$-contact structure on the manifold $\widetilde{T^{*k}M}$.

We have, also:

Theorem 10.8.6 *For a Cartan space $\mathcal{C}^{(k)n}$ the following properties hold:*

1^0 The pair $(\overset{\vee}{\mathbb{G}}, \overset{\vee}{\mathbb{F}})$ is a Riemannian almost $(k-1)n$-contact structure determined only by the canonical nonlinear connection N and the fundamental tensor g^{ij}.

2^0 The associated 2-form is

$$\theta = \delta p_i \wedge dx^i$$

and if the coefficients N_{ij} of N are symmetric then θ is the canonical presymplectic structure

$$\theta = dp_i \wedge dx^i.$$

Concluding, the space $(\widetilde{T^{*k}M}, \overset{\vee}{\mathbb{G}}, \overset{\vee}{\mathbb{F}})$ is the *geometrical model* of the Cartan space of order k, $\mathcal{C}^{(k)n}$.

Chapter 11

Generalized Hamilton and Cartan Spaces of Order k. Applications to Hamiltonian Relativistic Optics

On the total space of the dual bundle $(T^{*k}M, \pi^*, M)$ there exist some geometrical structures defined by a general d-tensor field $g^{ij}(x, y^{(1)}, ..., y^{(k-1)}, p)$, symmetric and nondegenerate, which are suggested by the Relativistic Optics. Generally the d-tensor g^{ij} is not the metric tensor of a Hamilton space $H^{(k)n}$ or of a Cartan space $\mathcal{C}^{(k)n}$. In this case the pair $GH^{(k)n} = (M, g^{ij}(x, y^{(1)}, ..., y^{(k-1)}, p))$ defines a 'Generalized Hamilton space of order k'. If g^{ij} is 0-homogeneous on the fibres of the bundle $T^{*k}M$ we say that the pair $GC^{(k)n}$ is a 'Generalized Cartan space of order k'.

We study, in this chapter, the geometry of these spaces and apply it to the theory of Hamiltonian Relativistic Optics.

11.1 The Space $GH^{(k)n}$

Definition 11.1.1 *A generalized Hamilton space of order* k *is a pair* $GH^{(k)n} = (M, g^{ij}(x, y^{(1)}, ..., y^{(k-1)}, p))$, *where*

1° g^{ij} *is a d-tensor field of type* $(2, 0)$, *symmetric and nondegenerate on the manifold* $\widetilde{T^{*k}M}$:

$$rank \left\|g^{ij}\right\| = n \tag{11.1.1}$$

2° The quadratic form $g^{ij}X_iX_j$ *has a constant signature on* $\widetilde{T^{*k}M}$.

The tensor g^{ij} is called fundamental for the space $GH^{(k)n}$.

In the case when the base manifold M is paracompact then $\widetilde{T^{*k}M}$ is paracompact. Thus on $\widetilde{T^{*k}M}$ there exists the d -tensor g^{ij}, with the property that $GH^{(k)n} = (M, g^{ij})$ is a generalized Hamilton space of order k.

If g^{ij} is positively defined on $\widetilde{T^{*k}M}$, then the conditions (11.1.1) is verified.

Definition 11.1.2 *The space $GH^{(k)n} = (M, g^{ij})$ is called reducible to a Hamilton space of order k, if there exists an Hamiltonian $H(x, y^{(1)}, ..., y^{(k-1)}, p)$ such that the following equality holds:*

$$g^{ij} = \frac{1}{2}\, \dot{\partial}^i \dot{\partial}^j\, H \tag{11.1.2}$$

Let us consider the d-tensor field:

$$C^{ijh} = -\frac{1}{2}\, \dot{\partial}^h\, g^{ij} \tag{11.1.3}$$

We have:

Proposition 11.1.1 *A necessary condition for a generalized Hamilton space $GH^{(k)n} = (M, g^{ij})$ be reducible to a Hamilton space of order k is that the $d-$tensor field C^{ijk} be totally symmetric.*

Indeed, if (11.1.2) holds, then $C^{ijh} = -\frac{1}{4}\, \dot{\partial}^i \dot{\partial}^i \dot{\partial}^i\, H$ is totally symmetric.

Example 1. 1° Let $g^{ij}(x, y^{(1)}, ..., y^{(k-1)}, p) = \gamma^{ij}(x, y^{(1)}, ..., y^{(k-1)})$ be a d-tensor on the $\widetilde{T^{*k}M}$ determined by the fundamental tensor γ_{ij} of a Finsler space of order $k-1$.

The space $GH^{(k)n} = (M, g^{ij})$ is reducible to the Hamilton space $H^{(k)n} = (M, g^{ij}p_i p_j)$.

2° Consider the fundamental tensor $\gamma_{ij}(x, y^{(1)}, ..., y^{(k-1)})$ of a Finsler space of order $k-1$, $F^{(k-1)n} = (M, y^{(1)}, ..., y^{(k-1)}))$. The d-tensor field

$$g^{ij}(x, y^{(1)}, ..., y^{(k-1)}, p) = e^{-2\sigma(x, y^{(1)}, ..., y^{(k-1)}, p)}\gamma^{ij}(x, y^{(1)}, ..., y^{(k-1)}) \tag{11.1.4}$$

with the property $\sigma \in \mathcal{F}(\widetilde{T^{*k}M})$ and γ_{ij} is a fundamental tensor of $F^{(k-1)n}$, determines a generalized Hamilton space of order k, $GH^{(k)n} = (M, g^{ij})$.

This space is reducible to a Hamilton space $H^{(k)n}$ only if $\dot{\partial}^i\, \sigma = 0$.

Let g_{ij} be the covariant tensor of the fundamental tensor g^{ij} of the space $GH^{(k)n} = (M, g^{ij})$. Then we have:

$$g_{ih}g^{hj} = \delta_i^j \tag{11.1.5}$$

Also, we consider the d-tensor field

(11.1.6) $$C_i^{jh} = -\frac{1}{2} g_{is}(\dot{\partial}^j g^{sh} + \dot{\partial}^h g^{js} - \dot{\partial}^s g^{jh})$$

Evidently:

(11.1.7) $$S_i^{jh} = C_i^{jh} - C_i^{hj} = 0.$$

If the space $GH^{(k)n}$ is reducible to a Hamilton space then:

(11.1.6') $$C_i^{jh} = -g_{is} C^{shj}.$$

The d-tensor field C_i^{jh} are the coefficients of the w_k-covariant derivatives. Indeed, we have:

(11.1.8) $$g^{ij}|^h = \dot{\partial}^h g^{ij} + C_s^{ih} g^{sj} + C_s^{jh} g^{is} = 0.$$

11.2 Metrical N-Linear Connections

If we consider a generalized Hamilton space of order k, $GH^{(k)n} = (M, g^{ij})$, in general we can not determine a nonlinear connection only by means of the fundamental tensor g^{ij}.

But there are some particular cases when this is possible. For instance, in examples 1° and 2° we can consider the canonical nonlinear connection $\overset{\circ}{N}$ of the Finsler space $F^{(k-1)n} = (M, F(x, y^{(1)}, ..., y^{(k-1)})$ with the coefficients $(\underset{(1)}{N}{}^i_j, ... \underset{(k-1)}{N}{}^i_j)$. Then the system of functions $(\underset{(1)}{N}{}^i_j, ... \underset{(k-1)}{N}{}^i_j, N_{ij})$ with $N_{ij} = \frac{\delta}{\delta y^{(1)j}}(\underset{(1)}{N}{}^h_j p_h)$, (if $\underset{(1)}{N}{}^i_j(x, y^{(1)})$, does not depend on $y^{(2)i}, ..., y^{(k-1)i}$) determines a nonlinear connection only by means of the fundamental tensor g^{ij} of the space.

Now, let us consider an apriori fixed nonlinear connection N, with the coefficients $(\underset{(1)}{N}{}^i_j, ... \underset{(k-1)}{N}{}^i_j, N_{ij})$ on the manifold $\widetilde{T^{*k}M}$. We will study the geometry of the space $GH^{(k)n}$ endowed with the nonlinear connection N.

The adapted basis to the direct decomposition (6.2.9), $(\frac{\delta}{\delta x^i}, \frac{\delta}{\delta y^{(1)i}}, ..., \frac{\partial}{\partial y^{(k-1)i}}, \frac{\partial}{\partial p_i})$ is expressed in (6.2.10) and the adapted cobasis $(dx^i, \delta y^{(1)i}, ..., \delta y^{(k-1)i}, \delta p_i)$ is written in (6.3.2).

Now, as usual, we can prove:

Theorem 11.2.1 *1) A generalized Hamilton space of order k endowed with a nonlinear connection N, has an unique N-linear connection* $C\Gamma(N) = (H^i_{jh}, \underset{(1)}{C}{}^i_{jh}, ..., \underset{(k-1)}{C}{}^i_{jh}, C_i^{jh})$ *satisfying the following axioms:*

1° The nonlinear connection N is apriori given.

2° $g^{ij}{}_{|h}=0,\ g^{ij}\overset{(\alpha)}{|}{}_{h}=0,\ (\alpha=1,...,k-1),\ g^{ij}|^{h}=0.$

3° $T^{i}{}_{jh}=0,\ \underset{(\alpha)}{S}{}^{i}{}_{jh}=0,\ (\alpha=1,...,k-1),\ S_{i}{}^{jh}=0.$

2) The coefficients of $C\Gamma(N)$ *are the following generalized Christoffel symbols:*

$$H^{i}_{jh}=\frac{1}{2}g^{is}\left(\frac{\delta g_{sh}}{\delta x^{j}}+\frac{\delta g_{js}}{\delta x^{h}}-\frac{\delta g_{jh}}{\delta x^{s}}\right),$$

$$\underset{(\alpha)}{C^{i}_{jh}}=\frac{1}{2}g^{is}\left(\frac{\delta g_{sh}}{\delta y^{(\alpha)j}}+\frac{\delta g_{js}}{\delta y^{(\alpha)h}}-\frac{\delta g_{jh}}{\delta y^{(\alpha)s}}\right),\ (\alpha=1,...,k-1),\tag{11.2.1}$$

$$C^{jh}_{i}=-\frac{1}{2}g_{is}(\dot{\partial}^{j}g^{sh}+\dot{\partial}^{h}g^{js}-\dot{\partial}^{s}g^{jh}).$$

The previous connection $C\Gamma(N)$ is called canonical metrical N -connection.

More general, one proves

Theorem 11.2.2 *1) A generalized Hamilton space* $GH^{(k)n}$, *endowed with a nonlinear connection* N, *has an unique* N- *linear connection* $\overline{D}\Gamma(N)=(\overline{H}^{i}_{jh},\underset{(1)}{\overline{C}^{i}_{jh}},...,\underset{(k-1)}{\overline{C}^{i}_{jh}},\overline{C}^{jh}_{i})$ *satisfying the axioms:*

1° N is apriori given on $\widetilde{T^{*k}M}$.

2° h-, v_{α}*- and* w_{k}*- covariant derivation of* g^{ij} *vanish:*

$$g^{ij}{}_{|h}=0,\ g^{ij}\overset{(\alpha)}{|}{}_{h}=0,\ g^{ij}|^{h}=0,\tag{11.2.2}$$

3° The skewsymmetric tensors of torsion

(11.2.2')

$$\overline{T}^{i}{}_{jh}=\overline{H}^{i}_{jh}-\overline{H}^{i}_{hj},\ \underset{(\alpha)}{\overline{S}}{}^{i}{}_{jh}=\underset{(\alpha)}{\overline{C}^{i}_{jh}}-\underset{(\alpha)}{\overline{C}^{i}_{hj}},(\alpha=1,...,k-1),\ \overline{S}_{i}{}^{jh}=\overline{C}^{jh}_{i}-\overline{C}^{hj}_{i}$$

are apriori given.

2) This connection has the following coefficients:

$$\overline{H}^{i}_{jh}=H^{i}_{jh}+\frac{1}{2}g^{is}(g_{sr}\overline{T}^{r}_{jh}-g_{jr}\overline{T}^{r}_{sh}+g_{hr}\overline{T}^{r}_{js}),$$

$$\underset{(\alpha)}{\overline{C}^{i}_{jh}}=\underset{(\alpha)}{C^{i}_{jh}}+\frac{1}{2}g^{is}(g_{sr}\underset{(\alpha)}{\overline{S}^{r}_{jh}}-g_{jr}\underset{(\alpha)}{\overline{S}^{r}_{sh}}+g_{hr}\underset{(\alpha)}{\overline{S}^{r}_{js}}),\ (\alpha=1,...,k-1),\tag{11.2.2"}$$

$$\overline{C}^{jh}_{i}=C^{jh}_{i}-\frac{1}{2}g_{is}(g^{sh}\overline{S}^{jh}_{r}-g^{jr}\overline{S}^{jh}_{r}+g^{hr}\overline{S}^{js}_{r}),$$

where $(H^{i}_{jh},\underset{(1)}{C}{}^{i}_{jh},...,\underset{(k-1)}{C}{}^{i}_{jh},C^{jh}_{i})$ *are the coefficients of the canonical metrical N-connection* $C\Gamma(N)$.

If we are interested on the all N-linear connections which verify the equations (11.2.2), we can prove:

Theorem 11.2.3 *In a generalized Hamilton space $GH^{(k)n} = (M, g^{ij})$ the set of all N -linear connections $\overline{D}\Gamma(N)$ which satisfy the equations (11.2.2) is given by*

$$\overline{H}^i_{jh} = H^i_{jh} + \Omega^{is}_{rj} X^r_{sh},$$

$$(11.2.3) \qquad \underset{(\alpha)}{\overline{C}^i_{jh}} = \underset{(\alpha)}{C^i_{jh}} + \Omega^{is}_{rj} \underset{(\alpha)}{X^r_{sh}}, \ (\alpha = 1, ..., k-1)$$

$$\overline{C}^{jh}_i = C^{jh}_i + \Omega^{js}_{ri} X^{rh}_s$$

where $\Omega^{is}_{rj} = \frac{1}{2}(\delta^i_r \delta^s_j - g_{rj} g^{is})$ *are Obata's operators and* $C\Gamma(N) = (H^i_{jh}, \underset{(1)}{C^i_{jh}}, ..., \underset{(k-1)}{C^i_{jh}}, C^{jh}_i)$ *is the canonical metrical N-connection and* X^r_{sh}, $\underset{(\alpha)}{X^r_{sh}}$, $(\alpha = 1, ..., k-1)$, X^{rh}_s *are arbitrary d-tensor fields.*

Corollary 11.2.1 *The mappings $D\Gamma(N) \to \overline{D}\Gamma(N)$ determined by (11.2.3), together with the composition of these mappings is an abelian group.*

Now, we can repeat all considerations from the section 7 of the chapter 8. So, we have:

Proposition 11.2.1 *The curvature d-tensor fields of the canonical metrical N-connection $C\Gamma(N)$, (11.2.1) satisfy the identities (7.6.8).*

Proposition 11.2.2 *The tensors of deflection of $C\Gamma(N)$, (11.2.1) satisfy the identities (7.6.9), with* $T^i{}_{jk} = 0$, $S_i{}^{jh} = 0$, $\underset{(\alpha)}{S^i{}_{jh}} = 0$, $(\alpha = 1, ..., k-1)$.

Let ω^i_j be the 1-forms connection of $C\Gamma(N)$,

$$(11.2.4) \qquad \omega^i_j = H^i_{js} dx^s + \sum_{\alpha=1}^{k-1} \underset{(\alpha)}{C^i_{js}} \delta y^{(\alpha)s} + C^{js}_i \delta p_s$$

and a curve $\gamma : t \in I \to \gamma(t) \in T^{*k}M$ expressed by (7.7.1).

Thus, we have (cf. Th. 7.7.1):

Theorem 11.2.4 *In a space $GH^{(k)n}$ endowed with the canonical metrical N -connection $C\Gamma(N)$ a vector field* $X = X^{(0)i}\frac{\delta}{\delta x^i} + ... + X^{(k-1)i}\frac{\delta}{\delta y^{(k-1)i}} + X_i \frac{\partial}{\partial p_i}$

is parallel along curve γ if and only if $(X^{(0)i}, ..., X^{(k-1)}, X_i)$ are the solutions of the system of differential equations

(11.2.5)

$$\frac{dX^{(0)i}}{dt} + X^{(0)s}\frac{\omega_s^i}{dt} = 0, ..., \frac{dX^{(k-1)i}}{dt} + X^{(k-1)s}\frac{\omega_s^i}{dt} = 0, \ \frac{dX_i}{dt} - X_s\frac{\omega_i^s}{dt} = 0.$$

Theorem 11.2.5 *The space $GH^{(k)n}$ endowed with the N-linear connection $C\Gamma(N)$ is with absolute parallelism of vectors if and only if all curvature d - tensor of $C\Gamma(N)$ vanish.*

A smooth parametrized curve $\gamma : t \in I \to (x(t), y^{(1)}(t), ..., y^{(k-1)}(t), p(t)) \in T^{*k}M$ is an autoparallel curve of $C\Gamma(N)$ if $D_{\dot\gamma}\, \dot\gamma = 0$.

Thus, applying Theorem 7.7.3 we have:

Theorem 11.2.6 *The curve $\gamma : I \to \widetilde{T^{*k}M}$ is autoparallel for the space $GH^{(k)n}$, with respect to the canonical N-linear connection $C\Gamma(N)$ if and only if the following system of differential equations is verified:*

$$\frac{d^2x^i}{dt^2} + \frac{dx^s}{dt}\frac{\omega_s^i}{dt} = 0,$$

(11.2.6)
$$\frac{d}{dt}\left(\frac{\delta y^{(\alpha)i}}{dt}\right) + \frac{\delta y^{(\alpha)s}}{dt}\frac{\omega_s^i}{dt} = 0, (\alpha = 1, ..., k-1),$$

$$\frac{d}{dt}\left(\frac{\delta p_i}{dt}\right) - \frac{\delta p_s}{dt}\frac{\omega_i^s}{dt} = 0.$$

Recall that γ is an horizontal curve if and only if:

$$x^i = x^i(t), \frac{\delta p_i}{dt} = 0, \frac{\delta y^{(\alpha)i}}{dt} = 0, (\alpha = 1, ..., k-1).$$

Therefore, we have:

Theorem 11.2.7 *For a generalized Hamilton space of order k, $GH^{(k)n}$ endowed with the canonical metrical N-connection $C\Gamma(N)$ the following properties hold:*

1° The horizontal paths are characterized by the system of differential equations:

$$\frac{d^2x^i}{dt^2} + H^i_{jh}\frac{dx^j}{dt}\frac{dx^h}{dt} = 0, \ \frac{\delta y^{(\alpha)i}}{dt} = 0, \ \frac{\delta p_i}{dt} = 0, (\alpha = 1, ..., k-1).$$

2° The v_α- paths in a point $x_0 \in M$ are characterized by

$$x^i = x^i_0, \frac{dy^{(\beta)i}}{dt} = 0, (\beta \neq \alpha), \frac{dp_i}{dt} = 0,$$

$$\frac{d}{dt}\left(\frac{\delta y^{(\alpha)i}}{dt}\right)+\underset{(\alpha)}{C^{i}_{sj}}\frac{dy^{(\alpha)s}}{dt}\frac{dy^{(\alpha)j}}{dt}=0$$

3° The w_k-path are characterized by

$$\frac{dx^i}{dt}=0,\frac{dy^{(1)i}}{dt}=...=\frac{dy^{(k-1)i}}{dt}=0,$$

$$\frac{d^2p_i}{dt^2}-C_i^{jm}(x_0,0,...0,p)\frac{dp_j}{dt}\frac{dp_m}{dt}=0.$$

Finally, we remark:

The structure equations of the canonical metrical N-connection $C\Gamma(N)$ of the space $GH^{(k)n}$ are given by Theorem 7.8.1.

where $T^i{}_{jh}=0$, $\underset{(\alpha)}{S^i{}_{jh}}=0$, $(\alpha=1,...,k-1)$, $S_i{}^{jh}=0$.

11.3 Hamiltonian Relativistic Optics

In the book: 'The Geometry of Higher Order Lagrange Spaces' - Kluwer FTPH, Vol. 82 is given a generalized Lagrange metric, formula (10.5.6) of the Relativistic Optics of order k. It is rather complicated. In the dual spaces $T^{*k}M$ it can be introduced much more simple.

Consider a semidefined Finsler space of order $k-1$, $F^{(k-1)}M = (M,F(x,y^{(1)},...,y^{(k-1)}))$ and $\gamma_{ij}(x,y^{(1)},...,y^{(k-1)})$ its fundamental tensor field.

The projection $\pi^{*k}_{k-1}:(x,y^{(1)},...,y^{(k-1)},p)\in T^{*k}M\to(x,y^{(1)},...,y^{(k-1)})\in T^{k-1}M$ allows to consider d-tensor $\gamma_{ij}\circ\pi^{*k}_{k-1}$ on $T^{*k}M$. It will be denoted by γ_{ij}, too. Its contravariant γ^{ij} will be considered defined on the manifold $\widetilde{T^{*k}M}$.

Let us consider a differentiable function n on $T^{*k}M$ with the property $n>1$. It will be called a refractive index.

Some notations:

$$\text{(11.3.1)}\qquad\begin{cases}\overset{\vee}{p}{}^i=\gamma^{ij}p_j,\ p_i=\gamma_{ij}\overset{\vee}{p}{}^j,\ p_i\overset{\vee}{p}{}^i=\gamma^{ij}p_ip_j=\|p\|^2,\\[2mm] \dfrac{1}{n(x,y^{(1)},...,y^{(k-1)},p)}=u(x,y^{(1)},...,y^{(k-1)},p).\end{cases}$$

Now, we define on $\widetilde{T^{*k}M}$ the d-tensor field

(11.3.2)

$$g^{ij}(x,y^{(1)},...,y^{(k-1)},p)=$$

$$=\gamma^{ij}(x,y^{(1)},...,y^{(k-1)})+\left(1-\frac{1}{n^2(x,y^{(1)},...,y^{(k-1)},p)}\right)\overset{\vee}{p}{}^i\overset{\vee}{p}{}^j$$

Theorem 11.3.1 *1° g^{ij} is a symmetric d-tensor field of type (2, 0) on the manifold $\widetilde{T^{*k}M}$.*

2° rank $\|g^{ij}\| = n$.

Indeed, 1° g^{ij} is a sum of two symmetric d-tensor of type (2,0).

2° Consider the d-tensor

$$g_{ij} = \gamma_{ij} - \frac{1}{a^2}\left(1 - \frac{1}{n^2}\right) p_i p_j, \tag{11.3.3}$$

where

$$a = 1 + \left(1 - \frac{1}{n^2}\right) \|p\|^2 . \tag{11.3.3'}$$

It is easy to verify the following equality:

$$g^{ih} g_{hj} = \delta^i_j . \tag{11.3.4}$$

Consequently the pair $GH^{(k)n} = (M, g^{ij})$ is a generalized Hamilton space of order k.

Theorem 11.3.2 *The space $GH^{(k)n} = (M, g^{ij})$ is not reducible to a Hamilton space of order k.*

Proof. The tensor field C^{ijk} from (11.1.3) is as follows

$$-C^{ijh} = \dot{\partial}^h \sigma \overset{\vee}{p^i} \overset{\vee}{p^j} + \sigma(\gamma^{ih} \overset{\vee}{p^j} + \gamma^{jh} \overset{\vee}{p^i}), \ \sigma = 1 - u^2$$

If we assume that $C^{ijh} = C^{ihj}$, we obtain

$$\left(\dot{\partial}^h \sigma \overset{\vee}{p^j} - \dot{\partial}^j \sigma \overset{\vee}{p^h}\right) \overset{\vee}{p^i} + \sigma \left(\gamma^{ih} \overset{\vee}{p^j} - \gamma^{ij} \overset{\vee}{p^h}\right) = 0 \tag{*}$$

Contracting by p_i we have

$$\dot{\partial}^h \sigma \overset{\vee}{p^j} - \dot{\partial}^j \sigma \overset{\vee}{p^h} = 0.$$

Substituting in (*) we get $\gamma^{ih} \overset{\vee}{p^j} - \gamma^{ij} \overset{\vee}{p^h} = 0$. A new contraction with γ_{ih} leads to $(n-1) \overset{\vee}{p^j} = 0$. Consequently $\overset{\vee}{p^j} = 0$, i.e $p_j = 0$. But this is impossible and the assumption we made is false. q.e.d.

The space $GH^{(k)n}$ will be called the generalized Hamiltonian space of order k of the Hamiltonian Relativistic Optics.

Let us consider a local d-vector field $V^i(x)$ and a local d-covector field $\eta_i(x)$ on the manifold M. It is not difficult to see that the mapping $S_{V,\eta} : M \to T^{*k}M$ defined locally by

$$\begin{aligned} &x^i = x^{i'}, \\ (11.3.5)\qquad &y^{(1)i} = V^i(x), ..., y^{(k-1)i} = \frac{1}{(k-2)!}\frac{d^{k-2}V^i}{dt^{k-2}}, \\ &p_i = \eta_i(x) \end{aligned}$$

is a cross- section of the canonical projection $\pi^{*k} : T^{*k}M \to M$. It follows that $S_{V,\eta}(M)$ is a local embedding of M in the manifold $\widetilde{T^{*k}M}$.

The restriction of the fundamental tensor g^{ij} of the space $GH^{(k)n}$ to $S_{V,\eta}$ will be called the Synge metric of the *Hamiltonian Relativistic Optics.* The restriction of function $n(x, y^{(1)}, ..., y^{(k-1)}, p)$ to $S_{V,\eta}(M)$ is the refractive index of the dispersive optic medium:

$(M, V(x), \eta(x), n(x, V(x), ..., \frac{1}{(k-2)!}\frac{d^{k-2}V}{dt^{k-2}}, \eta(x)).$

Therefore, we say that the geometry of the generalized Hamilton space of order k, $GH^{(k)n} = (M, g^{ij})$, with the fundamental tensor g^{ij} in (11.3.2) is the geometrical theory of the previous optic medium, endowed with Synge metric.

If the refractive index n depend on $x \in M$ only, then the optic medium is called nondispersive.

Let us consider the canonical nonlinear connection $\overset{\circ}{N}$ with the dual coefficients $\underset{(1)}{M^i_j}, ..., \underset{(k-1)}{M^i_j}$ of the Finsler space $F^{(k-1)n} = (M, F)$. For simplicity we assume that $\underset{(1)}{M^i_j} = \underset{(1)}{M^i_j}(x, y^{(1)})$, (see ch. 8, §8). In this case $(\underset{(1)}{M^i_j}, ..., \underset{(k-1)}{M^i_j}, N_{ij})$, with:

$$(11.3.6)\qquad N_{ij} = \frac{\delta}{\delta y^{(1)i}}(\underset{(1)}{M^h_j})p_h$$

define a nonlinear connection N of the space $GH^{(k)n} = (M, g^{ij})$ determined only by the fundamental tensor g^{ij}.

Let $(\frac{\delta}{\delta x^i}, \frac{\delta}{\delta y^{(1)i}}, ..., \frac{\delta}{\delta y^{(k-1)i}}, \frac{\delta}{\delta p_i})$ and $(dx^i, \delta y^{(1)i}, ..., \delta y^{(k-1)i}, \delta p_i)$ be the adapted local basis and adapted local cobasis corresponding to the nonlinear connection N.

We can determine the canonical metrical N-linear connection $C\Gamma(N)$ of the space $GH^{(k)n}$ starting from the expressions (11.3.2) and (11.3.3) for the fundamental tensor g^{ij} and its covariant tensor field g_{ij} and applying the usual tehniques of calculus.

The connections N and $C\Gamma(N)$ allow to study the geometry of the space $GH^{(k)n}$. Taking the restriction of its main geometrical object field to the section $S_{V,\eta}(M)$ we obtain the main notions and properties of the Hamiltonian Relativistic Optics.

Evidently, there are some particular cases, as:

a) the nondispersive media

b) the Finsler space $F^{(k-1)n}$ is the prolongation of order $k-1$ of a Finsler space $F^n = (M, F(x, y^{(1)}))$.

Let us consider the absolute energy of the space $GH^{(k)n}$, [115]:

$$\mathcal{E} = g^{ij} p_i p_j \tag{11.3.7}$$

Taking into account the expression (11.3.2) of g^{ij} and the function a from (11.3.2'), we get:

$$\mathcal{E} = a\,\|p\|^2 \tag{11.3.7'}$$

Consequently, we can say that $\mathcal{E}(x, y^{(1)}, ..., y^{(k-1)}, p)$ is a differentiable Hamiltonian uniquely determined by the fundamental tensor g^{ij} of the space $GH^{(k)n}$. It allows to determine the Hamilton - Jacobi equations of the space. These equations are given by (5.1.17), (5.1.17'). The energies of order $k-1, ..., 1$ are expressed in (5.3.1) with $H = \mathcal{E}$ and the low of conservation is mentioned in Theorem 5.3.2. Also, Theorem 5.5.3 gives the Nöther symmetries for the Hamiltonian $H = \mathcal{E}$ from (11.3.7).

11.4 The Metrical Almost Contact Structure of the Space $GH^{(k)n}$

The generalized Hamilton space of order k, $GH^{(k)n} = (M, g^{ij})$, endowed with an apriori given nonlinear connection N determines a metrical almost contact structure on the manifold $\widetilde{T^{*k}M}$.

The N-lift of the fundamental tensor field g^{ij} is

$$\check{\mathbb{G}} = g_{ij} dx^i \otimes dx^j + \sum_{\alpha=1}^{k-1} g_{ij} \delta y^{(\alpha)i} \otimes \delta y^{(\alpha)j} + g^{ij} \delta p_i \otimes \delta p_j \tag{11.4.1}$$

Evidently:

1° $\check{\mathbb{G}}$ is a pseudo-Riemannian structure on $T^{*k}M$.

2° The distributions $\underset{0}{N}, \underset{1}{N}, ..., \underset{k-1}{N}, V_{k-1}$ and W_k are mutual orthogonal with respect to $\check{\mathbb{G}}$.

3° $\mathbb{G}^H = g_{ij} dx^i \otimes dx^j$, $\mathbb{G}^{V_\alpha} = g_{ij} \delta y^{(\alpha)i} \otimes \delta y^{(\alpha)j}, (\alpha = 1, ..., k-1)$,

$\mathbb{G}^{W_k} = g^{ij}\delta p_i \otimes \delta p_j$ are d-tensor fields.

4° $\mathbb{G}^H, \mathbb{G}^{V_\alpha}, (\alpha = 1, ..., k-1)$ and $\mathbb{G}^{W_k}$ are covariant constant with respect to the canonical metrical N -connection $C\Gamma(N)$.

The geometrical object fields N and g^{ij} determine an almost contact $(k-1)n$-structure $\overset{\vee}{\mathbb{F}}$,(given in (6.8.2)):

(11.4.2)
$$\overset{\vee}{\mathbb{F}}\left(\frac{\delta}{\delta x^i}\right) = -g_{ij}\frac{\delta}{\delta p_j},\ \overset{\vee}{\mathbb{F}}\left(\frac{\delta}{\delta y^{(\alpha)i}}\right) = 0, (\alpha = \overline{1, k-1}),\ \overset{\vee}{\mathbb{F}}\left(\frac{\delta}{\delta p_i}\right) = g^{ij}\frac{\delta}{\delta x^j}$$

Theorem 6.8.1, from ch. 6 is valid:

Theorem 11.4.1 *1° The structure $\overset{\vee}{\mathbb{F}}$ is defined only by N and g^{ij}.*

2° $\overset{\vee}{\mathbb{F}}$ is the following tensor field of type (11.1.1) on $T^{*k}M$:

(11.4.3)
$$\overset{\vee}{\mathbb{F}} = -g_{ij}\frac{\delta}{\delta p_j} \otimes dx^i + g^{ij}\frac{\delta}{\delta x^i} \otimes \delta p_j.$$

3° $Ker\ \overset{\vee}{\mathbb{F}} = \underset{1}{N} \oplus ... \oplus \underset{k-1}{N},\ Im\ \overset{\vee}{\mathbb{F}} = \underset{1}{N} \oplus W_k$.

4° $rank\ \overset{\vee}{\mathbb{F}} = 2n$.

5° $\overset{\vee}{\mathbb{F}}^3 + \overset{\vee}{\mathbb{F}} = 0$.

Consequently $\overset{\vee}{\mathbb{F}}$ is an almost $(k-1)n$- contact structure on the manifold $\widetilde{T^{*k}M}$.

The condition of normality of the structure $\overset{\vee}{\mathbb{F}}$ is given by (see (6.6.5)):

(11.4.4)
$$\mathcal{N}_{\overset{\vee}{\mathbb{F}}}(X,Y) + \sum_{i=1}^{n}\left[\sum_{\alpha=1}^{k-1} d(\delta y^{(\alpha)i})(X,Y) + d(\delta p_i)(X,Y)\right] = 0,\ \forall X, Y \in \mathcal{X}(T^{*k}M)$$

where $\mathcal{N}_{\overset{\vee}{\mathbb{F}}}$ is the Nijenjuis tensor of $\overset{\vee}{\mathbb{F}}$.

Theorem 8.9.4 is valid for spaces $GH^{(k)n}$.

Theorem 11.4.2 *For any generalized Hamilton space of order k, $GH^{(k)n} = (M, g^{ij})$ endowed with a nonlinear connection N the following properties hold.*

1° The pair $(\overset{\vee}{\mathbb{G}},, \overset{\vee}{\mathbb{F}})$ is a pseudo-Riemannian almost $(k-1)n$-contact structure determined only by N and g^{ij}.

2° The associated 2-form is

$$\theta = \delta p_i \wedge dx^i$$

3° If the coefficients N_{ij} of N are symmetric, then

$$\theta = dp_i \wedge dx^i$$

is the canonical presymplectic structure on the manifold $\widetilde{T^{*k}M}$.

4° The conditions of normality of the structure $\overset{\vee}{\mathbb{F}}$ *is expressed by (11.4.4).*

Finally, taking into account $\mathbb{G}$ from (11.4.1) and $\mathbb{F}$ from (11.4.3) it follows.

Theorem 11.4.3 *With respect to the canonical metrical connection* $C\Gamma(N)$ *of the space* $GH^{(k)n}$ *we have*

$$D_X \overset{\vee}{\mathbb{G}} = 0, D_X \overset{\vee}{\mathbb{F}} = 0$$

As such the geometry of the pseudo-Riemannian almost $(k-1)n$-contact space $(T^{*k}M, \overset{\vee}{\mathbb{G}}, , \overset{\vee}{\mathbb{F}})$ can be studied by means of the canonical metrical N-linear connection $C\Gamma(N)$ of the generalized Hamilton space of order k, $GH^{(k)n}$.

11.5 Generalized Cartan Space of Order k

Definition 11.5.1 *A generalized Cartan space of order* k, *is a Generalized Hamilton space of order* k, $GH^{(k)n} = (M, g^{ij})$ *in which the fundamental tensor* g^{ij} *satisfies the axioms:*

1° g^{ij} *is positively defined on* $\widetilde{T^{*k}M}$.

2° g^{ij} *is 0-homogeneous on the fibres of the dual bundle* $(T^{*k}M, \pi^{*k}, M)$.

We denote by $GC^{(k)n} = (M, g^{ij})$ a generalized Cartan space of order k.
¿From the axiom 2° it follows

Proposition 11.5.1 *The following identities hold:*

1° g^{ij} *being 0-homogeneous, we have*

$$\mathcal{L}_{\overset{k-1}{\Gamma}+kC^{*}} g^{ij} = 0 \tag{11.5.1}$$

or, developed:

$$y^{(1)i}\frac{\partial g^{jh}}{\partial y^{(1)i}} + \ldots + (k-1)y^{(k-1)i}\frac{\partial g^{jh}}{\partial y^{(k-1)i}} + kp_i \dot{\partial}^i g^{jh} = 0 \tag{11.5.1'}$$

2° The absolute energy

$$\mathcal{E} = g^{ij}p_i p_j \tag{11.5.2}$$

is 2k-homogeneous on the fibres of $T^{*k}M$.

3° $\mathcal{L}_{\overset{k-1}{\Gamma}+kC^{*}} \mathcal{E} = 2k\mathcal{E}$.

4° $\mathcal{L}_{\overset{k-1}{\Gamma}+kC^{*}} C^{ijh} = -kC^{ijh}$.

Example 2. Let $\overset{\circ}{g}{}^{ij}$ be the fundamental tensor of the Cartan space $\mathcal{C}^{(k)n}$ and $\sigma \in \mathcal{F}(\widetilde{T^{*k}M})$ with the properties:

a) σ is 0-homogeneous;

b) $\dot{\partial}^i\ \sigma$ nonvanishes.

The pair $GC^{(k)n} = e^{-2\sigma}\ \overset{\circ}{g}{}^{ij}$ is a generalized Cartan space of order k. In particular we can consider

$$\sigma = \frac{p_i y^{(1)i}}{\sqrt{\overset{\circ}{g}{}^{ij}\ p_i p_j}\sqrt{\overset{\circ}{g}_{ij}\ y^{(1)i} y^{(1)j}}}.$$

The previous example shows the existence of the spaces $GC^{(k)n}$ are not reducible to a Cartan space of order k.

Let N^* be a nonlinear connection on $T^{*k}M$, having the coefficients $(\underset{(1)}{M_j^{*i}}, ..., \underset{(k-1)}{M_j^{*i}}, N_{ij})$ homogeneous of degree $k-1, ..., 1, k$ respectively.

Theorem 11.5.1 *There exists an unique canonical metrical N^*-connection of the space $\mathbb{G}C^{(k)n}$. Its coefficients are given by (9.5.5).*

Now, the geometry of $\mathbb{G}C^{(k)n}$ can be studied like the geometry of $GH^{(k)n}$ spaces.

The N^*-lift to $\widetilde{T^{*k}M}$ of the fundamental tensor g^{ij} of $GC^{(k)n}$, given by (11.3.2) is not homogeneous on the fibres of the bundle $T^{*k}M$. Therefore, taking into account the following Hamiltonians:

(11.5.3)

$$\mathcal{E}_1 = g_{ij} \overset{(1)i(1)j}{z\ \ z}, \mathcal{E}_2 = g_{ij} \overset{(2)i(2)j}{z\ \ z}, ..., \mathcal{E}_{k-1} = g_{ij} \overset{(k-1)i(k-1)j}{z\ \ \ \ z}, \mathcal{E} = g^{ij} p_i p_j,$$

we can define the following tensor field:

$$\overset{\circ}{\mathbb{G}} = g_{ij} dx^i \otimes dx^j + \sum_{\alpha=1}^{k-1} \frac{1}{\mathcal{E}_\alpha} g_{ij} \delta y^{(\alpha)i} \otimes \delta y^{(\alpha)j} + \frac{1}{\mathcal{E}} g^{ij} \delta p_i \otimes \delta p_j. \tag{11.5.4}$$

Evidently, $\mathcal{E} > 0$.

Theorem 11.5.2 $\overset{\circ}{\mathbb{G}}$ *is a Riemannian structure on $T^{*k}M$, determined only by the fundamental tensor g^{ij} of the space $GC^{(k)n}$ and the nonlinear connection N^*.*

Consider the $\mathcal{F}(T^{*k}M)$-linear mapping $\overset{\circ}{\mathbb{F}}$: $\mathcal{X}(\widetilde{T^{*k}M}) \to \mathcal{X}(\widetilde{T^{*k}M})$ defined by:

(11.5.5)

$$\overset{\circ}{\mathbb{F}}\left(\frac{\delta}{\delta x^i}\right) = -\mathcal{E} g_{ij} \frac{\delta}{\delta p_j},\ \overset{\circ}{\mathbb{F}}\left(\frac{\delta}{\delta y^{(\alpha)i}}\right) = 0, (\alpha = 1, ..., k-1),\ \overset{\circ}{\mathbb{F}}\left(\frac{\delta}{\delta p_i}\right) = \frac{g^{ij}}{\mathcal{E}} \frac{\delta}{\delta x^j}$$

We obtain:

Theorem 11.5.3 *We have:*

1° $\overset{\circ}{\mathbb{F}}$ *is a tensor field on* $T^{*k}M$ *of type (1,1).*

2° $\overset{\circ}{\mathbb{F}}$ *is expressed in the adapted basis by*

$$\overset{\circ}{\mathbb{F}} = -\mathcal{E}g_{ij}\frac{\delta}{\delta p_j} \otimes dx^i + \frac{g^{ij}}{\mathcal{E}}\frac{\delta}{\delta x^i} \otimes \delta p_j \tag{11.5.6}$$

3° *rank* $\overset{\circ}{\mathbb{F}} = 2n$

4° $\overset{\circ}{\mathbb{F}}{}^3 + \overset{\circ}{\mathbb{F}} = 0$

5° $\overset{\circ}{\mathbb{F}}$ *is determined only by* g^{ij} *and* N^*.

Theorem 11.5.4 *For a generalized Cartan spaces of order k,* $GC^{(k)n} = (M, g^{ij})$ *endowed with a nonlinear connection* N^* *the following properties hold:*

1° *The pair* $(\overset{\circ}{\mathbb{G}}, \overset{\circ}{\mathbb{F}})$ *is a Riemannian almost* $(k-1)n$*-contact structure determined only by* N^* *and* g^{ij}.

2° *The associated 2-form is*

$$\overset{\circ}{\theta} = \mathcal{E}g_{ij}\delta p_i \wedge dx^j$$

The proofs are made by usual methods.

Concluding, the space $(\widetilde{T^{*k}M}, \overset{\circ}{\mathbb{G}}, \overset{\circ}{\mathbb{F}})$ is the geometrical model for the generalized Cartan space of order k.

Bibliography

[1] Abate, M. and Patrizio, G. (1994) *Finsler Metrics- A Global Approach with applications to geometric functions theory*, Lecture Notes in Mathematics **1591**, Springer-Verlag, Berlin.

[2] Abraham, R. and Marsden, J. (1978) *Foundations of Mechanics*, (second edition), Reading Mass: Benjamin-Cummings.

[3] Anastasiei, M. (1996) A historical remark on the connections of Chern and Rund, *Contemporary mathematics*, Finsler Geometry **196**, 171-176.

[4] Anastasiei, M. (1996) Certain generalizations of Finsler metrics. *Contemporary Mathematics*, **196**, 161-170.

[5] Anastasiei, M. and Antonelli P.L. (1996) The differential Geometry of Lagrangians which Generate Sprays, *Lagrange and Finsler Geometry*, P.L.Antonelli and R. Miron (eds.), Kluwer Academic Publishers, 15-34.

[6] Anastasiei, M. and Shimada, H. (2000) Deformation of Finsler metrics, *Finslerian Geometries*, Kluwer Academic Publishers, FTPH. **109**, 53-66.

[7] Antonelli, P. L., Ingarden, R.S. and Matsumoto, M. (1993) *The Theory of Sprays and Finsler spaces with Applications in Physics and Biology*, Kluwer Academic Publishers.

[8] Antonelli, P. L. and Hrimiuc, D. (1996) A new class of spray- generating Lagrangians, *Lagrange and Finsler Geometry. Applications to Physics and Biology*, Kluwer Academic Publishers **76**, 81-92.

[9] Antonelli, P. L. and Lackey, B. C. (eds.) (1998), *The Theory of Finslerian Laplacians and Applications,* Kluwer Academic Publishers, **459**.

[10] Antonelli, P. L. and Miron, R. (eds.) (1996), *Lagrange and Finsler Geometry. Applications to Physics and Biology*, Kluwer Academic Publishers, **76**.

[11] Antonelli, P. L. and Zawstaniak, T.J. (1999) *Fundamental of Finslerian Diffusion with Applications*, Kluwer Academic Publishers, **101**.

[12] Arnold, V.I. (1978) *Mathematical Methods of Classical Mechanics,* Springer, New York.

[13] Arnold, V.I., Gelfand, I.M. and Retakh V.S., (eds), (1997) Birkhäuser, Boston, 77-92.

[14] Asanov, G. S. (1985) *Finsler Geometry, Relativity and Gauge Theory,* Reidel-Dordrecht.

[15] Bacso, S. and Matsumoto, M. (1997) On Finsler Spaces of Douglas type, *Publ. Math.* Debrecen **51/3-4**, 385-406.

[16] Balan, V. and Stavrinos, P.C. (1996) Deviations of Geodesics in the Fibred Finslerian Approach, *Kluwer Academic Publishers*, **76**, 65-75.

[17] Bao, D., Chern, S.S. and Shen, Z. (2000) *An Introduction to Riemann-Finsler Geometry*, Springer, GTM **200**.

[18] Bejancu, A. (1990) *Finsler Geometry and Applications*, Ellis Harwood Ltd.

[19] Bejancu, A, (2000) On the theory of Finsler submanifolds, *Finslerian Geometry*, Kluwer Academic Publishers, FTPH. **109**, 111-131.

[20] Bejancu, A. and Farran, H. R. (2000) *Geometry of Pseudo-Finsler Submanifolds*, Kluwer Academic Publishers.

[21] Beil, R. G. (1992) Finsler Gauge Transformations and General Relativity, *International Journal Theoretical Physics*, **31**, 1025-1044.

[22] Blair, D. (1987) *Contact Manifolds in Riemannian Geometry*, Lecture Notes in Mathematics, **509**, Springer Verlag, Berlin.

[23] Blănuţă, V. and Yamata, M. (1994) Infinitesimal transformations of the $2-\pi$ structures on the tangent bundle, *N.S. Tensor*, Japan, **55**, 43-52.

[24] Blănuţă, V. (2001) Natural $2-\pi$ Structures in Lagrange Spaces of Higher Order, *Journal of Egyptian Mathematical Society*, Cairo, vol. **9(2)**, 151-164.

[25] Bryant, R. L. (1996) Finsler Structures on the 2-sphere satisfying $K = 1$, *Finsler Geometry, Contemporary Mathematics,* **AMS 196**, 27-42.

[26] Bucataru, I. (1996) Sprays and homogeneous connections in the higher order geometry, *Studii şi Cercetări Matematice*, **50** 5-6, 253-258.

[27] Bucataru, I. (1996) Horizontal lifts in the higher order geometry, *Publ. Math.* Debrecen **56**, 1-2(2000), 21-32.

[28] Busemann, H. (1955) *The Geometry of Geodesics*, Academic Press.

[29] Bernd, R. (2001) An introduction to Symplectic Geometry, Graduate Studies in Math., vol. **26**, AMS.

[30] do Carmo, M. (1992) *Riemannian Geometry*, Birkhäuser, Boston.

[31] Cartan, E. (1934) *Les espaces de Finsler*, Hermann, Paris.

[32] Chaki, M.C. (2001) On generalized quasi Einstein manifolds, *Publ. Math.* Debrecen **58**, no.4, 683–691.

[33] Chern, S.S. (1996) *A Mathematician and his Mathematical Work*, (Selected papers of S. S. Chern), Edited by S. Y. Cheng, P. Li and G. Tian, World Scientific.

[34] Crampin, M. (1995) *Jet Bundle Techniques in Analytical Mechanics*, Quaderni del Cons. Naz. delle Ricerche, Gr. Naz. di Fizica Matematica, no.**47**.

[35] Crampin, M., Sarlet, W. and Cantrijn, F., (1986) Higher Order Differential Equations and Higher Order Lagrangian Mechanics, *Math. Proc. Camb. Phil. Soc.* **99**, 565-587.

[36] Crâşmăreanu, M. (2000) Conservation laws generated by pseudosymmetries with applications to Hamiltonnian systems, *Italian Journal of Pure Application in Mathematics* **8**(2000), 91-98.

[37] Cruceanu, V. (1993) Une classe de structure géométrique sur fibré cotangent, *Tensor N.S.*, **53**, 196-201.

[38] Cruceanu, V. (1986) Sur la theorie de sous-fibre vectoriels, *C. R. Acad. Sci. Paris*, **302**, 701-708.

[39] Cruceanu, V., Furtuny, P. and Gadea, P.M. (1996) A survey on Paracomplex Geometry, *Rockey Mountain J. of Math.* **26**, No. 1, 83-115.

[40] Craioveanu, M. and Puta, M., (1991) Geometric Quantization of Some Constrained Mechanical Systems, *Proceedings of 22-th Conference on Differential Geometry and Topology*, Bucharest, 87-91.

[41] Dombrowski, P. (1962) On the geometry of tangent bundles, *Journal Reine Angew Math.*, **210**, 73-88.

[42] Dragomir, S. (1986) Submanifolds of Finsler Spaces, *Conf. Sem. Math. Univ. Bari*, **217**, 15 p.

[43] Eliapoulos, H.A. (1965) A generalized metric space for electromagnetic theory, *Acad. Roy. Belg. Bull Sci.*, **(5)51**, 986-995.

[44] Gábos, Z., Mangeron, D. and Stan, I. (1962) *Fundamentele Mecanicii*, Ed. Academiei Române.

[45] Guillemin V. and Sternberg S. (1984) *Symplectic Tehniques in Physics*, Cambridge Univ. Press.

[46] Grifone, J. (1972) Structures Presque Tangente et Connexions, I, *An. Inst. Fourier*, Grenoble **22**, 287-334.

[47] Grifone, J. (1972) Structures Presque Tangente et Connexions, II, *An. Inst. Fourier*, Grenoble **22-3**, 291-338.

[48] Gromov, M. (1983) Filling Riemannian Manifolds, *Journal of Differential Geometry*, **18**, 1-147.

[49] Haimovici, M. (1934) Formulas fundamentales dans la théorie des hypersurface d'un espace de Finsler, *C. R. Acad. Sci. Paris*, **198** , 426-427.

[50] Haimovici, M. (1939) Variétés totalment extremales et variétés totalment géodésiques dans les espaces de Finsler, *Ann. Sci. Univ. Iassy,* **25**, 559-644.

[51] Hashiguchi, M. (1975) On Wagner's generalized Berwald spaces, *Journal of Koreean Mathematical Society*, **12**, 51-61.

[52] Hashiguchi, M. (1976) On conformal Transformations of Finsler Metrics, *Journal of Math.*, Kyoto Univ. 16, 25-50.

[53] Hassan, B.T. (1994) The Theory of Geodesics in Finsler Spaces (*"Ph. D. Thesis"*), Southampton Univ., 57-65.

[54] Hrimiuc, D. (1994) Hamilton Geometry, *Pergamon Press*, Math. Comput. Modelling **20**, no. 415, 57-65.

[55] Hrimiuc, D. and Shimada, H. (1996) On $\mathcal{L}$-duality between Lagrange and Hamilton Manifolds, *Nonlinear World*, **3**, 613-641.

[56] Hrimiuc, D. (1996) On some special problems concerning the $\mathcal{L}$- duality between Finsler and Cartan spaces, *Tensor N.S.*, **58**, 48-61.

[57] Ichijyo, Y. (1976) Finsler manifolds modeled on a Minkowski space, *J. Math. Kyoto Univ.*, **16**, 639-652.

[58] Ichijyo, Y. (1988) On the Finsler group and an almost symplectic structure on a tangent bundle, *J. Math Kyoto Univ.* **28**, 153-163.

[59] Ikeda, S. (1955) *Advanced Study in Applied Geometry*, Seizansha, Kanagawa.

[60] Ingarden, R. S. (1987) On Physical interpretations of Finsler and Kawaguchi spaces, *Tensor N. S.* **46**, 354-360.

[61] Ishikawa,H. (1981) Note on Finslerian relativity, *Journal of Mathematical Physics*, **22 (5)**, 995-1004.

[62] Izumi, H. (1991) On the geometry of generalized metric spaces I. Connections and identities, *Publicationes Mathematicae,* Debrecen **39**, 113-134.

[63] Kawaguchi, A. (1961) On the vector of Higher Order and the extended affine connection, *Ann. di. Matem. Pura ed Appl.* **IV (55)**,105-118.

[64] Kawaguchi, M. (1962) An Introduction to the Theory of the Higher Order Space, I: The Theory of Kawaguchi spaces, *RAAG Memoirs*, **3**.

[65] Kern, J. (1974) Lagrange Geometry, *Arch. Math.*, **25**, 438-443.

[66] Kikuchi, S. (1979) On the conditions that a space with (α, β)-metric be locally Minkowskian, *Tensor N.S.*, **33**, 242-246.

[67] Klein, J. (1962) Espaces Variationnels et Mécaniques, *Ann. Inst. Fourier* (Grenoble), **12**, 1-124.

[68] Kobayashi, S. and Nomizu, K. (1969) *Foundations of Differential Geometry*, John Wiley & Sons, Inc., New-York **I** (1963), **II**.

[69] Kondo, K. (1963) On the Physical Meanings of the Zermelo Conditions of the Kawaguchi Space, *Tensor N.S.*, **14**, 191-215.

[70] Kondo, K. (1991) *Epistemological Foundations of Quasi-Microscopic phenomena from the stand-point of Finsler's and Kawaguchis higher order Geometry*, Post RAAG Reports **241**, **242**, **243**.

[71] Krupka, D. and Janyska, J. (1990) *Lectures on Differential Invariants*, Univ. Brno.

[72] Krupkova, O. (1997) *The Geometry of Ordinary Variational Equations*, Lecture Notes in Math. nr. 1678, Springer.

[73] de Leon, M. and Rodrigues, P.R. (1985) *Generalized Classical Mechanics and field Theory*, North-Holland.

[74] de Leon, M. and Rodrigues, P. R. (1989) *Methods of Differential Geometry in Analytical Mechanics,* North-Holland.

[75] Liberman, P. and Marle, Ch.M. (1987) *Symplectic Geometry and Analytical Mechanics*, D. Reidel Publ. Comp.

[76] Matsumoto, M. (1986) *Foundations of Finsler Geometry and Special Finsler spaces*, Kaiseisha Press, Otsu, Japan.

[77] Matsumoto, M. (1992) Theory of Finsler Space with (α, β)-metric, *Rep. Math. Phys.* **31**, 43-83.

[78] Matsumoto, M. and Miron, R. (1977) On an invariant theory of the Finsler spaces, *Periodica Math. Hungar*, **8**, 73-82.

[79] Matsumoto, M. and Shimada, H. (1978) On Finsler spaces with 1-form metric, *Tensor N.S* **32**, 161-169.

[80] Matsumoto, K., Mihai, I. and Miron, R. The almost Kählerian structures determined by Riemannian structures, (to appear).

[81] Mazetis, E. (1998) Geometry in Cartan spaces, *Lithuanian Math. Jour.* vol. **38**., no. 2., 172-180.

[82] Mihai, I. (1995) *Capitole Speciale de Geometria Varietăţilor Complexe*, Ed. Univ. Bucharest, (in Romanian).

[83] Mihai, I. (1995) Certain Submanifolds of a Kähler manifold, *Geometry and Topology of Submanifolds,* World Scintific **VII**, 186-188.

[84] Mihai, I., Rosca, R. and Verstaelen, L. (1996) *Some Aspects of the Differential Geometry of Vector Fields*, PADGE, KU Leuven, KU Brussel.

[85] Miron, R. (1989) Hamilton Geometry, *Analele ştiinţifice ale Univ. 'Al.I. Cuza'*, Iaşi, **35**, 33-85.

[86] Miron, R. (1987) Cartan Spaces in a new point of view by considering them as duals of Finsler Spaces, *Tensor N. S.* **46**, 330-334.

[87] Miron, R. (1988) The geometry of Cartan spaces, *Prog. in Math.*, India **22** (1&2), 1-38.

[88] Miron, R. (1990) Subspaces in Hamilton spaces, *Indian J. Pure Appl. Math.*, **21(3)**, 211-233.

[89] Miron, R. (1996) General Randers Spaces, *Lagrange and Finsler Geometry*, Ed. by P. L. Antonelli and R. Miron, Kluwer Academic Publ. FTPH no. **76**, 123-140.

[90] Miron, R. (1993) Space with Higher Order Metric Structures, *Tensor N.S.* **53**, 1-23.

[91] Miron, R. (1994) Lagrange Geometry, *Math. Comput. Modelling* **20(4/5)**, 25-40.

[92] Miron, R. (1995) Nöther Theorems in Higher Order Lagrangian Mechanics, *International Journal of Theoretical Physics* **34**, 1123-1146.

[93] Miron, R. (1996) Some problems in the Geometries of Finsler type, *Contemporary mathematics, Finsler Geometry* **196**, 187-196.

[94] Miron, R. (1997) *The Geometry of Higher Order Lagrange Spaces. Applications to Mechanics and Physics*, Kluwer Acad. Publ. FTPH no. **82**.

[95] Miron, R. (1998) *The Geometry of Higher-Order Finsler Spaces*, Handronic Press, Inc., USA.

[96] Miron, R. (2000) Hamilton spaces of order k grater than or equal to 1, *International Journal of Theoretical Physics*, **39**, 9, 2327-2336.

[97] Miron, R. (1996) Some Problems in the Geometries of Finsler type, *Finsler Geometry,* Contemp. Math., AMS **196**, 187-195.

[98] Miron, R. (2002) Variational Problems of Higher Order Hamiltonians, *Proc. non-Euclidian geometry in modern Physics*, Debrecen, Hungary.

[99] Miron, R. and Anastasiei, M. (1994) *The Geometry of Lagrange Space: Theory and Applications*, Kluwer Acad. Publishers, FTPH no. **59** .

[100] Miron, R. and Anastasiei, M. (1997) *Vector Bundles and Lagrange Spaces with Applications to Relativity*, Geometry Balkan Press, Bucharest.

[101] Miron, R., Chaki, M. C. and Barua, B. (2000) Symmetries of the Synge metric in the relativistic optics., *Bull. Calcutta Math. Soc.* **92**, no. 3, 219–224

[102] Miron, R., Ianuş, S. and Anastasiei, M. (1989) The Geometry of Dual of Vector Bundle, *Publ. de l'Inst. Math.* **46 (60)**,145-162.

[103] Miron, R. and Zet, Gh. (1993) Post-Newtonian Approximation in Relativistic Optics, *Tensor N.S.* **53**, 92-95.

[104] Miron, R. and Hashiguchi, M. (2003) The Generalized Hamilton Spaces with Regular Metic, *Proc. of workshop "Finsler Geometry"*, Debrecen (Hungary).

[105] Miron, R. and Kawaguchi, T. (1991) Relativistic Geometrical Optics, *International Journal of Theoretical Physics*, **30**, 1521-1543.

[106] Miron, R. and Sabău V. S. (1999) On the foundations of Higher Order Finsler Spaces, *Algebras, Groups and Geometries*, Hadronic Press, Vol. **16**, No. **1**, 83-90.

[107] Miron, R., Sabău V. S. and Shimada, H. (1998) Abstract from the Higher Order Finsler Spaces and examples, *Proceedings of Hokkaido Tokai Univ.* **9**, 1-12.

[108] Miron, R., Watanabe, S. and Ikeda, S. (1987) Some Connections on Tangent Bundle and their Applications to General Relativity, *Tensor N.S.* **46**, 8-22.

[109] Miron, R. and Bejancu, A. (1984) A non-standard theory of Finsler subspaces, *Collq. Math. Soc. J. Bolyai*, vol. **46**, 815-851.

[110] Miron, R. and Klepp, F.C. (1984) Almost complex Finsler structures on the tangent bundle, *Collq. Math. Soc. J. Bolyai*, 4-Debrecen, 853-881.

[111] Miron, R. and Matsumoto, K. The almost Hermitian structures determined by the Riemannian structure on the tangent bundle, (to appear).

[112] Miron, R. and Radivoiovici-Tatoiu, M. (1989) A Lagrangian theory of electromagnetism, *Rep. Math. Phys.* **27**, 49-84.

[113] Miron, R., Roşca, R., Anastasiei, M. and Buchner, K. (1992) New Aspects of Lagrangean Relativity, *Foundations of Physics Letters* **5(2)**, 141-171.

[114] Miron, R., Tavakol, R. K., Balan, V. and Roxburgh, L. (1993) Geometry of Space-Time and Generalized Lagrange Gauge Theory, *Publ. Math. Debrecen* **42(3-4)** (1993), 215-224.

[115] Miron, R., Hrimiuc, D., Shimada, H. and Sabău, V.S. (2001) *The Geometry of Hamilton and Lagrange Spaces*, Kluwer Acad. Publ. FTPH, no. **118**.

[116] Modugno, M. (1987) *Systems of vector valued forms on forms on a fibered manifold and applications to gauge theories*, Proc. Conf. Diff. Geom. in Math. Phys. Salamanca, L. N. in Math., Springer Verlag **1251**.

[117] Morimoto, A. (1968) Prolongations of Geometric Structures, *Nagoya Math. Jour.* **32**, 67-108.

[118] Munteanu, Gh. (1988) Lagrange geometry of Second Order, *Proceedings of the Fifth National Seminar on Finsler and Lagrange Spaces*, Braşov, 285-295.

[119] Munteanu, Gh. (2000) Complex Hamilton Spaces, *Algebras, Groups and Geometries*, vol. **17**, nr. 3, 293-302.

[120] Nagano, T. (2000) Singular Finsler Spaces, *Algebras, Groups and Geometries*, vol. **17**, nr. **3**, 303-311.

[121] Nicolescu, L. (1981) Champs presque principaux dans l'algébre de déformation, *Rev. Roum. Math. Pures Appl.* **26**, 611-616.

[122] Nicolescu, L. (1994) *Grupuri Lie*, Ed. Univ. Bucharest, (in Romanian).

[123] Numata, S. (1975) On Landsberg spaces of scalar curvature, *Journal of Korean Math. Soc.* **12**, 97-100.

[124] Okada, T. (1982) Minkowskian product of Finsler spaces and the Berwald connections, *Journal of Math. Kyoto Univ.* **22**, 323-332.

[125] Opriş, D. and Butulescu, I. (1997) *Metode Geometrice in Studiul Sistemelor de Ecuaţii Diferenţiale*, Ed. MIRTON, Timişoara, (in Romanian).

[126] Pitiş, Gh. (1984) Pontriagin Classes in Finsler Geometry, *Analele ştiinţifice ale univ.'Al. I. Cuza'*, Iasi, mat. **XXX**, **s. I. a**, 61-64.

[127] Popescu, L. (2000) On the Homogeneous lift G in the Cotangent Bundle (II), *Differential Geometry, Dynamical Systems*, Balkan Press, vol. **2**, no.**1**, 32-35.

[128] Postolache, M. (1993) Computational Methods in Lagrange Geometry, *Kluwer Academic Publishers*, FTPH. **76**, 163-177.

[129] Randers, G. (1941) On an asymmetric metric in the four-space of general relativity, *Phys. Rev.* **(2)**, **59**, 195-199.

[130] Rund, H. (1959) The differential geometry of Finsler spaces, *Grundlehr. Math Wiss.* Springer Verlag **101**.

[131] Rund, H. (1966) The Hamilton-Jacobi Theory in the Calculus of Variations, *O. van Nostrand*, London.

[132] Rizza, G.B. (1962) Structure di Finsler sulle varieta quasi complesse, *Leicei Rend.* **(8) 33**, 271-275.

[133] Ros, A., Verstraelen, L. (1984) On a conjecture of K. Ogiue, *Journal of Differential Geometry*, **19**, 561-556.

[134] Roman, M. (2000) On the homogeneous hyperbolic lift of a Finslerian metric, *Algebras, Groups and Geometries*, vol.**17**, nr. **3**,323-330.

[135] Sabău, V.S. and Shimada, H. (2000) Finsler Geometry, *Kluwer Academic Publishers*, P. L. Antonelli (ed.),**109**, 15-24.

[136] Sabău, V.S. and Shimada, H. (1996) The Electromagnetic Field in the Second Order Relativistic Optics, *Memoriile Academiei Romane* (1996) (to appear).

[137] Sabau, V.S. and Shimada, H. (1996) Some remarks on the Geometry of Higher Order Osculator Bundle, *Proceedings of Hokkaido Tokai Univ.* **9**, 47-58.

[138] Sakaguchi, T. (1988) Subspaces in Finsler space, *Mem. Nat. Def., Acad.*, **28**, 1-37.

[139] Santilli, M.R. (1978) Foundations of Theoretical Mechanics, I: The Inverse Problem in Newtonian Mechanics, Springer-Verlag, (1981) *Foundations of Theoretical Mechanics: Birkhoffian Generalization of Hamiltonian Mechanics II,* Springer-Verlag.

[140] Santilli, M.R. (2001) Foundations of Hadronic Chemistry with Applications to New Clean Energies and Fuelnes, Kluwer Academic Publishers, FTPH **120**.

[141] Şandru I. (1995) *Conexiuni metrice în varietăţi de elemente liniare. Aplicaţii în teoria relativităţii generale*, Ed. Didactică şi Pedagogică, Bucureşti,(in Romanian).

[142] Şandru O., Intrinsic geometrization of the Hamiltonian variational calculus, (to appear).

[143] Sasaki, S. (1958) On the Differential Geometry of Tangent Bundle of Riemannian Manifolds, *Tohoku Math. Jour.* **10**, 338-354.

[144] Saunders, D.J. (1989) *The Geometry of Jet Bundles*, Cambridge Univ. Press.

[145] Souriau, J-M. (1997) Structure of Dynamical Systems. A Symplectic View of Physics, *Prog. Math.* vol. **149**, Birkhäuser, Boston.

[146] Shen, Z. *Differential Geometry of Spray and Finsler Spaces*, (book to appear).

[147] Shibata, C. (1978) On Finsler space with Kropina metric, *Rep. on Math. Phys.* **13**, 117-128.

[148] Shibata, C., Shimada, H., Azuma, M. and Yasuda, H. (1997) On Finsler space with Randers metric, *Tensor N. S.* **33**, 365-372.

[149] Shimada, H. and Sabău, V.S. (1998) Higher Order Miron-Tavakol metric, *Revue Roumain des Mathematiques Pures et Appliques* **XLII (9-10)**, 813-822.

[150] Shimada, H. (1979) On Finsler space witk the metric $L =^m \sqrt{a_{i_1 \ldots i_m}(x) y^{i_1} \ldots y^{i_m}}$, *Tensor N. S.* **33**, 365-372.

[151] Stavre, P. (1988) The *d*-linear connections, *Proc. Nat. Sem on Finsler and Lagrange Spaces*, Braşov, Romania, 375-382.

[152] Stavre, P. (1999) On the integrability of the structures $(T^*M, \overset{*}{G}, \overset{*}{F})$, *Algebras, Groups and Geometries*, **16**, 107-114.

[153] Stavre, P. and Popescu, L. The Homogeneous lift $\overset{*}{G}$ on the Cotangent Bundle, *Novi-Sad Journal of Mathematics*, (to appear).

[154] Stavrinos, P.C. and Manouselis, P. (1995) The non-localised field theory: Poicaré gravity and Yang-Mills fields. *Rep. Math. Phys.*, vol. **36**.

[155] Synge, J. L. (1966) *Relativity: General Theory*, North-Holland, Amsterdam.

[156] Szabo, Z.I. (1981) Positively Define Berwald Spaces, *Tensor N.S.*, **35**, 25-39.

[157] Szilasi, J. and Kozma, L. (1984) Remarks on Finsler-Type Connections, *Proc. Nat. Sem. on Finsler and Lagrange Spaces*, Braşov, Romania, 181-197.

[158] Tamassy, L. (1996) Affine Connections inducing a Finsler Connection, *Coll. Math. Soc. Janos Bolyai*, North-Holland **46** , 1185-1193.

[159] Tamim, A. (1998) Submanifolds of a Finsler Manifold, *Journal of Egyptian Mathematical Society*, **6(1)**.

[160] Tamim, A. and Youssef, N.L. On Generalized Randers Manifolds, *Algebras, Groups and Geometries* (Hadronic Press) **16**, 115-126.

[161] Tonnelat, M.A. (1995) *Les théories unitaires de l'élecromagnetisme et de la gravitation*, Gauthier-Villars, Paris.

[162] Tarină, M. (1987) Invariant Finsler Connections on Vector Bundles, *Analele Ştiinţifice ale Univ. Al. I. Cuza*, Iasi, **sI, a. Mat. 33**, 87-94.

[163] Thompson, A.C. (1993) Minkowski Geometry, *Encicl. of Math. and its Appl.* **63**.

[164] Tsagas, Gr. and Sourlas, D. (1995) Isomanifolds and Their Isotensor Fields. Isomappings between Isomanifolds, *Algebras, Groups and Geometries,* vol. **12**, no. 1, 1-89.

[165] Tulczyjew, W.M. (1980) The Euler-Lagrange Resolution, *Lecture Notes in Math.* 836, Diff. Geom Methods in Math. -Phys., Springer, Berlin, 22-48.

[166] Vacaru, S. (1988) Interactions, Strings and Isotopies in Higher Order Anisotropic Superspaces, *Handronic Press*, Palm Harbor, USA.

[167] Vaisman, I. (1994) Lectures on the Geometry of Poisson manifolds, Birkhaser, Basel.

[168] Vaisman, I., (1987) Symplectic Geometry and Secondary Characteristic Classes, Birkhäuser, Basel.

[169] Verstaelen, L. (1990) Curves and Surfaces of finite Chen type, *Geometry and Topology of Submanifolds*, Word Scintific **III**, 304-311.

[170] Vrănceanu, Gh. (1969, 1971, 1973) *Opera Matematică*, vol. **I, II, III**, Ed. Academiei, Bucureşti, (in Romanian).

[171] Vrănceanu, Gh. and Mihăileanu, N. (1978) *Introducere în teoria relativităţii*, Ed. Tehnică, Bucureşti, (in Romanian).

[172] Watanabe, S., Ikeda, S. and Ikeda, F. (1983) On a metrical Finsler connection of generalized Finsler metric $g_{ij} = \exp(2\sigma(x,y))\gamma_{ij}(x)$, *Tensor, N.S.* **40**, 97-102.

[173] Zet, Gh. (1994) Lagrangian Geometrical Models in Physics, *Math. Comput. Modelling* 20(4/5), 83-91.

[174] Zet, Gh. (1998) *Unitary Symmetries and Gauge Theories*, (Romanian), Ed. Gh.Asachi, Univ. Tehnică, Iaşi.

[175] Yano, K. and Ishihara, S. (1973) *Tangent and Cotangent Bundles*, Marcel Dekker.

[176] Yasuda, H. and Shimada, H. (1977) On Randers spaces of scalar curvature, *Report on Mathematical Physics* **11**, 347-360.

[177] Yawata, M. (1996) Killing equations in tangent bundle, *Lagrange and Finsler Geometries*, (P.L. Antonelli and R. Miron (eds.)), Kluwer Academic Publishers FTPH, vol. **76**, 189-195.

Index

MIX
Papier aus verantwortungsvollen Quellen
Paper from responsible sources
FSC® C105338

If you have any concerns about our products,
you can contact us on
ProductSafety@springernature.com

In case Publisher is established outside the EU,
the EU authorized representative is:
Springer Nature Customer Service Center GmbH
Europaplatz 3, 69115 Heidelberg, Germany

Printed by Libri Plureos GmbH
in Hamburg, Germany